Ruby Free is an award-wi[nning] Conservation Biologist base[d ...] about saving nature and r[...] worked and volunteered in the environmental conservation sector since she was 16 and has been named one of the 100 most influential Environmental Professionals in the UK on the ENDS Power List.

Since living on Rathlin Island and moving to Ballyconnelly Farm, Ruby has pursued an MSc in Ecological Management and Conservation Biology, which took her to another island off the coast of Canada where she conducted her research thesis.

Back home in Ireland, Ruby has been working away at Ballyconnelly Farm with her partner Craig, setting up a community growing space and rewilding areas of the farm for nature. Alongside her passion of nature-writing, these days any spare opportunities to unwind are spent adventuring with her dog Isla, surfing, growing food or illustrating.

PRAISE FOR *RATHLIN, A WILD LIFE*

John Barry, Professor of Green Political Economy and Co-Chair Belfast Climate Commission

An honest account and unapologetic appreciation of the beauty of and necessity for us to care for nature.

With childlike joy and curiosity, Ruby Free brings a fresh pair of eyes and an open heart to Rathlin Island. I have been going there for twenty years but Ruby brings to life its vitality, vibrancy and unpredictability in all its natural glory. The book can be read as a 'love poem' to the island in gratitude for all it provided her. But more than that, the book not only offers an eloquent and unaffected experience of the more than human world, but it is a young woman's experience.

The book charts how in 'digging where she stands', and in connecting with people and place, she also connects with herself. She looks at nature and knows the scientific name and ecological function of what she sees, but what makes this book unique is that she also feels what she sees, as well as knowing the stories and meaning of nature. *Rathlin, A Wild Life* conveys the how and what of nature, but also the why.

The book is a tremendous achievement and all the more welcome for being a small but important ray of local light in the dimming and globally turbulent times we live in. If what we need to do in relation to our planetary crisis is to 'save the whole by saving the parts', Free's *Rathlin* is a beautifully written example of a place-based way to start. After all, we cannot save that which we neither know about nor love, and *Rathlin, A Wild Life* shows us how to do both.

For peace, people and planet.

Claire Barnett, RSPB NI Area Manager (East)
Rathlin Island is a very special place. Every corner turned you encounter something magical – the views, the wildlife, the history, the people. It's a place where I feel grounded. Ruby's writing, her storytelling, brings the essence and beauty of Rathlin direct to the reader.

I so enjoyed reading this book. Ruby's energy, passion and charm is, and always will be, contagious.

She is an incredible advocate for nature, a reasoned, knowledgeable and poetic voice and I can't wait to see what she does next.

Rathlin
A Wild Life

Island Living, Seabirds
and Extraordinary Gifts
from Nature

Ruby Free

This book is memoir. It reflects the author's present recollections of experiences over time. Some names have been changed for privacy and some events have been compressed.

First published in 2024 by Blackstaff Press
an imprint of Colourpoint Creative Ltd
Colourpoint House
Jubilee Business Park
21 Jubilee Road
Newtownards BT23 4YH

© Text, Ruby Free, 2024
© Foreword, Megan McCubbin, 2024
© Illustrations, Ruby Free, 2024

All rights reserved

Ruby Free has asserted her right under the Copyright, Designs and Patents Act 1988 to be identified as the author of this work.

Printed and bound by CPI Group UK Ltd, Croydon CR0 4YY

A CIP catalogue record for this book is available from the British Library

ISBN 978 1 78073 386 9

www.blackstaffpress.com

Dedicated to my mum, Sarah Ward,
for sailing right behind.

In loving memory of Lizzie Holmes.

Contents

Foreword by Megan McCubbin xi
Introduction xiii
Map of Rathlin. xvi
Chapter 1 Arriving 1
Chapter 2 Landing 21
Chapter 3 Understanding 45
Chapter 4 Discovering 65
Chapter 5 Flourishing. 83
Chapter 6 Gliding. 105
Chapter 7 Surviving. 129
Chapter 8 Thriving. 155
Chapter 9 Reflecting 175
Chapter 10 Returning. 199
Act Now for Nature 221
What does Rathlin mean to you? 227
Acknowledgements 236

Foreword

Megan McCubbin, Wildlife Broadcaster and Zoologist

Clutched within the Irish Sea, a small body of water that separates Northern Ireland with Great Britain, is a small unassuming L-shaped island which lays unnoticed by most. It's a place where time seemingly stands still, where the worries of the world slip away like sand through your fingers, and where nature is making a last stand. And an important one at that...

Rathlin Island is six miles long, and whilst some might deem it quiet and dreary with a population of only 150 people, Ruby Free has clearly found her sanctuary nestled within its wild, rocky landscape. To uproot your life from Dorset in the south of England to this small island takes courage. Generally, people do this due to a strong sense of purpose – that could be a purpose for adventure, for work passions, for freedom or even simply to find that special place that makes you happy. Whilst this transition was slightly unexpected for Ruby, you can read along and feel her whole world shift as this tiny little rock changes her life forever.

This is an authentic love-letter to Rathlin and to the wildlife that lives on and around its coastline. Based on diary entries, drawings and records that Ruby has noted during her time working for RSPB West Light Seabird Centre, this book highlights the pure joy of experiencing nature but is expertly balanced with the reality that we could, and are, losing the battle to protect it. The pages are filled with important information about the state of our environment, but it was refreshing to read how Ruby approached these sensitive topics being so devoted to implementing the solutions. I was left feeling empowered and motivated to stand up alongside her, to fight for what I believe in. And I believe in puffins – like Puff, Finn and Busy Lizzie (three great individual birds we get to know throughout the pages), gannets, seals, corncrakes, thresher sharks and all the other species Ruby found herself living alongside during her time there.

I personally adore seabirds, so I knew from the offset I was going to love this book and I genuinely could not put it down (and this, coming from a severe dyslexic is quite rare)! Ruby's ability to describe her surroundings, not only visually but through scent and sound, is captivating. It's like you're stood with her overlooking the cliff tops where vast numbers of breeding razorbills and guillemots hastily go about rearing their young chicks. The way she describes her pure emotions and connections to the land and the animals was really beautiful to read, too. Her honesty and willingness to express herself is refreshing. As someone who has faced her fair share of challenges as well as being diagnosed with ADHD as an adult, it's clear to see why Rathlin Island has become her home away from home. Nature's ability to support you when you need it can never be underestimated – it's the ultimate escapism.

This brilliantly written book is the adventure everyone dreams of, but very few are able to chase. It's a testament to Ruby's character – her kind and loving nature leaps off every page as she shares her experiences getting to know an unfamiliar place, reconnecting to nature and finding her happy place.

Introduction

In 2021, I was offered a dream job with the Royal Society for the Protection of Birds (RSPB), which meant living and working on Rathlin Island, a Special Area of Conservation, renowned for its diverse wildlife, beautiful scenery and rich heritage. With internationally important seabird populations, I was immersed in the colony's smelly, busy, brilliant world for one season and through this book, I hope to take you there with me.

Positioned off the north coast of Ireland, Rathlin is the wildest, most biodiverse place I have ever lived; the first place with a larger biomass of wild species than human life. What a beautiful, rare thing. But for most of human history, our species lived amongst the natural world, not against it nor apart from it. Ten thousand years ago – which in the scale of earth's history is just a speck of time – animal biomass existed as 1 per cent humans and 99 per cent wild animals. Today, it's 32 per cent humans, 1 per cent wild animals and 67 per cent livestock. Earthly systems are out of kilter, we are disconnected from wild places and are suffering as a result.

I came to Rathlin to protest against this reality. To not just love nature but reside alongside it. I wanted to wake up every day and know I was contributing to something good. I longed to inhabit wild spaces, alongside endangered species experiencing steep decline, and live in a way that would not hinder their existence but help to conserve it. I'm so glad I did, because it changed me forever.

Nature underpins all of us – the healthier it is, and the closer we can live to it, the more we thrive. Through the process of evolution, because of our close relationship with nature, human brains formed positive chemical reactions to the sights and sounds that signposted survival. The dawn chorus signals that the harsh winter is over and the reassuring sound of trickling water brings us comfort because it tells us we aren't to dehydrate. This is one of the many reasons that nature gives us such joy – it belongs in our lives because we are nature, not some separate entity.

Rathlin Island gave me so much; amazing nature encounters, but more than that, it connected me to a way of living and thinking I'd not experienced before.

It taught me important truths that remain with me now.

Truths we cannot and must not supress.

Currently there's an ecological 'elephant in the room'. While two thirds of the population worry about climate change, most of us do not speak about it, causing a phenomenon called 'climate silence.' This silence has come about for several reasons, with one of the main drivers being how politicised and polarising the conversation surrounding the climate and ecological crisis has become. But it's so important we have these conversations because the more informed and empowered we all feel, the better collective actions we can make.

We do not need any particular qualification to understand what's happening to the natural world. Conservation doesn't have to be our vocation in order for us to discuss it with those around us. There's a misconception that we have to do 'environmentalism' perfectly or fit into a label or category – be a scientist, green-fingered hippie or young person – to be vocal and care. This isn't true, not least because it's in all of our interests to live in a safe and habitable world. It's going to take a wealth of diverse individuals to save nature, and also to get the transition from where we are now to where we need to be right. But of course, being a human right now can be mentally exhausting. The size of the burden that we consciously or sub-consciously take on to do right for our planet has often made it easier to disengage. It's hard to know where to start – or begin again – to care for nature and, even better, make a positive impact to it, but I hope this book can help.

Whether you're a nature nerd struggling with eco-anxiety or a nature novice not yet aware of the issues facing biodiversity, the best first step is to take a physical one into a wild landscape; to understand it, to feel fascinated by it and to cherish it. Once you have renewed that connection with nature – whether it be the mushrooms lining your garden path, the geese at your local pond or a pod of dolphins dancing off the coast – then advocating for it will come much more easily.

Changing your neighbours' hearts and minds is of equal value to a scientific paper that predicts nature's decline. Knowing the facts is great, but what good are they if we don't use them? We have the power to make a difference and communication is our greatest tool. On Rathlin, narrating the nature in front of me to visitors was my day-to-day job, and I soon found it was an incredibly important one. I had a duty to educate others about wildlife, to make visitors fall utterly in love with nature, to encourage them to pay closer attention to the intricacies of the natural world; the little, the large, the soil, the seaweed. Because to protect something with all our might, we must love it.

I hope that this story – not of my life, but of a period of time that changed my life – gives you hope but also highlights the need for action on the nature and climate crisis and empowers you to find your role in this movement.

WILD LIFE

Chapter 1

Arriving

Not all those who wander are lost.

Late April: milky sea mist, blue skies, summer on the tip of everyone's tongues. With my feet planted on the concrete slipway of Ballycastle harbour, I looked across to Rathlin Island. The smell of seaweed made my nose tingle as nerves knotted in my stomach. It was time to board the *Spirit of Rathlin*, the ferry to my new life.

A smiley ferryman guessed I was the new RSPB person and guided me to the level on the boat that was the best for seeing cetaceans and seabirds. 'Right, I'm in a rush, have to be off. Enjoy!' he said. 'Thank you very much,' I called as he left, noticing how embarrassingly English I sounded amongst the surrounding crowd.

Not long after we left Ballycastle harbour, Fairhead started to dominate the horizon and gannets swept past my eyeline. It felt as though the moaning seals at Rue Lighthouse were drawing us closer to Church Bay and the enamel white cottages on Rathlin started to shine in the sun. A forty-five minute, relatively wobbly journey across the sound was alleviated by the scenes that surround me. I clung to my suitcase and clutched my camera as we neared the turbulent waters surrounding the island; so green, ragged and rocky. We docked.

Rachele, my soon-to-be manager and occasional housemate, was standing by the harbour wall frantically waving. I had only met Rachele through a Zoom call and I wanted to make an excellent first impression, so I let the boat empty before me. A deep inhale, exhale, one foot in front of the other – and I had arrived. I had made it to Rathlin. Wow.

Rachele greeted me warmly and led me to our cottage, situated at the heart of Church Bay. After a walk of no more than a minute, we arrived to a garden of overgrown rapeseed that looked as though it had sown itself some time ago. It was clear we were the first inhabitants in a while. The RSPB had recently bought the property to house whoever would be willing to stay for a six-month stint, working as an officer at the West Lighthouse Seabird Reserve. The rippling teal water was so close to the cottage that it reflected glassy light patterns onto the cracked yellowy walls. There was a window shaped like a porthole in the centre of the building – it was a house built like a ship, ready for anything Rathlin's weather could throw its way. Inside there was no furniture, not even a bed. The emptiness felt daunting. It was just Rachele and me, a mattress and a ridiculous amount of tinned food.

I unpacked my leaking toiletries and then gazed out of my new bedroom window in shock. Although I'd known for months that I was coming, the reality smacked me in the face. The length of time ahead of me seemed overwhelming, but it was reassuring to realise that this breathtaking view would be the last thing I'd see at night and the first in the morning.

Given that I was going to be here a while, mostly on my own, I wanted to make my room feel like home. Beside my mattress, I put a whale tail statue that I'd bought in Cornwall just days after seeing my first humpback whale – it's a little good luck charm. There was also a book – Rachel Carson's *Silent Spring* – a sketch pad, and, most importantly, a diary.

Arriving in a new place, most folk would scout it out using maps or books, but I decided to set off and let the island surprise me. I convinced a tired Rachele to come with me just before nightfall. We headed west, to a hill just beyond where we'd docked. Mesmerised by flickering lights on the mainland, miles away, we contemplated the remoteness of our location. 'What have we done?' joked Rachele.

The sun slipped away past the headland so we headed back. At the cottage I made a vat of curry that would last a week. Before going to bed, I collected my thoughts by the kitchen window, taking in the peace of the harbour. My mum texted to see how I was doing as the waves lapped the shore outside. I usually slept badly so had downloaded an app with wave sounds to soothe me to sleep. Now I had the real thing.

The next morning, I woke with the sun. My plan was to leave my cottage at dawn and walk in a straight line to see where the island would take me. Within the space of an hour, I had walked to the far west of the island, passing the Viking settlements I'd read about, hidden caves, and a wildflower meadow rich in buttercups, foxgloves and clover. Cowlease is rare – I'd seen very few fields beaming with life like this before, which isn't surprising – in the UK, we've lost 97 per cent of our wildflower meadows since 1930.

I started documenting what I saw in my diary. Within a few miles I had noted six habitats and sixteen species I'd never seen before. My list included five kinds of insect, one being a rose chafer, a fluorescent green beetle bright enough to blind someone. There were ten new birds, including a greenfinch that I spotted perched on a planter in a wonderfully wild garden. I also saw a donkey, who gave me lots of attention.

I sat down for a well-earned tea break on a headland just beyond the West Light Seabird Centre, my soon-to-be place of work, and heard a stonechat chatting, as if congratulating me on the completion of my first proper hike. Darts of black and white shot across the sky – toing and froing, seabirds were collecting beakfuls of sticks, seaweed and grass to make their nests. For many of them, it would be their first year making their precious nest. Seabirds exhibit strong site fidelity and return to the location they once hatched when they reach sexual maturity, three to five years after they first fledge. It blows my mind that they can find their way home to Rathlin after spending so much time away from it.

A raven interrupted my thoughts with a loud squawk. There was a disturbance on the cliffs below, forcing hundreds of seabirds to rocket from their roosts. Another call echoed above, so loud and so different to the raven's that I soon realised I was in the presence of something much bigger ... white-tailed sea eagles!

At first, I was slightly intimidated since these birds' wingspans are much wider than I am tall – but thankfully I wasn't on their menu. After my heart finished racing, I sat in astonishment for thirty minutes as the pair glided around the sea cliffs ahead. For all the years I'd lived in Dorset I'd hoped to see a sea eagle but I never had. I could hardly believe that early on my first day on Rathlin I'd seen two. With brown body plumage and a conspicuously pale tail, it's like this raptor has been dunked into a pot of white paint. In flight it has broad wings with 'fingered' ends that look like a hand waving. This species was driven to extinction in the UK during the twentieth century – the present population is descended from reintroduced birds. Knowing this, I waved to the sky with immense joy.

White-tailed sea eagles are flexible, opportunistic hunters, often stealing food from other birds. The pair in front of me were attempting to snatch the ravens' catch directly from their claws, hence the almighty display and the disturbance on the cliffs.

If my time on Rathlin was due to play out like my first morning, I was in for a treat.

* * *

Rathlin has a fluctuating population of 150. Many people have a strong family history on the island going back generations, but a large proportion have arrived from elsewhere in search of a different lifestyle. I think the tonic that draws people to the island or makes them stay is the quality of life.

My mum, who is a carer for my younger sister, called to tell me that a friend of hers from a Facebook support group lives on Rathlin. What a coincidence. This friend, Jane, had seen my mum's gushingly proud social media posts about me securing a dream job on an island and had got in touch to say it was the island she lives on. My mum forwarded me Jane's number and told me to reach out, so I did. 'If you ever need anything, don't hesitate to ask; if your friends want to stay, I have spare accommodation to the east; if you run out of food, my husband commutes on and off the island most days, so, just reach out if you need anything! I can be your island mum.'

It was comforting to know I already had people there for me, so close by.

During my first few days on Rathlin, I introduced myself to a woman called Charlie. She has the most beautiful, old red boat called *Family's Pride* that she'd restored herself. Charlie is also the Rathlin harbour master, which I found incredibly badass, given that the conditions are often so challenging.

There were sometimes impromptu music sessions on her boat. The first time I heard the music and edged closer, Charlie handed me a beer and told me to hop on board. Wendy, a local folk singer, was even kind enough to lend me her guitar for a few songs.

Charlie is a member of a dipping group on the island – I met them all that first week and they immediately made me feel so welcome. One of the most outgoing was Kirsty, who had worked at the Seabird Centre when she was younger. Living here she developed a deep love for the island, and for a local fisherman, that led to her staying and raising a family.

Being a lover of an Atlantic swim myself, I spent time scouting out dipping spots. One day, after five minutes of walking, I found Charlie's dog, Meg, and Kirsty's dog, Toby, chilling at Mill Bay, a beach to the left of the harbour. The midday sun was beaming down; for these island dogs, who resembled sleepy seals, life was blissful and free.

Aoife, who moved to Rathlin several years ago, works in the community's Co-op shop and dips almost every day. I met her by Mill Bay. She had clocked my fresh face during that first week and often asked if I had enough supplies, occasionally leaving something yummy on my doorstep.

Halfway through that first week, as I carted flatpack IKEA units between the ferry and my cottage, I bumped into Ali and Liam, who were instantly welcoming. They are like a Rathlin power couple. Liam is a nature-friendly farmer and the RSPB Warden for Rathlin. He grew up on the island and his history here goes back generations. Like me, Ali is from Dorset – a lovely coincidence – and like Kirsty, she had come to Rathlin, worked at the Seabird Centre, fallen in love with the island (and then with Liam), so stayed. It's a trend that has happened multiple times to various generations of islanders.

The island has this magnetic quality, a pull like no other, and I understood this from just a few days here. It's like the instinct that tells seabirds to fly home, calling them hundreds of miles back to the place they belong.

As I settled into life on the island, I got to know Rachele, my boss, better. We became good friends quickly, bonding over art and spending time showing each other our drawings from previous lives. She could only stay a few nights a week as she had to manage another reserve on the mainland – Belfast's Window on Wildlife – so I soon found that I spent most of my time alone. At first, I didn't know if I'd cope, but soon I came to love it. For years I thought the worst thing a person could be is alone; on Rathlin I realised that it wasn't black and white. I may have been apart from other people but I was surrounded by life.

The next Monday came quickly after a week of building my home, and it was finally time to explore the Seabird Centre. Rachele, her partner Eddie and I hopped into the work van and made our way to West Light to prepare the centre for opening. Foot tapping, hands sweating, I couldn't wait to get there. With nothing breaking the horizon until America, it's the best place on the island to watch seabirds nest and feed. Rachele drove us up the path I had hiked the days before – the journey felt a lot shorter on wheels…

The door was covered in cobwebs and the guttering above was dripping. Not the best start, but walking in still felt surreal to me – I'd studied the pictures of the building online, counting down the days until I'd arrive. Eddie helped us carry gift shop stock, COVID safety signs and beverages from the van to the building.

We had a week to prepare for opening and a lot needed sorting. Because of the pandemic, the building hadn't been entered in yonks. Sunbeams highlighted dust particles as they drifted through the stagnant air. The place felt so still.

The centre might have been closed throughout lockdown but the wildlife on the island carried on. In fact, the pandemic was of enormous benefit to nature. It was a conservationists' point proven: that if we treat a crisis like a crisis, we can make change happen. Our systems could do a lot to help wildlife and sustain people, but they choose not to. Capitalism is of course at fault here – but how did we even get to capitalism? A huge factor is that, as a modern species, especially in western cultures, we do not value nature as we should because we are not connected to it, we somehow now feel above it, and most of us are apart from it. I restocked the Seabird Centre gift shop and dreamt of a day when GDP didn't represent the destruction of our planet, a day when all decision makers must have some level of understanding of ecological systems before they are allowed to govern a population. Preventing further declines in nature, even reversing much of the damage we've caused, is still within reach and I believe everyone has a part to play. I resolved then to use the summer to inspire as many people as possible and to connect them with nature – something that I could do to try to avert the crisis.

First we tackled the visitor centre, a small building perched at the top of the cliff overlooking the West Lighthouse. It comprises only a few toilets, a small shop and a refreshment area but I underestimated the work that would need to be done – it felt like the first time in decades that the building had had a deep clean. I was a few minutes into a rummage and soon found a treasure trove of material about Rathlin, from recent seabird handbooks written by previous staff, Ric and Hazel, to pocket-sized walking guides that covered every inch of the island. I took pictures of all the pages from various booklets and maps that I thought would come in handy.

After spending a few hours sorting the visitor centre, Rachele asked me to clean the West Lighthouse and neighbouring seabird platform. Eager to get closer to the birds, I hurried down the eighty-nine steep sea-cliff steps and two paths that lead to the seabird platform. I was met by a symphony of sound as I came to the first bend in the stairway. Then some of the 250,000 seabirds that migrate here every year came into view and the first sea stack appeared. Here, on the edge of the Atlantic Ocean, beside one of the largest seabird colonies in the world, the wildlife was tossing and turning, hustling and bustling. The cliffs were alive with life. Descending the penultimate path to the seabird viewing platform, the cove opened up further and my sensations were overwhelmed. The smell of guano punched me in the face. It was something I would eventually become nose blind to after countless hours on these cliffs, but in that moment it took my breath away.

Readjusting to the wildness surrounding me, standing in awe, I felt my puffy face well up and a tear roll down my cheek. It hadn't been easy deciding to come here, so arriving to this view felt like a sigh of relief. Just a few months earlier, I'd been living in Cornwall with my boyfriend Craig. We loved it and life was good, but suddenly Craig's granny became unwell. Granny Lizzie is everything to Craig – she is the connection to nature that inspired him to become all the things he is today. So we decided to move to Northern Ireland so that he could be with her.

It was hard leaving our Cornish idyll – a beautiful cottage perched at the top of a hill on the outskirts of the quaintest village. We had a horticultural centre close by and just minutes away, down luscious rolling hills, was a world class surf break. Our friends were amazing, there was a fabulous pasty shop and my job working in a seal sanctuary gave me such purpose. I didn't think life could get any better. The end of our time in Cornwall was like wiping out – falling off a board and smacking my head on the sandy floor, having just been riding the cleanest wave. I put Craig's needs before my own. He knew how I felt about leaving, but I tried not to make a fuss. Cornwall lit the fire inside me for conservation, it's like a best friend. Every flowering blue cove is a hug waiting to be embraced, every windswept ledge is a salty kiss, every well-trodden path is a trip down memory lane, every twist and turn is as ingrained in my brain as the gorse is embedded on the steep cliff below. I owe a lot to it.

Before we'd left, Craig explained to me that there would be plenty of opportunities in Ireland, that the wildlife was amazing and that we could start to make our mark on his family farm since he would be inheriting it at some stage. At that point in time, even though he was trying to be kind, I could feel nothing but loss for the home we had made for ourselves.

Now, facing hundreds of miles of sheer Atlantic wilderness, I was already glad we'd made the decision. It was as if someone had placed me in the middle of the climax shot of a wildlife documentary. Senses readjusted, clinging to the safety bars that lined the top of the sheer seabird platform, I set out to identify everything and anything I could. Cleaning the centre ready for reopening was the last thing on my mind.

A magical mayhem: fulmars, everywhere, circling the bay with a stiff upper wing, asserting their dominance. Kittiwakes, squawking at neighbours, competing for the best patch and partner on the cliff. Razorbills, high above my head, with razor-sharp white eyeliner that stands out from a mile away. Guillemots, so exposed but with safety in numbers – there were so many of them, one-hundred thousand to be exact, lining every square inch of the stacks. Adrenaline coursing through my blood, I stood alone on the edge of this island, the wildest place I'd ever been.

I realised that I hadn't yet seen the bird that Rathlin is best known for. Using my grandad's old binoculars, I scanned from side to side before realising the puffins were hiding in plain sight. Everything else was happening above but the puffins were below. Lining the grassy floor there were tons of burrows, and beside every burrow, PUFFINS! I saw more appear, and pretty soon, I could see hundreds. This was brilliant – but I knew that there should have been many more. Puffins have declined massively in recent decades and if trends carry on the way they are, we could lose 89 per cent of them by 2050.

Like everything in nature, seabirds have a vital role to play in a functioning, healthy ecosystem. When seabirds' poo, guano, washes into the ocean, it acts like a fertiliser, stimulating the growth of phytoplankton, the organism that produces over half the oxygen we breathe. Even this species' poo, something that most of us wouldn't know to be important, is vital for our survival.

I watched as the puffins waddled and tumbled around. Clowns of the sea and collectively called a circus, these brightly billed beauties brought another tear to my eyes. They have patterns of life, millions of years old, ingrained into their psyche. Their ancestors were from here, they were born here so they return here, and so will their young. They have thrived on land, air and sea for centuries but now humans are messing things up. Extreme weather patterns, depleted fish stocks, pollution, disease and invasive species all affect their ability to live in the way they have for centuries.

A puffin flew up the cliff and landed by my side. I noticed that it had a distinguishable scratch on the left side of its beak.

As I studied it, I thought about how hard it had been for me to leave home and come here – but that's exactly what this puffin had done, with a million more obstacles to face than me. Side by side with this creature who was no more than twenty-five centimetres tall, already emotional, I knelt down and whispered, 'I'm trying to help' to the intrigued little bird. I watched it launch off the platform as it free-fell fell to a vertical world, before catching wind under its wing to circle the lighthouse.

I felt compelled to follow. Fiddling with old, clattering keys to unlock the lighthouse door, I ran down hundreds of steps that would take me to a lower level of the cliff – closer to the puffin. The Rathlin West Lighthouse is famously upside down and it didn't take me long to reach the light room on the ground floor, which glows a fluorescent red colour. A big wooden door confronted me. When I opened it I saw a view so vast, I'd never felt more exposed.

Closer to the sea stacks, everything was louder and smellier. The blue of the ocean below was more intense, and the wind was so wild it drew sea spray up the entire cliff face, slamming it into the hundred-year-old lighthouse. Getting drenched with salty ocean water, I felt like I was back on the wildlife boats I used to work on. The horizon was bigger and more daunting down here, and I'd never felt more joyfully insignificant. Leaving the safety of the lighthouse, I too had launched into the puffins' vertical world, with towering cliffs surrounding me.

I found the puffin with a scratch on its face on a grassy section of cliff. With a tilted head and eyes so curious, it hopped down a slope, and of course, I followed. I crawled across the top of a grassy bank to keep it in view but halted after a few seconds – there was a sixty-foot drop and I felt weak with vertigo. I saw the puffin's burrow just feet away, and then the puffin's partner emerged. I watched as a curious courtship played out – the birds tapped beaks to reaffirm their bond. This was spectacular. Using no imagination, I named them Puff and Finn, Puff being the adventurous bird with the chipped bill that I'd first met. Swapping roles – they are very progressive birds – Puff waddled to her burrow while Finn swooped down to the sea below to start fishing.

In that moment, I noticed another puffin, waddling about with a huge catch of sand eels, sharing the same ledge. It's unusual to see a puffin on its own and without a burrow, and I watched this bird for another hour or so as it flew back and forth to sea. Every time it came back to land, it sat among an open stretch of sea pinks gobbling up its catch, keeping it all to itself. With its peculiar behaviour, this third puffin stood out to me, so I named it Busy Lizzie. Intrigued to find out more about her, I knew I'd be venturing here often. A noise chimed in the lighthouse, knocking me out of my wildlife trance.

The public weren't allowed to access this part of the reserve when I was on Rathlin – there had been maintenance work carried out on the lighthouse and it was a bit of a tip. Worried

I'd get caught there evading my cleaning responsibilities, I hurried back up the hundreds of steps of the lighthouse and the further eighty-nine steps to the visitor centre. I felt a ball of exhausted excitement in my chest when I got back. I was beginning to connect to this place, all these miles away from any previous normality. Watching the puffins was like looking through a window into another wild life.

Rachele asked me what I thought of the reserve and I stumbled over my words. 'Incredible, astounding, smelly – there really isn't any word that does it justice,' I replied. She found my stunned face funny, and said we had to go since there was some parcels that needed collecting from the harbour. I looked behind me as we left the Seabird Centre car park – the dark blue sea turned to green hills, then the hills turned to blue loughs, the loughs turned to green bogs and the bogs turned to the blues of Church Bay.

When I got home, I opened my camera roll and brought up the photographs that I'd taken from the guides earlier in the day. Dinner eaten, plates washed, I shot out of the house to find a place every guide had directed me towards – Rue Point, on the south-east tip of the island. Heading into the unknown felt scary but so invigorating and I tried to trust my intuition. The anxious teen who often worried about leaving the house was nowhere to be seen. This state of self-acceptance came with time and it's still not perfect. But there are many things I want to tell young girls growing up now, lessons I wish I knew sooner – like how exercise should make you stronger, not skinnier; self-worth is not defined by the way you look; and lips are much more beautiful belly-laughing than pouting.

I care about this so much because my younger sister – part of an online generation that is drowning in beauty standards – nearly died from anorexia. Chances are the way you look naturally will come in and out of fashion during your lifetime. Beauty standards change but you should never have to. I was teased for my curly hair and thick eyebrows in school, but now these features are trendy. Walking to Rue reminded me of drag star RuPaul, who said, 'If you can't love yourself, how the hell you gonna love somebody else?' That also applies to nature – if you can't look after yourself, how are you going to allow yourself the capacity to care for anything. We have to maintain our physical and mental health to have the most positive impact on the world around us. We need to feel joy and being close to nature gives you that. When I'm outside the birds don't judge my bare face, nor do the seals hate my clothes. I exist, so do they, and that's all that matters.

The path to Rue was twisty and hilly. Just as I thought I'd arrived, another slope appeared. I'd stopped for a break behind Ushet Lough when I saw my first golden hare. A glimmer of gold shot across a mottled landscape, faster than lightning. I'd brought my camera, but the hare moved far too fast for me to capture it. The golden hare is unique to Rathlin. It has a genetic variation, like human red hair, which gives it piercing blue eyes with a bright blonde coat. It's funny that I saw my first golden hare so early during my time on Rathlin. I never saw one again, and in that moment, didn't realise just how rare they are. I resolved to be quicker with my camera next time, annoyed that I'd missed my opportunity to snap the white-tailed sea eagles and now the golden hare.

Approaching yet another incline, with the sound of seals getting louder, I thought I must be getting close to the summit. When I finally got to the top of the hill at Roonivoolin that overlooked the east side of the island, my mind was blown – Scotland was so close, and the Mull of Kintyre and Islay looked stunning lit up in the evening glow. The crumbled ruins of cottages lining Rue's coast, the seals' moans and the waders' calls combined to create a creepy atmosphere. I could see why there are so many folk tales from this part of the world. One of the information booklets I'd read earlier that day had explained how people often mistook the curlew's call for a banshee's scream. I was not surprised. A group of curlews is called a curfew, but this group were not in any rush to quiet down for the night. Famously a poet's muse, inspiring Vernon Watkins and W.B. Yeats, the Eurasian curlew is recognisable by its long legs and distinctive down-curved bill. The UK's breeding population of Eurasian curlews is of national importance, estimated to represent over 30 percent of the west European population, yet we've lost 65 per cent since 1970. Changing farm practices have meant they are now Britain and Ireland's highest conservation priority bird species. I sat in awe.

Peering through swards of long grass, I watched on, feeling like a spy. The tide was coming in and so were the common seals as they hauled out after hunting for fish. In trying to find a good patch to rest, they got too close to the oystercatchers' nests. The more the seals intruded, the more the oystercatchers objected, their calls clear even over the crashing waves.

Rathlin was the first place I'd lived with a population of common seals – which have unfortunately become a lot less common. When I'd worked at the seal sanctuary, most of our patients were grey seals. The way I distinguished between the two species was remembering that the greys look like dogs and the commons like cats. I find the commons extraordinarily cute, often holding their body in an upright, curved banana shape. The UK lost half its common seal population a few decades ago due to the phocine distemper virus. It is now listed as a priority species but is still at risk from threats such as disease, pollution and overfishing. The greys are feistier and harder to rescue, wriggling and biting any chance they get, while the commons are so docile that I could carry them in a towel like a baby.

After hours on the plains of low tideline, the curlews were full and ready for rest, but sheep ran up and down the valley, ruining everyone's quiet night. The sunset painted the island a deeper red, baking the cliffs and decorating the sea in dappled warm light. I found huge peace sitting here. As a human animal I wondered what my place was in this world and how an ancient me would have lived.

Nature is the one constant in life, maybe that's why I take so much comfort in it. The tide receding, sun rising, moon waning – these patterns keep me connected. The leaves growing, flowering, falling, rotting – these cycles keep me grounded. The birds calling at dawn and foxes hunting by dusk – these sights give form to the days, months and seasons.

Sitting in this wild landscape, I recognised myself as part of it.

Sláinte Rathlin.

Chapter 2

Landing

Home isn't where you're from, it's where you belong.

The next morning, I was woken by a buzzing in my pocket. Tired from all the walking the previous day, I'd gone to bed in my clothes without brushing my teeth or washing my face. It was Craig calling, and so early in the morning – what could he need at 7 a.m.? I could've guessed it wouldn't be good news. It tuned out that his granny had fallen in the night and had needed to go to hospital for a check-up. Because of COVID restrictions Craig couldn't go with her. His voice wobbled as he told me what happened. I felt awful not to be there with him.

It was hard to comfort Craig – his granny is one of the people in his family that he's closest to, and I felt upset about what had happened myself. Following previous scares like this, I knew that I shouldn't try to fix things as I often attempt to do, or distract him, or tell him that everything would be fine. I felt doing these things would deny Craig the opportunity to face his emotions. Instead, I asked him, 'What would make Lizzie happy?' 'Collecting a vase of flowers,' he replied. 'Well, go and pick a bunch of the most beautiful, fragrant flowers you can find, ready for when she's home and leave it beside her armchair.' 'That's a lovely idea,' he replied.

After the call ended I sat in a pool of worry. My memories of last night's incredible sunset hike at Rue were dampened by the news; I felt awful but so far away and as though there was so little I could do to help. A chap sang in his garden as he watered the wildflowers that surrounded his house and provided some much-needed nectar for the island pollinators. His singing brightened my mood a little and prompted me to get on with my day, but Craig and his granny were never far from my mind.

I walked the ten steps from my door to the sea and splashed the salty solution to my face, feeling lucky to have the ocean on my doorstep. I was due at the centre for 9 a.m. to carry on finishing preparations for the grand reopening. After several postponements due to the ever-changing COVID restrictions, the centre was scheduled to reopen in late May, which wasn't far away. I spent hours mopping and cleaning, all the time mentally running through bird facts that I could tell future visitors.

When I wasn't at work, all the daily domestic tasks that used to feel laborious were renewed with excitement – I was hanging clothes out to dry to the sound of sand martins and washing dishes with a view of sea cliffs. The mundane was now marvellous.

I felt my new island life slowly knitting together. Days of endless sorting and cleaning had flown by and it was time to welcome some more seasonal staff and get ready to open West Light to the public.

I was eating breakfast the morning that the new staff were due to arrive when my phone lit up with an incoming call from Craig. Apart from the odd text to make sure Craig was ok, I'd not spoken to him properly since he told me about Lizzie's fall. I let it ring a few times, preparing what I'd say if it was bad news. To my surprise, he told me that Lizzie was now home and doing well, even gardening again like the determined woman she was. The prognosis wasn't good though. 'Sepsis,' he said, a new addition to an already long list of ailments. 'And while she's stable now, the doctors said it will get worse.' 'Right, well, are you okay? You sound calm considering ...' 'I'm enjoying our time together. Lizzie is old, she will pass sometime soon, I know that now, and I've cried enough. I want to enjoy what time we have left together.'

I felt reassured to hear Craig sounding happier than he had during our last conversation, having already made peace with the prognosis – but the news hit me like a bullet to the chest. It was the first time that a doctor had said she wasn't going to get better, just worse. I tried to stay strong for Craig but my voice wobbled as I said goodbye.

Work was a good distraction, and I was excited to see the new faces arrive at the centre. The team was a diverse mix: we were all from different places and had different stories. Caitlin, who was a similar age to me, came from a nearby coastal town called Portstewart and Benedict, a student originally from the island, who'd come back to work on Rathlin for a month of the summer. I then met David, a young lad from Ballymena, Eleanor, a Scot living in Ballycastle, and Jean, a local who probably knew Rathlin better than any of us. We sat down for an introductory session, getting to know each other better before learning about the conservation aims of the RSPB from Liam, the warden. The rest of the team left on the 5 p.m. ferry every day. I felt so lucky that I got to both live and work here.

Finally came opening day. Rachele and I took the small blue RSPB van down to the harbour to collect our colleagues. We were all so excited to welcome people to Rathlin and give them a brilliant wildlife experience they wouldn't forget.

We were all comparing weird and wonderful nature facts on the drive to the centre but were soon ground to a halt. Hundreds of cows lined the small, windy road ahead, blocking the only way in. No amount of honking moved them. Time was of the essence, so I got out of the van and tried to move them by waving and clapping, which was scary and funny all at once. The cows eventually cleared from the road and we all thought the worst was over until we arrived at the centre to find the car park covered in cow pats. Rachele and I had accidentally left the gates leading to the Seabird Centre car park wide open when we'd headed home the night before. It had all looked so beautiful and clean when we left – how could we have been so silly?

Each of us grabbed a shovel and, by the time the first Puffin bus pulled up, carrying tonnes of tourists, the car park was clear. I joked that it was the best team building activity we could possibly have done. I welcomed crowds of people as I wiped away the sweat and poo from my forehead. The day had only just begun and I was already knackered.

The Puffin bus is a Rathlin institution, bringing visitors from the harbour up to the centre. The driver Francis is well known for wearing a different flamboyant shirt every day. He commutes from the mainland to Rathlin so would bring a supply of goodies to the Seabird Centre for us all. This bag consisted of tray bakes, biscuits, cupcakes and sometimes, to add a touch of health, the odd banana. I don't eat dairy so couldn't eat most of these treats, except for the bananas and some oatie shortbreads, which didn't have dairy in them.

Visitors charged in through the doors, out through the next set and down towards the birds. Some were regulars, who couldn't wait to see the colony again after two years of reserve closure, but for most, it was their first visit. I had been mentally preparing to meet crowds of knowledgeable bird nerds, but was surprised to find that most people hadn't taken any previous interest. I spoke to crowds of people about our conservation work, why the wildlife here is so under threat and what they could do in their own lives to help the cause. I watched as people fell in love with the beauty around them and took comfort in seeing these connections form. At intervals throughout the day, I pointed down towards the puffins, showing them off like I was some sort of proud mother. 'Down there you can find Puff, Finn and Busy Lizzie!' My colleagues looked over, surprised that I had already named three birds. 'Is this what living on an island does to you?' Caitlin asked.

The crowds gasped with excitement as Puff circled above our heads, launching off the lower platform to greet everyone. What a performer, I thought. Caitlin was stunned – it was the first puffin she'd seen relatively close-up here. Every hour, a new busload of visitors arrived, and the previous group returned to the harbour. During these changeovers, I had ten precious minutes to rest, before narrating the seabird scenes again to a new group of people. Although I had a rough script to stick to, something new and unexpected happened that I could commentate on every hour. It was hard to keep up with the colony's antics.

As the day drew to a close, I realised I hadn't eaten, and my voice had nearly disappeared. I'd thrown myself into the role, making sure I spoke to everyone, but I was suffering for it. I needed to pace myself better.

When Francis was picking up the last visitors, he said he'd overheard conversations on the bus journeys back to the harbour and that everyone had enjoyed themselves. He told me that a man had mentioned my name, saying he'd be sure to come back again. I was delighted. Maybe losing my voice had paid off.

That night, armed with a cossie, skateboard and some paints, I headed back to Mill Bay for a swim. When I got out, I started to sketch the seabird species I'd met on the cliffs, with the idea of making an interactive sighting leaflet for kids, or anyone who'd use it! I'll call it The Big Five tick sheet, I thought.

Skating back, I came across a young girl in a school uniform, zooming up and down the only smooth road on the island. She was enjoying an evening skate too. We had a chat about boards and skating, and she had a go on mine. 'Wow, it's so fun,' she said, picking up speed, before nearly running over a cat mid swivel. 'That's our pet cat Marmalade,' she said. I promised to watch out for her when I was skating next. After ten minutes of chatting I headed back to the cottage.

I could faintly hear a tin whistle from the direction of the harbour as I got closer, and a silky shiver ran through my body. The air was so still, and a gentle mist started to roll in. The noise, maritime and primal, was like a call from the sea. I didn't want to miss a minute so I sat outside to finish my Big Five leaflet, sketching into the night. As my eyes grew weary, I heard a gentle purr, purr at my side. Marmalade had found her way to my cottage and started begging for cuddles. I've had pets all my life and the cottage felt lonely without one, so I was very tempted to take her inside with me, but I walked her back to where she belonged and she scurried away.

At the centre the next day, before any of my colleagues arrived, I scanned my artwork for the Big Five tick sheet and printed hundreds of copies. Rachele loved the idea and laid them out on the entrance desk ready for people to take. The beautiful morning had brought hundreds of people to the centre. A young girl called Edith, in purple dungarees and with plaited hair, was the first to walk through the door and pick up a sheet.

As I guided Edith and her mum around the reserve, Edith asked if she could use my fancier, older-looking binoculars: 'Yours look better than mine.' She wasn't wrong. I handed over my grandad's precious possession, and Edith smiled as she gazed through the finder towards a flock of fulmars. 'Yes, yours are much, much better,' she proclaimed. A budding young naturalist, Edith was clearly so excited to be here and her mum looked so proud.

I left Edith to welcome others. The sheets worked well: once people found the species, they could tick it off, turning wildlife watching into a fun and competitive game of 'who can spot it first?' Edith tugged my fleece, interrupting a highly intellectual conversation I was having with an old man about nest building, and returned my binoculars.

The day was braw but breezy – I shivered so added another layer. Later that morning, looking through my camera, I could see some guillemots laying eggs. What a moment! I didn't want any of the visitors to miss this, so showed them where to look. Apart from the kittiwakes, the seabirds found on Rathlin lay just one precious egg each year, and have to be on guard constantly to protect it from strong winds and unpredictable predators.

I walked back and forth along the platform edge, making sure everyone had binoculars and a pencil – and then I felt a crunch under my foot. It was a razorbill egg that had fallen from the nests above the platform. It wasn't too broken, so I decided to keep the specimen in my pocket to show to visitors. A bit later I noticed another egg lying broken on the floor, this time a guillemot's. When I looked more closely, I was amazed by the bright vibrant blues and the brown speckles. This egg has adapted over time so that it's less likely to fall from a cliff nest – it's shaped like a cone and rolls a full 360-degree circle when it's nudged. Given these adaptations, the egg seemed unlikely to have fallen, so I wondered if it might have been stolen by a raptor and discarded.

The bright afternoon rolled past and eventually the centre fell quiet. I stayed into the later part of the afternoon while everyone else left in good time for the last ferry back to the mainland. Just as I started to slow down, the seabird stacks livened up. Predators knew egg-laying had begun – they could smell broken shells from miles away. From out of the corner of my eye, I glimpsed a peregrine. I'd never seen one before so ran to the platform edge with my camera, and watched as it caused chaos and commotion among the five colonies.

Peregrine falcons are the fastest animal in the world, capable of reaching 200 miles per hour, and are the most agile predator on the cliffs, over any raven, black backed gull or skua that roams here. Rather adaptable birds, they can be found outside what an ecologist would class as their 'natural environment'. As long as they have prey nearby, and a high vantage point where they can nest, they thrive. In fact, the Empire State Building in New York was the highest place on earth they've been found to nest. Adaptable as they are, they face many threats: persecution, food decline and fewer available nesting sites, to name a few.

I was thrilled to see such a magnificent creature soar

around the cove, but felt worried for the puffins I had been watching over. They were all at risk, especially Busy Lizzie, who wandered around so openly. She would have been easy pickings for the peregrine.

The hunt played out in front of my eyes with a razorbill getting snatched from its nest. It seemed the peregrine would take anything it wanted with its razor-sharp claws and swooping dives. I leant over the platform railing and saw – to my relief – Puff and Finn in their burrow, but Busy Lizzie stood out on her usual patch of grass, exposed. I was beside myself and bit my nails until the peregrine left. I understood why the peregrine needed to hunt, but also felt for the seabirds for the fright they'd endured. I saw the dark and light in the situation. Everyone needed to eat, but it did not seem fair to the puffins, already so endangered and outnumbered ... but the peregrines were too. All the species in this scenario are declining as a consequence of human behaviour. Every death in this cove felt unfair, but then every death always does.

As I glanced down towards Busy Lizzie, carrying on as though nothing had happened, I thought about Craig's granny, who cracked on with life in her usual calm way despite the reality of death being so close by. The farm back on the mainland had its own, metaphorical peregrine circling above but – like the puffins – Lizzie's only choice was to carry on.

The activity on the cliffs was more dramatic that an episode of Eastenders. I felt emotionally exhausted and knew it was time for some rest.

* * *

Before I came to Rathlin I would lie in bed tired but wired, overthinking, and had been diagnosed with insomnia. But here

I slept like a baby, leaping out of bed every morning with an energy I'd never felt before. Although I have always loved the freeing feeling of rising early, mornings used to seem a chore and the days slipped past me so quickly. Night, morning, day, night. But it was different now and I slept better too – there was something almost medicinal about the salty air. My body was exhausted from daily adventures and my racing mind was hushed to sleep by the elements outside.

I didn't know what the next day would hold, what I'd see, who I'd meet, the wildlife I'd encounter. For the first time in my adult life, it was all unpredictable. Apart from work, there was no pressure to be anywhere or do anything, which is why … I wanted to do everything! My routine evolved around seeing and experiencing as much as I could, not working myself into the ground. Mind inspired, body re-energised, I spent time walking, swimming and drawing before getting ready for work. Collecting any extra staff from the harbour in the rickety blue van, I'd then head straight to the dramatic cliffs at West Light to find my birds at 9 a.m.

One morning, the first real warmth of the season shone through. Mist rose off the cold sea and black guillemots, otherwise known as tysies, lined the harbour wall. This species of guillemot is different from the common guillemot. Black guillemots are like cool, emo cousins that chill to the side of the drama, keeping themselves to themselves. Unlike the common guillemots – whose nesting territories are the smallest of any UK bird, extending only a beak's-length around their nest – black guillemot couples occupy much larger nesting areas, often sheltering in harbours.

Most of the Seabird Centre team enjoyed a catch-up in the office before we opened the doors to visitors, but I used this moment of quiet to check on the puffins at the lower lighthouse platform on my own, still aware I wasn't really meant to be down there. On cue, as I opened the lighthouse doors, Busy Lizzie flew down to her grassy verge with a mouthful of sand eels. Over recent days a blanket of white sea campion had carpeted the cliffs. This otherworldly place felt like a dreamscape. Puff and Finn were also there, standing at the mouth of their burrow looking like they were debating who would go fishing next. Making sure I wasn't disturbing them and their natural behaviours, I sat behind a mound of earth that concealed me but allowed me to see all the action. Sipping my coffee contentedly, I watched them closely.

I felt a strong connection to each of the three birds. Gazing down at Finn fiddling with sticks, Lizzie staring out to sea, and Puff prancing around the lower lighthouse ledge, I soon realised why. Puff, an adventurous little bird, reminded me of myself; Finn, an efficient nest guarder and caring partner, of Craig; and of course, Busy Lizzie, of Lizzie, Craig's granny. Watching the puffins' unique behaviours unfold hour by hour, day by day, I became glued to this patch of cliff, and found huge joy in telling visitors all about my new friends.

After my first month on Rathlin I'd run out of food.

Lesson number one already learnt: pack more food than you think you'll need! I also had to go back to the mainland to collect my car and drive it onto the island. In the next week, we would be welcoming some residential volunteers, so we needed another vehicle to transport the team up to the centre every day.

The night before I left, I listed everything I was going to buy. Staples such as rice, pasta and cereals came first, then tins, followed by fresh veg, frozen food and snacks. If I needed to head back to the mainland at any stage, I had to book my ferry ticket at least two days in advance of leaving ... so preparation was key.

There are some off-grid systems that greatly benefit the residents of Rathlin, though. For example, the Co-op shop stocks very basic essentials and homemade goods. In emergencies, islanders can be airlifted to the nearest hospital meaning anyone calling 999 from Rathlin could get a much quicker response than they would on the mainland. The 150 islanders also have their own fire service, coastguard and GP surgery!

I will never forget the feeling of returning to Ballycastle after moving to Rathlin. Weeks had felt like months – time goes so wonderfully slowly with little distraction. I didn't realise how much I'd adjusted to the island's slow pace of life until I left. Everything felt louder, crazier and more hectic – even though I'd been at the seabird platform in peak season!

Craig was there to pick me up from the ferry port. I was so excited to see him, especially after everything he'd been through these past weeks. We hugged tightly. He asked me how my first few weeks on Rathlin had been. Just as I'd found it hard to explain to Rachele how I'd felt after visiting the seabird platform for the first time, I couldn't find the words to explain how I felt about island life to Craig, but he could tell how happy I was from my beaming smile.

Apart from my mum, Craig knows me better than anyone – he's been there for me through thick and thin. He is also a passionate naturalist and conservationist, so we share many of the same values. We met at a training weekend in Scotland, doing work for an environmental charity on the Isle of Cumbrae, which is tucked away behind the offshore Scottish islands that face Rathlin. That night there was a music group playing traditional Scottish music and teaching us to dance, called the Céilidh Minogues! Like a couple in an old-fashioned fairy tale, we were instantly besotted by each other, dancing after a few too many pints of Guinness, and the rest is history.

Our relationship has stood some tests, including long-distancing for over a year before we moved to Cornwall together. Before moving to Ireland, I'd spent a month in Dorset, sorting out my possessions, coming to terms with this big upheaval. It was nice to have some quality time in my cradlehood home, catching up with friends and family.

After the sadness of leaving Cornwall had subsided, I started applying for new jobs, to ensure I could carry on living a life full of purpose, and soon came across the advertisement for the 'Visitor Experience Officer' job at Rathlin West Light Seabird Centre. I knew instantly it was what I wanted to do. I was so stubborn I didn't even apply for anything else. Weeks passed with no news and I began to feel slightly silly, but the week I left Dorset and set sail for Ireland, I found out that I had been successful in securing the job. I'd felt bad that I was moving to Craig's home to then leave him and live on an island, but we both knew it was what I needed. A new adventure called.

As we drove back to the farm from Ballycastle, I asked after Lizzie. 'She's just fine. I mean, you can tell she's worse, but she's a tough strong lady who will never stop.' That was true. When we arrived back to the farm after collecting some food from the shops, we could see Lizzie out in her garden, tending to the begonias and daisies. Craig and I went over to help and sure enough, still the queen of Ballyconnelly Farm, she had plenty of jobs for us to be getting on with. Just like the flower (and now the puffin), Craig's granny is known as Busy Lizzie and you can see why.

We planted agapanthus, worked on the veg bed and listened to garden birds tweeting over the rustling trees Lizzie had planted herself at least fifty years ago. We identified the birdsong – 'The starlings!' Lizzie said. She was happy to hear them in such numbers, with their electrifyingly techno calls. Next Craig heard a goldcrest – its light drumming sounds as if it is playing the cymbals. This bird is a rockstar in its own right and the lightning yellow crest on its head matches the sound it makes. Then I heard the drumming of a woodpecker, a 'teacher, teacher' of a great tit and a melodic song from a robin. I joked that we had our own five-piece band playing to us while we worked. 'Listen! We have an electric piano, drums, a tambourine, and a lead singer – what a symphony, and for free.' With a quarter of Northern Ireland's garden birds at risk of extinction, this place felt like a much-needed haven for them. The sound was hundreds of decibels lower than the seabird calls at West Light, and with a slightly prettier tone – I enjoyed the soothing waves of song humming in.

It was getting dark. The stars began to glow and the bats circled around the trees above Lizzie's bungalow before roosting for the night along the canopy of ancient oaks, pines, alders, ash and sycamore. The farm is perched high on 'Long Mountain Ridge' and looks across half of Northern Ireland, with breathtaking views of the Glens of Antrim, Slemish, the Sperrins and even the Mournes on a clear day. Pretty as it is, this view shows just how nature-depleted the landscape has become. What many people see as rolling green fields, I see as ecological deserts, devoid of any wildlife and with very few trees. Across the island of Ireland, only 8 per cent of the land area is home to woodland, and of that only 2 per cent is native – the worst cover in Europe. Of the land in Northern Ireland, 75 per cent is farmed in some way, predominantly for meat and dairy, which has caused huge declines in biodiversity and an unprecedented rise in greenhouse gas emissions.

Craig and I talked about how we could change the ways we eat and farm, as we often do, feeling inspired to create a change, but still unsure what this would be. The sun dipped below the farm's highest hill and sunk us into shadows of dark blue. The birds we had just been listening to dispersed to their nests for the night. Feeling almost as if I had just landed back to my own nest, I felt content that this place, so full of nature, was here for me.

With hands washed, the fire lit and veggies picked, Lizzie and I got comfy in her lounge. I showed her photos of some the wildlife I'd seen on Rathlin, telling her about some of the most exciting encounters. Lizzie had gone there herself when she younger, before there were proper ferries, just little boats big enough to take a family and a cow! A wildlife lover, she visited the seabirds and enjoyed the spectacular views of Church Bay, where I now lived. Sharing this moment was special as I didn't know how many more times I'd get to have one like it. It reminded me to cherish everyone in my life.

As caring and considerate as any Irish granny, Lizzie asked me if I was eating enough. I admitted that I had run out of food, and she insisted that I should take any veggies that I wanted from her patch. A true farmer, she asked me how the weather was on the island and if I was keeping warm enough. Despite being adamant that I was nice and toasty in my RSPB gilet, Lizzie handed me a scarf that she had recently knitted. It was beautiful with a rainbow pattern and would ensure I never got lost on any hike. I could see she was tired so we decided to call it a night.

Before I left the room, I looked back one last time to see Lizzie happily sitting in her armchair by the flickering fire.

With all the awfulness that goes on in this world, it felt comforting to see an old lady live out her days in the place she loved, by the fire, after an afternoon of gardening, now knitting something new. Although I'd only known her the few years I'd been with Craig, I found Lizzie an amazingly inspiring person.

Lizzie met Craig's grandad Jimmy in the early 1960s and moved to his farm, which had already been in the family for hundreds of years. Jimmy tragically died from sudden death in 1989, leaving Lizzie to take on the entire farm. A custodian and warden to Ballyconnelly Farm almost her whole life, she had seen the seasons come and go, probably over sixty times. I couldn't imagine how connected to the living world around her she was. Growing millions of potatoes, seeing thousands of sunrises and counting hundreds of birds here in that time, I wonder if this nature that surrounded her helped through the grief of losing a husband.

Seven of Lizzie and Jimmy's eight children went their separate ways, leaving the youngest, Alan – Craig's dad – to take on the farm. As Lizzie got older, Alan carried on working the land for years by her side. Through this hardship, Lizzie also lived through the Troubles. She was implacably opposed to the conflict and the hate it brought – she just wanted peace and togetherness.

When I looked at her, I felt a shiver up my spine for the influence she'd had in this world. She made the farm what it is today, brought Alan up who then made Craig, and she never once lost sight of the important simplicities of this world: nature, homegrown food, flowers, wildlife and ... knitting. 'Goodnight, see you soon, I'm off again in the morning,' I said gently. 'Night, night,' said Lizzie.

Craig was very touched that she'd given me her rainbow scarf. 'You better keep that safe,' he told me. I promised I would. Full of the mashed potatoes and veggie sausages we'd had for dinner, I lay in a food coma beside Craig. 'I wonder what our home will look like one day,' I said. It was then that Craig told me about the plan he'd been making. He wanted to convert a caravan into a permanent home for us both, here on the farm. I was so excited.

I rolled over and wrapped my arms around Craig, knowing I was leaving bright and early the next day. Sensing that I felt guilty to be leaving him and his beloved granny, he told me, 'Go live.' So I decided that I would, for all of us.

The next morning I headed back to the ferry port – I was taking over my little Nissan Micra for the first time. Although he was busy helping to care for Lizzie, Craig said he'd visit Rathlin whenever he possibly could. I'd packed my life into my tiny car as well as enough food to get me through an apocalypse. There was a mountain bike, which was split into two sections to fit in my boot, my wetsuit, flippers and other swimming gear. I crammed in some furniture and bedding to make the house feel a little less empty.

Reversing my car onto the ferry was hilarious – the boat wiggled from side to side, and I kept losing my bearings. The ferrymen attempted to guide me in the right direction, no doubt remarking to each other that I needed more practice.

As the ferry set off, I got excited about heading back to my island cottage, to carry on living my new life, exploring more, and getting closer to the birds. Just five minutes into the journey, Tom McDonnell, who is both a member of the ferry crew and a wildlife photographer, rushed over to me and said, 'Dolphins!' Running up the echoey stairs to the top deck, I looked over the side and saw over twenty bottlenoses bow-riding the ferry. They stayed and played in the boat's wake for the entirety of the journey. I was close enough to hear their playful clicks and some of them started to breach so high that I felt they were looking at me. What a welcome home.

During the first few days of early June, I spent my early mornings sorting the new things I'd carted over; my days on the cliffs obsessing over eggs; and my evenings walking new sections of coast path that I hadn't explored before. Every day I ate my lunch at the lighthouse's lower platform, always watching out for Puff, Finn and Busy Lizzie. Eating sand eels or wandering around on the open grass, Busy Lizzie was always doing something. Her mannerisms became so familiar. One day as I sat there alone, the horizon looked particularly huge; it made me feel small, and so far away from my family back home. When I lived in Cornwall, I could drive back to my mum inside two hours. Feeling a wave of homesickness, I called her. Somehow her comforting words, even though they came through the phone and from hundreds of miles away, had the same effect on me as being hugged.

My mum, Sarah Ward, is the strongest person I know and she's also the kindest. It takes a particular kind of person to work for a homeless charity for thirty years, and then leave employment to become a carer when her daughter's illness got bad. When my sisters and I were children, my mum would buy us hot chocolates to take to the homeless men and women she worked with. She encouraged a conversation, then, as we left, explained their stories, teaching us from a young age that no matter what happens in life, what someone may look like, or where they are from, we all deserve to be respected equally. 'Any of us could end up needing help one day' are words imprinted on my brain. My values were instilled early: treat others as you wish to be treated, and always be compassionate.

These values stretch out beyond human boundaries – I aim to treat all species with the same respect I would wish to receive. My admiration and appreciation of nature comes in large part from the way I was brought up. If I can be anywhere near as remarkable as my mum in my life, I will be happy. And that's what pushes me on every day – doing crazy things like moving to Rathlin are down to the strength and resilience I learnt from her. I truly feel like I can do anything because of her, and I know she got that strength from her dad, Brian. I never got to meet him but I feel his spirit somehow living through my own. They both loved the lyrical genius of Simon & Garfunkel. There's a part of their song 'Bridge Over Troubled Water' that's been passed down from my grandad to my mum, and it strikes a chord inside my soul each time I undertake something that scares me:

'Sail on silver girl, sail on by,
your time has come to shine,
all your dreams are on their way,
see how they shine,
if you need a friend,
I'm sailing right behind.'

And my mum is: through my phone, across the sea, in spirit, always.

I left the seabirds and went back to work, feeling slightly lighter than I was before. When I got home that evening, I opened my camera roll to look at some family photos, attempting to feel closer to everyone at home in Dorset. As I sat in my muddy front garden, smiling back at pictures of little me covered head to toe in dirt, I realised that the trick to being happy is to be a child. The curiosity, the not caring about what others think, the imagination that we are born with … it's all what we strive to crawl back to after the awkward stages of adolescence, when society teaches us to lose ourselves for the benefit of a system.

I kept scrolling. And as I stared down at my phone, I saw that little girl...

– on her mum's shoulders, mesmerised as she watched the red squirrels leap through the canopy on Brownsea Island,

– sitting on the sand, meticulously identifying every shell she came across, holding shiny ones in her hand like newly found treasure,

– standing by the shoreline watching the waves endlessly crash, wondering what lies beneath, hoping a whale would leap out in the distance,

– building tents in the garden made of the bamboo that grew there, tied together with hairbands, covered with old blankets, determined she and her sisters would be allowed to sleep out all night,

– feeling ecstatic to be camping to the sound of heavy rain,

– making sandcastle villages with her dad,

– painting and drawing until her fingers were sore,

– playing in the autumn leaves until it was too cold,

– swimming until she was too tired.

That little girl is very much still here. I hope she'd be proud of where I am now, that I chose to keep her favourite pastimes over beauty standards, money and status quo. These simplicities would never have to leave me as long as I looked to find them. Pulling a few of my favourite childhood snaps into a collage, I made a screensaver so that I would see a slice of home every time I opened my phone.

I left the cottage for a swim in nothing but a cossie. Walking down the coastal road out of Church Bay with my flippers in hand, I came across some tourists setting up camp at the old Kelp House along the shore. 'All right, love. Nice ... flippers,' one of them said. Gosh, you can get catcalled even on an island, I thought.

I arrived at the pristine waters of Mill Bay, gasping for a swim. The water was a shimmering teal colour in the afternoon sun. Applying some sun cream, I watched seals haul out on to almost every rock, disappearing into their surroundings. I could also see Charlie and some other ladies swimming together, so decided to join them. At first, I thought there were kids swimming too, but I soon realised it was in fact a line of young, playful seals trailing behind them.

I'd swum with seals at my local beach in Cornwall often and I'd missed these encounters so I felt excited to dive in. After not too long I had the attention of a young, female grey seal. It was fascinated by my movement in the water. I splashed around to find the seal beneath my feet.

Charlie asked if I felt scared, but I'd never once felt scared swimming with seals! It comes from a place of understanding their behaviour, and I knew how to act respectfully around them. Having rescued over fifty seals and tended to a hundred wild animals in distress through wildlife medic work, I'd got to sense the preciousness of each encounter. I would never approach a seal, not on the land or in the sea; wildlife has to come to you and it's even more meaningful when that happens.

The local ladies left the sea after twenty minutes or so but I stayed and swam for a while longer. Taking a deep breath, I dived to the forest of kelp beds that thrive here – islanders even make seaweed pesto from the stuff! In my haste to enter the water I'd forgotten to put on my flippers and goggles so decided to challenge myself to swim as if I were at one with the elements, like a mermaid. The young seal swam close to my face, before jolting in a different direction. My head turned upwards and I was almost blinded by the streams of burnt sunset rays dazzling into the sea. The seal then swam overhead, looking like nothing more than a fuzzy black blob to my irritated eyes. Eager to take a breath, I pushed off the sandy floor, rising upwards towards the seal. Before I could get much closer, she darted off again, this time beneath me. Checking out my silly human feet, her whiskers tickled my ankle and I giggled, letting out an explosion of bubbles that rose to the surface.

Back to reality: I looked to the shore to see that I'd floated out quite far and decided to head back. Swimming, or by this stage of exhaustion, doggy paddling, I struggled in the intense currents. I just wish I'd put those flippers on, I thought. It took me ten minutes to finally reach the shore. I was sure the seal had left me long ago, but as I reached for the towel that I'd placed on a rock, she popped up as if to say goodbye.

I stood snug in my towel on the shoreline and as I let my feet sink into the wet sand, I felt a sense of belonging to this land.

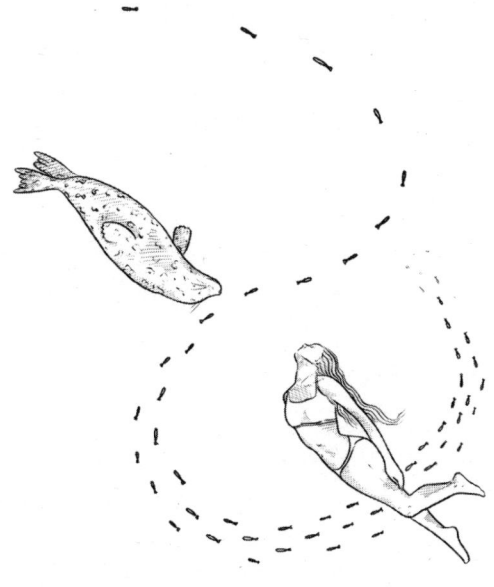

Chapter 3

Understanding

Some people only know the cost of everything and the value of nothing.

Floating out past the pebbles of Church Bay, I lay back, letting the water hug me. Buoyant and blissful, I faced the wide sky and replayed memories of my previous week here over and over like a cassette tape. I went back to the house and made a Sunday roast just for me, feeling proud of what I had achieved since I'd got here: starting a new job, moving country, moving to an island, and immersing myself in this new life. I was doing okay.

The next morning, the sun cracked through my blinds and the piercing blue ocean water lit up my room. It was another glorious day. As I drew the blinds and wiped sleepy dust from my eyes, I saw movement in the water behind the harbour wall. Ripples rolled in from the flat, calm sea but I couldn't spot anything, so put them down to the strong undercurrents that encircle the island. I headed downstairs, dishevelled and half asleep in some underwear and one of Craig's baggy old tops. Coffee poured and toast half-eaten, I looked out of the large kitchen window to see more of these ripples. I ran to the front door and opened it to get a better look, determined to crack the case. A few moments later, a dolphin breached straight ahead and in my astonishment I dropped my coffee cup, which broke on the flint garden path. I rubbed my face, amazed to see a dolphin so close to shore.

I ran out of the door with no shoes on and headed straight to the harbour wall to watch not one but ten dolphins play! I was nothing short of mesmerised. Reality came crashing in when some neighbours joined me – fully clothed. I was still in the pants and T-shirt I'd woken up in, looking like I'd been dragged through a bush. I hadn't met this family yet and felt worried I wasn't making a great first impression, but they told me they admired my eager nature. I stayed with the family for half an hour more, until the dolphins headed back out to sea.

I raced home, got dressed and went to the centre. Holding coffee number two, I walked down the stairs to the colony before we opened up for the day. I spent time admiring the four seabird species I'd slightly neglected up until now – kittiwakes, guillemots, razorbills and fulmars – and pinpointed some nesting pairs, ready to follow their journeys.

Some naturalists don't agree with naming wildlife, but I think it makes for great storytelling, which is vital for connecting people, especially children, with nature. Naming and following certain birds was a way for me to get closer to the wildlife.

There was a particular pair of fulmars that caught my eye, and I named them Freddie and Fanny. Almost gull-like, these charismatic birds are related to the albatross. They fly low over the sea on rigid wings, with shallow wingbeats, gliding and banking to show their white underparts. At their breeding sites they fly high up the cliff face, keeping a watchful eye on their nests. If any potential predator gets too close, fulmars defend their nests by spitting out a foul-smelling oil. This gives them their name – 'ful' meaning foul and 'mar' meaning gull.

A kittiwake couple to the left of the platform stood out to me because of their squabbling – they were always getting into fights with neighbouring birds. I called them Kitti and Kat. They were hilarious to watch. The name kittiwake comes from the way they call – it sounds like someone shouting 'kitti-wakee, kitti-wakee'.

My colleague Jean arrived at the cliffs. I told her what I was doing and she decided to join in, naming a razorbill couple nesting to the right, above our heads, James and Michelle.

'Very civilised,' I said. The razorbills looked satisfied with their nest, which was deep within a nook of a cliff. They were so attentive and loving towards each other; it was beautiful to watch. Their name 'razorbill' comes from the distinctive, sharp-edged bill, which resembles the shape of a razor blade. There is a beautiful white line that darts down it too. I like to think they apply fresh eyeliner every morning.

The guillemot couple I named were rather far away. Orla and Ottie sat just beside a tidal pool on a sharp basalt ledge. There were hundreds of thousands of other guillemots so it was hard choosing who to focus on. Orla and Ottie stood out because they always had big white splodges of poo on their heads from the nesting birds above. Unfortunate for them but handy for me.

Guillemots have evolved to nest near each other to provide safety in numbers, because they raise their chicks on such exposed slabs of rock. What I love the most about these birds is that they can dive to depths of 180 metres, enduring crushing oceanic pressures. Deep sea divers have been known to mistake a guillemot for a penguin – they don't expect to see the guillemot at such depths. There is no other bird of a similar size or weight that can fly *and* dive to such depths. They really are a true seabird. Like the duck-billed platypus, a mammal that lays eggs, or the bat, a mammal that flies, seabirds push the boundaries of their class's usual characteristics.

So there we had it, five pairs from five species, all with their own stories to follow. It was exciting getting to know them. All day, when I wasn't chatting to visitors, my eyes rotated around the cove like the hands of a clock, watching each of the pairs until it was time to close the centre. I loved getting to know my new friends, but my attention was always drawn back to the puffin trio at the lighthouse ledge. Looking over my shoulder to Puff and Finn, I watched them preen one another in the afternoon sun. Although I was unable to see it, knowing that they almost certainly had an egg in their burrow felt extremely exciting. I anticipated their little puffling appearing one day.

Every day visitors came in their droves and buses toed and froed. I watched closely and got to know the couples I had picked out. Each morning, after checking on Puff, Finn and Busy Lizzie, I walked up from the lower lighthouse perch to the visitor platform to find the other four of my big five.

One of the fulmar couple, Freddie and Fanny, was always away at sea, spending hours at a time to catch fish, so it was rare to find them side by side. However, early one Tuesday morning, they were both perched on their mossy patch, directly under my nose. It was a beautiful surprise to see them together and – even better – turning their precious egg. Every time a feathery bum lifted I got a glimpse of it. I could see just how careful they were being: one fulmar spread its wings to protect the egg from the colony while the other squatted and used the tip of its beak to gently turn it, ensuring every part of the developing embryo was incubated. Just as the egg-turning came to an end, a sneaky black-backed gull flew in from behind, coming dangerously close to their nest. The fulmar used its wings to create a broad shield and let out an almighty cry in an attempt to deter the gull … but it wasn't working. The gull came closer. It was time for plan B – release the puke. A surge of fulmar stomach acid shot into the gull's eyes, causing it to spiral back down the cliffs, away from Freddie and Fanny. Usually, I'd feel a level of empathy for both sides, but in that moment, I felt that the gull deserved it.

The fulmars squabbled, starting to peck at each other. It all looked quite aggressive. When I got back to the office, I consulted a seabird guide to try to find out why this couple was fighting. After a thirty-second read on fulmar breeding displays, I felt rather silly. I had projected human emotions on to these birds when in fact this behaviour translated to a seabird 'I love you so much'. That afternoon, every time I heard what sounded like fulmars fighting, I was touched – thanks to my newfound understanding.

During the first week of June it became warmer, so most mornings Rachele and I set off to the harbour in shirts and shorts. One day, as we collected our colleagues Eleanor and

Caitlin from the ferry, eider ducks were making themselves known, quacking loudly to keep us away from their tiny sooty chicks.

When we neared the car park gate, I spotted a colony of rats amongst the bins we had left out the day before. I'd known there was a problem with non-native species here, but I didn't understand just how bad it was until that morning.

Rats are not native to Rathlin, or to the UK and Ireland for that matter. They came on boats that arrived from central Asia in the late eighteenth century and have caused havoc since. As an animal lover, I adore all of them – rats included – but I hate what this species has done to native wildlife populations. It was never their fault they arrived here – it's ours. It's the same story with the ferrets. Years ago, ferrets were introduced to control rabbits on Rathlin but they have instead wiped out many populations of ground-nesting birds. Island avifauna hasn't evolved alongside predators such as rats and ferrets, so is particularly vulnerable to them. Ground nesting species that once dominated grassy wet areas, such as curlew, snipe and lapwing, have experienced a severe decline here. Puffins go a step further than ground nesting – I call it 'underground nesting' – meaning they are severely at risk because rodents can easily take eggs from their burrows, and sometimes even chicks. Just a few decades ago, puffins on Rathlin used to nest on the tops of the grassy cliffs but due to predatory pressures they've been pushed further down. Some species, such as the Manx shearwater, have quit breeding on the island entirely, but there is hope they will come back.

Ferrets and rats are both incredibly intelligent species that deserve to live, eat and survive like any other animal on this planet, just not here. As someone who follows a plant-based diet, I am against animal cruelty, so the species-control aspect of conservation has always troubled me. However, I understand that it's vital that we undo some of the things we have done so that we can improve the lives and survival of native wildlife populations. If wildlife conservation charities such as the RSPB didn't control species like rats and ferrets on Rathlin, puffins would decline a whole lot faster, further impacting a wider, global ecosystem that, in turn, affects all of us. Apart from the rats and ferrets, Rathlin felt like a safe haven for wildlife – I had already encountered so many species I'd not seen anywhere else, which is why we must fight for what native wildlife remains.

I know many animal lovers that work on species eradication. The act of culling pains them, but the consolation is the greater benefit that is species recovery. It's one of the weirdest dynamics a conservationist can face. Animal lovers at heart, people enter the field with the hope of saving species, often not realising a huge part of that could entail … killing.

Rachele told me about a project to eradicate the rats and ferrets that would soon launch. 'Why haven't I heard about this?' I asked, still in shock. Rachele explained, 'It's secret at the moment. It's a very complex topic and we need the entire island to be on board for the eradication to go ahead. It's called LIFE Raft. It's a four-year, four-million-pound project that the RSPB hopes will remove all rats and ferrets from the island, in turn bringing many ground-nesting species back from the brink of local extinction. It starts next year.' I was excited for the puffins but sad it ever had to get to this stage.

I found the situation bitterly ironic. Humans are helping puffins, among other species, by controlling the population of rats when arguably, on a global scale, humans are the ones having the most impact. If there's one species that should be controlled, it's us, I thought – feeling guilty just for existing as I often did.

We would have never purposefully set out to destroy nature – which provides us with everything – but in a world that is increasingly lacking abundant green spaces, in a society that prioritises an 'economy' over ecology, it is no surprise that we forget the value of nature. And when I say we are forgetting the value, I do not mean the cost, I mean the life it supports.

I got to watch natural systems play out before my eyes on the cliffs every day, and it made me think deeply about the time in history when we allowed ourselves to become separate. Was it religion that made us think we were superior to the species around us, or class systems? I wasn't sure but I longed to know. Working in such a wild place, I felt equal to the flowers, bugs and birds, yet everything society had taught me allowed me to think I was worth more. It made me laugh that I'd been jokingly called 'bird brain' as a child, when these seabirds I'd been connecting with could navigate their way across oceans. I couldn't even find my way home without Google maps.

There was a point in many civilisations, about ten thousand years ago, when humans realised that if we created strong communities to share knowledge, food and resources, we would have a much stronger chance of survival. But a few thousand years into this framework, that structure of strength then turned into our very weakness. The moment we placed a form of monetary value on our existence was the moment human civilisation formed the concept of ever-growing, limitless capital in a finite world. We evolved from taking what was needed and living in balance with the natural world to a mindset that aimed to control it.

We may be entering the Anthropocene, a unit of geologic time used to describe the most recent period in Earth's history during which human activity has started to have a significant impact on the planet. But I would argue it's also a time when humans can realise this mistake, coin the term and attempt to undo our doings. Perhaps, the first step is to focus on where it all went wrong – the way we place value on the natural world.

I feel this is especially important now because ever more frequently, even when we are trying to prove a species' worth, we use its monetary value as a means of persuasion. The International Monetary Fund (IMF) estimated the value of a single great whale at more than $2 million – which meant that the value of the current 'stock' of great whales came to more than $1 trillion. The IMF based its figures on each whale's contribution to carbon capture, the fishing industry and the whale-watching sector, which is worth over $2 billion. This costing took place so that decision makers can view conservation efforts as more credible and economic. This is widely seen as a positive step in the right direction, but surely, while we are costing whales based on the economic benefit they provide to a capitalist system, we are not truly valuing their existence. If we did, we would ensure their world was free of noise, plastic and chemicals; we wouldn't deplete their food sources and trap them in nets. And I worry that overstating a whale's ability to mitigate anthropogenically induced changes may redirect attention from known methods of reducing greenhouse gases, such as ending fossil fuel expansion.

If we base the intrinsic value of something – human, wild animal or plant – on their economic cost we are not truly valuing that species at all, nor our planet, nor ourselves. Doesn't nature deserve the right to exist as we do?

In my relatively short time on Rathlin many of the people

I'd met seemed rich, not necessarily rich in monetary terms, but something a lot more valuable – happiness. The people who lived close to nature on Rathlin, who cared deeply about protecting its biodiversity, reminded me of the Indigenous communities spanning our planet that attempt to cling onto their wild existence. Despite the fact Indigenous peoples make up around 15 per cent of the world's extreme poor and just 5 per cent of the global population, they are protecting 80 per cent of the world's remaining biodiversity. If they had some form of power and say in the world, if human civilisation mirrored these minorities, I know nature would bounce back. We would all be so rich.

The longer I lived on Rathlin, the more I felt I was living as a human should, valuing the resources and rewards nature gave me. I was living as seasonally and naturally as I could. I ate the seaweed that grew in abundance, the bread neighbours baked and the vegetables that Lizzie grew in her garden. I had everything I needed.

I began to take my sketchpad to work. Whenever I could, I drew the puffins – eating, playing and resting – like they were my muses. One day, as I sketched Puff preening Finn, Busy Lizzie landed with a bundle of sand eels by my watching spot. I hid behind my mound of earth but Busy Lizzie was not dim – she could sense my presence and came to check me out. She squawked at me. 'Hello,' I answered, before she waddled back to her usual perch.

Puff went into the burrow with a feast and Finn flew off – it was his turn to fish. Every time I saw these three birds, I felt thankful for them, especially given all the obstacles in their way. I didn't ever take them for granted.

I was to return to the farm that coming weekend – it had been a few weeks since my last visit. Craig had started to make us a tiny house from an old caravan he'd found on Facebook. I was going to have my own nest to return to after Rathlin! The seabird season was due to become busier and more hectic, meaning I soon wouldn't be able to leave the island, so I took every opportunity to see Craig and Lizzie that I could.

The pollen from nearby oxeye daisies had made my nose run. I reached into my pocket for a tissue, and found a crunchy, sticky mess. The razorbill egg I had picked up on the cliffs a few days prior had broken, and the gunky remains covered my fingers ... It was disgusting. When I tried my other pocket, I found the guillemot egg, which was at least a bit drier and less gunky. But I was still a sticky, eggy mess. I raced up the stairs to the lighthouse office to wash my fingers under the shuddering old taps but no amount of soap and water could remove the smell. The day rolled on and I worried more and more that I stank. When all the staff came together for a short afternoon meeting, Caitlin asked if anyone could smell rotten fish ... so I clearly hadn't got away with it.

Break over, I sat on the far-left edge of the seabird platform pointing out what was going on with the kittiwakes to visitors. Kitti and Kat were defending their eggs from two ravens who were trying to steal them. As one of the kittiwake parents flew away, a raven swooped in and then, a second later, sped away again with an egg clutched in its black claws. I watched the raven land beside what must have been its nest. The two ravens then dug a hole to bury the egg using their claws and beaks. Among the most intelligent animals in the world, with the ability to think ahead, ravens store food for later in the year; in this case probably for the time when the seabirds wouldn't be here, and there would be less prey for them. Watching these scenes made some visitors cry. Many got angry, some even asked why we weren't doing anything to intervene, not understanding that what they were watching was natural behaviour.

At the end of the day I locked up the centre, bringing another eventful week to an end, and headed home. I never once took for granted how stunning my commute was. Stopping in a layby to chat with Liam, who was driving the other direction to meet a farmer, he told me how people here cut their crops to ensure no trampling of ground nesting birds. I drove off inspired, admiring the interconnected meadow of wildflowers for bugs, bees, moths, hares and owls. I'd realised that it takes a community effort to give nature a home.

This landscape-scale respect for nature was new to me. It's why there are so many untouched bogs, native Irish trees, waterways and beaches. Islanders have a deep respect and value for life here other than human ones, a sharing of the land that I'd never come across – scientists call it a mutualistic relationship, growers call it permaculture, conservationists call it symbiosis, but islanders call it … normal.

Before hopping on the ferry, I swam in the harbour in an attempt to get rid of the smell of the eggs. Returning to the mainland, I embraced Craig with the biggest hug, joking that this was my version of puffin courtship. Within moments he was crying with laughter, 'A few weeks on that island and you come home smelling like you haven't washed in years!'

That weekend the caravan shell was undergoing a major makeover. On the Saturday we pulled down old wallpaper, stripped the floor and demolished two entire walls, getting rid of the second bedroom to make an open-plan space. We were destroying the place, but all with the intention of creating something beautiful. It was a lot of fun! On the Sunday, Craig, his dad and his uncle Ivan insulated and repanelled the walls.

I couldn't do much of the insulation work and felt useless, so made endless cups of tea for everyone while they worked away. I hated standing on the sidelines watching others graft – I'm not that type of girl. I embrace challenge – when it's tasks that are stereotypically 'male,' I want to do them even more. Ever since I was little, the idea that boys could do things that I couldn't really got to me, even before I knew what feminism was. In school at the end of class, I got angry when the teacher asked the 'big strong boys' to stack the chairs. I opposed that thinking and got stuck into the chair stacking. Today I find myself in surf line-ups, attempting to catch more waves than male counterparts. This competitiveness and passion for dismantling misogyny, and all the experiences that come with it, has built a resilience inside me.

As Sunday evening drew to a close, Lizzie wandered up from her bungalow before dark to see what progress we'd made. She was delighted to see me wearing her rainbow scarf and gave me a hug. 'Where's mine?' joked Craig. When I first moved over, I realised Lizzie was never overly affectionate; neither were many members of Craig's family. He thinks it's a bit of an Irish thing, so receiving this hug was special. Craig had told me his family shows affection in different ways; in making sure you have everything you need, especially food. I grew up with 'I love yous' every bedtime. We hug a lot, and everyone talks about their feelings, which I realise as I get older seems to be quite a rare thing.

'That's a grand job,' Lizzie said. 'Thank you,' I replied, as if I had done much to help the build.

Craig, Lizzie and I chatted fondly about the puffins, and I updated her on all their antics. I found the courage to tell her I'd named one after her, not really knowing what she'd make of it. 'And there's one that reminds me of you – Busy Lizzie…' She laughed and said, 'That's brilliant.'

The sun fell lower in the sky and I felt the urge to find a sunset-lit adventure. Just after Craig and I had packed our tools away, I saw a clatter of swallows dancing in the evening light, feasting on clouds of insects. I grabbed Craig and insisted we follow them – we raced down the steep hills as if we were flying too.

The swallows had migrated from Africa and were nesting in one of the farm sheds. We kept up pretty well but stumbled to a halt when we saw six fox cubs playing in the long grass ahead. We watched in astonishment as the young pups played in the evening glow, unaware of our presence, content and thriving. 'How amazing is it to see these animals thriving on your farm,' I whispered. 'Our farm,' Craig replied.

We watched the cubs chasing the tails of their brothers and sisters before the mum came to round them up. I could imagine a fox version of 'but Mum I want to play longer' as the youngsters defied their mother's nudge to the hedgerow. By the time the foxes disappeared into the hedgerow it was later than we'd realised so we headed home to Lizzie's.

The next morning, Craig dropped me to the ferry terminal. This time I would be on Rathlin for a month but Craig promised he'd join me halfway through so I didn't get too lonely, crazy or … smelly. He waved me off, shouting, 'Don't miss me too much!' I jokingly cursed at him then blew a kiss. Running to the side of the boat, I pretended to be a seabird, flapping my arms like wings as the ferry picked up speed. Craig was disappearing out of view as we turned the corner of the harbour, but I could still hear him laughing.

When the ferry docked, I ran to the house and threw my food into the fridge, freezer and drawers before donning my uniform and dashing up to the Seabird Centre for opening time.

Francis arrived with the tourists and, as ever, a bag full of treats. To my surprise, he had gone out of his way to get me two extra packets of oaties, remembering I didn't eat dairy. I was touched by this gesture and delighted since I hadn't packed much for lunch that day.

We welcomed the visitors and led them down the stairs. It was raining quite heavily for the first time since I'd arrived. The place felt different – the rocks were drenched by the downpour and the smell of guano had slightly subsided. There were a lot less visitors. Usually the first Puffin bus would bring thirty guests, but this morning there was only ten. I guess people don't usually like Mondays or rain; on Rathlin I loved them both.

I used the quiet spell to find out what Puff, Finn and Busy Lizzie, Freddie and Fanny, Kitti and Kat, James and Michelle, and Orla and Ottie were doing on this fine, rainy morning. Not letting the rain hinder their hunting, it looked as though the seabirds were on a mission to eat as much as they possibly could before their chicks hatched next month.

It is incredibly hard to tell which seabird is male and which is female since they exhibit sexual monomorphism, meaning both sexes look alike in terms of plumage and size. They also share most breeding responsibilities. The only one I could confidently distinguish from their partner was Puff, because of her distinctive, confident behaviour and bright scar – and I still didn't know for sure that 'she' was female.

I started to chat with a couple at the platform who had travelled from America for a tour of Ireland. They were upset that they had come to Rathlin on such a damp day. They looked miserable, so I tried to cheer them up by telling them it was a good thing to have come in the rain because they had the entire platform to themselves. I also said the birds' poo was a lot less pungent and they laughed! I started to point out each of the nesting pairs. As I told them about the auks' characteristics and the fulmars' flying skills, the lady interrupted. 'Oh look, there, there's a seagull. I love birds but I really dislike seagulls – they steal my fries.'

'Did you know that there is no such thing as a seagull?' I asked. I explained that gulls are the overarching species – that the seagull she had disparagingly referred to was in fact a kittiwake, one of the few gulls that spends a significant time at sea and can be classified as a seabird. I also told the couple that kittiwakes are extremely endangered and threatened with extinction.

I then pointed to other gulls around the cove. 'Gulls are

predators but also scavengers, it's how they've adapted. This bird here is a greater black-backed gull. They are feisty predators and sometimes steal eggs. These ones are common gulls – they nest on this island using their sublime intelligence to make nests out of almost anything they can find. And these ones here are herring gulls, the ones you dislike, but really all they are doing is trying to survive. Fish populations are declining and hunting is becoming increasingly difficult. These birds have adapted to steal food from other species within their wild environment, at the coast. We see this same habitat as our seaside. When you think about it, we are in their habitat and it's within their nature to take from us!'

The lady was quiet after I'd finished my mini gull lecture. Had I said too much? Was I annoying, too smart, sarky? I wasn't sure. But then she smiled. 'I'd never thought about it that way,' she said. Well, that went okay, I thought.

To protect the climate we must protect nature; to protect nature, we must understand it and all its complexities. The idea of some aspect of nature being a 'pest' is a human construct, which has led to us demonising some species, and overlooking their positive attributes. A dandelion's pollen, a stinging nettle's habitat, a gull's nutrient-rich poo – these 'pests' are vital for the health of nature. They just don't always look neat and sometimes they steal our food, so we have found a way to put them into a negative box. This lady was so lovely, just disconnected, like most of society – but she was also an open book and soaked up my wildlife facts like a sponge.

I spent half an hour with her, pointing out the puffins' burrows, the guillemots' diving skills and the kittiwakes' finely balanced nests. There was a herd of seals rain-bathing on the rocks below and a gannet diving ahead. By the time the visitors had to leave, my throat was sore. 'Thank you so much,' the lady cried. 'Where I'm from in America, it takes ten hours to drive to the ocean. I had no idea we've discovered just 4 per cent of our seas, and no idea there wasn't such a thing as seagulls! I've never experienced this much wildlife, nothing like this. You've really changed my life. Thank you.'

During that one morning, in that half an hour of talking, the lady who turned up unimpressed and wet left with a beaming smile on her face, invigorated, empowered, happy – feeling as if her life had changed. It was because that was the day she'd not only encountered nature, but understood it differently. All I ever wanted was for anyone who came here to leave feeling inspired.

This encounter and many others I've had like it are why I don't dislike people who treat nature poorly, who hate certain species and view others as pests. It's very often because they have just never had the chance to feel part of it. The greatest tool that'll see us out of the climate and nature crisis is education, and real human connection.

That afternoon the rainclouds disappeared and visitors came by the dozen. The seabirds were noisier than ever – all striving to protect their nests, their partners and their eggs – while we humans watched in awe, looking more closely at the webs of life that sustain us, that amaze us.

Chapter 4

Discovering

If you travel far enough away, you'll meet yourself.

A few weeks into June, I dusted off my bike and cycled to the centre for the first time. Rachele was due to arrive a little later than usual in her blue van with some staff and a couple of new residential volunteers.

After a gruelling forty-minute journey, riding mainly uphill in the blazing morning sun, I arrived feeling invigorated. I got my breath back as I prepared the centre for opening with Jean. She was impressed that I'd decided to cycle instead of drive. 'So, you're cycling home after work, Ruby?' she asked. 'Yes,' I replied excitedly. 'At least it should be a lot easier going downhill.'

'Well, just beware of the fairies, bogs, bulls and the bus.' she said.

When I asked more about the fairies, thinking she was joking, she told me they were harmless, and resembled black shadowy figures, so not to freak out if I spotted one. I'd pictured fairies as looking like Tinker Bell and started to feel slightly concerned. Jean added, 'They appear when someone or something needs help,' and I felt a bit better.

I'd read about local folklore in one of the Rathlin history guides. I'd learned about selkies – creatures with the ability to shift between seal and human forms – and read stories about bog bodies, the banshee, mermaids and fairies. Until now I hadn't met someone who believed in them. Fascinated, I asked

Jean if the fairies had ever helped her. She told me about the time that there were plans to create a new slipway at the harbour on Rathlin, that would have ruined the seals' habitat. She'd asked the fairies for their help, and when the builders had arrived to start work on the site, they'd discovered tons of rock on the harbour floor and the project couldn't go ahead. She believed it was the fairies' doing.

I was fascinated by the story and honoured that Jean had told me something so personal. I thanked her and told her that I knew Rathlin was magical.

I don't believe in fairies or folklore, but I want to. I love hearing about it and admire people like Jean who are keeping these stories alive on the island. Even if we don't believe in folk stories, I think we should honour their messages, for humanity's sake. Often, these tales that have been passed down the generations contain an element of conservation. For example, in Ireland, farmers don't ever cut down a 'fairy tree' – usually a species of thorn or ash standing alone in a field – for fear the fairies will turn against them. This is why, even when you drive around the most deforested parts of Ireland, you see lone fairy trees standing tall in acres of open field.

I snapped out of dreamland and into work mode. Rachele arrived with Caitlin, David and the first residential volunteers of the season, Kevin and Meta, a loved-up retired couple who entered the building with a spring in their step. They were staying for the remainder of June at Kinramer Cottage. They couldn't have looked happier. Kinramer Cottage is a beautiful old building situated on the west of the island. I introduced myself and offered to show them around the seabird reserve, but Kevin said they wouldn't need a tour – it turned out that this was a familiar playground for them both. I spent the day getting to know them, admiring their immense enthusiasm for wildlife, excited to be working with such knowledgeable people.

Kevin had a massive camera that all the visitors wanted to know about. So he used it as a tool to engage with them, taking pictures of wildlife too far away to see with the naked eye and then, perched in the binocular hut, he uploaded the images to his laptop so that others could see them. Meta was incredible at chatting with the children; it seemed effortless for her. By the end of the day, we were doing bird impressions up and down the platform together for the kids, like we'd known each other for years. The couple were so open and kind. Meta told me about her brutal cancer battle and Kevin seemed like the most supportive partner. No matter what they'd been through, Rathlin had always been a place of solace and peace for them both.

As I watched guillemots getting splashed by sea foam through my binoculars, I saw Liam in a tiny rickety boat, bobbing around at the base of the caves counting seabirds. He was brave to be out at sea that day – the waves were large and the current looked strong.

It was nearly the end of the day. Kevin and Meta's arrival had made time fly. Rachele was heading off the island early that night – there had been a problem with deliveries, and we weren't getting everything we needed to run the centre. Just a few weeks in and the supplies of toilet roll, coffee, vending machine snacks and cleaning products were running low. I felt sorry for Rachele, constantly on and off the island, getting what was needed to ensure that everything ran smoothly. While I got to live here – having the time of my life – she hadn't been able to stay for longer than a night. 'When you're here for a few days longer, we'll go on an evening adventure,' I said. She looked delighted at the prospect.

Jean checked I was sure that I wanted to cycle; I thanked her and said I'd be fine. Kevin and Meta set off on foot as their cottage was close by and I set off on my bike. After a mile or so of racing down the road, I took the turn to Liam and Ali's place. Jane had mentioned that Liam and Ali used to run moth-trapping nights in their garden and I wanted to find out if I could do the same.

As I parked my bike, I saw a golden siskin dart overhead, welcoming me to Liam and Ali's. Ali offered me a cup of tea, but I declined, already shaky from the three coffees I'd had at West Light earlier that day. After half an hour of small talk, I cut to the chase, asking if I could use their famous moth trap. They were so delighted by the idea that they said they'd set up a trap for me that very night. Ali explained that nobody had used the trap for years so we toured around their old farm sheds, hunting everywhere for it – eventually Ali found it in a dark, dusty corner.

The design was pretty simple – it was a large plastic box with an inbuilt LED light to attract the moths and egg cartons inside where they can hide once they're in the trap. I don't know why but I was expecting something more high tech – but it was still a definite step-up from the torch and bed sheet set-up I'd previously used!

We turned the light on and placed the trap at the outer edge of Liam and Ali's front garden. It was a calm June evening, the perfect conditions for moth trapping. Liam found his old moth identification books and Ali posted on the Rathlin Facebook page to see if anyone wanted to join us the next morning to see what we'd discover.

It was getting dark. Ali asked where I had explored on the island so far and gave me some hiking recommendations. I told her I was excited to explore the woodland and heath around North Kinramer Nature Reserve some evening soon – but was surprised when she warned me not to stay out there too late as it was notoriously haunted.

Unable to tell if this was a joke or a real, folk-led concern like Jean's, I thanked Ali and shot off on my bike. Just ignore that, you know these stories aren't real, I told myself. But the long, hot heat of the day combined with the cold Atlantic Sea surrounding the island had created the largest, most misty land cloud I'd ever seen and it became harder to keep thoughts of ghostly fairies and haunted heaths out my head. Then it turned pitch black, and the visibility was so bad that I could barely make out a metre in front of me.

Fight or flight kicked in and I pedalled faster. With my heart racing and feeling highly disorientated, I nearly fell off my bike, spooked by a short-eared owl I'd accidentally flushed from the field perimeter. It darted in front of me so quickly and the lights from my bike lit it up like a ghost.

Pull yourself together, I whispered to myself. Keen as I was to get home, I realised cycling probably wasn't the safest activity in such bad mist so I walked and tried my best to enjoy the experience. After ten minutes, the path sank down towards the harbour and the mist quickly cleared. High on the rocks several seals scratched themselves and the moonlight glimmered on the dappled water.

I decided to call Craig. Hearing his voice calmed me. He laughed when he heard I'd almost fallen off my bike, and was fascinated to hear that Rathlin was so haunted. We planned to go on a night-time walk when he visited, although he wasn't sure when he'd be able to come. Lizzie wasn't great and – in seabird-style protection mode – Craig didn't want to leave her. We chatted for a while longer and then said good night. I headed to bed feeling slightly spooked, but sleepy.

The next morning I popped into Breakwater Studio, the little gift shop on Rathlin that also has a small art gallery. It was my mum's birthday that month and I wanted to buy her something nice. Yvonne, who runs the shop, showed me a handmade bag decorated with prints of puffins. It was perfect. I asked Yvonne if she also had anything I could send the parcel in, and she generously handed me an entire roll of wrapping paper and tape. 'Thank you!' I shouted, jogging off to the island Post Office to send my gift.

Next I headed to Liam and Ali's. When I arrived, there were four island kids, two teens and some other adults, all excited to see … moths. I couldn't imagine a crowd of strangers studying insects so early in the morning anywhere else.

Ali assumed I knew lots about moths because I'd been a wildlife guide for several years, but I didn't – I was excited to learn more myself. I opened the moth trap to discover an array of colours, shapes and sizes. Some moths were shaped like hearts, some camouflaged as leaves, some pink, some fluffy, some yellow. One by one, I pulled the egg cartons out and passed the tiny creatures around in awe. I let a colourful buff ermine rest at the tip of my nose. The moth vibrated its flight wings to keep warm – it tickled. I sat cross-eyed, captivated by its psychedelic under-wing pattern. Next out of the trap was a stunning garden tiger moth – the children became fascinated by the speckles on its hind wings, remarking that they looked like blue eyes staring up at them.

High numbers and a diverse range of moths are a brilliant indication of ecosystem health, making moths a great 'indicator species'. Elusive animals, they can be found throughout most ecosystems yet often go unseen. After marvelling at them for as long as I could, I gently placed them onto leaves and branches, conscious I was late for work. I was captivated and couldn't wait to do this again.

When I returned the next morning, Ali told me to take the moth trap home and keep it for a while – she could tell I was hooked. After work that day, I put the trap in my garden, excited to see what I would find. Every morning that week I hopped out of bed as if it was my birthday and hurried down to the garden to unbox my little gift from nature.

I counted 32 moth species and 109 individual moths by the end of the week – my favourites were a bright pink elephant hawk moth and a fluffy white puss moth. It was such a beautiful week of nights and mornings, with different islanders joining me at different points. On Rathlin, it was the smaller aspects of nature I soon found so fascinating, not least because they are often the pieces in the ecological puzzle that underpin everything else.

That week, Meta and Kevin settled in nicely and days on the platform were full of thrills. I pointed out Puff, Finn, Busy Lizzie and the other seabird couples to them. One day Puff circled high around the cliffs, practically skimming our heads – she must have been fond of Kevin and Meta too. I joked that Meta and Kevin were my lucky charms as we had been spoilt with incredible wildlife encounters every day since their arrival. Kevin also became our number one skua spotter. Twice a day these feisty birds would swoop past the lighthouse in search of prey. 'Bonxie, bonxie,' Kevin would shout to alert us. Bonxie is a nickname for the great skua, originating from Norse and commonly used in Scotland, that means 'scruffy woman' – a good description. I felt a newfound connection to the skua as I was turning into a rather scruffy woman myself. Their scarcity made them exciting to spot. Causing great commotions on the cliffs and getting into flight fights with ravens, they were like noisy neighbours who wouldn't quieten down.

The next week, multiple visitors complained about them. It was mid-June – aka peak skua breeding season – and walkers had been getting too close to the pair that nests on a public footpath beside the Seabird Centre. The skuas had stolen at least three hats and would sky-bomb anyone who came within half a mile of their nest.

The water got warmer and the ocean was full of jellyfish. One morning, as I reached the lower platform to greet Puff, Finn and Busy Lizzie, the sea appeared more orange than blue, with blooms of barrel jellyfish floating in the currents at the surface.

Our sightings chart had never been more packed – it kept filling with more and more incredible encounters as the week passed by. During the Friday of the following week, three minke whales circled the waters surrounding West Lighthouse for half an hour. We could practically smell their blow – they don't call them stinky minkes for nothing, I learned. The water was full of nutrient-rich guano, promoting planktonic species and enticing large schools of fish. No wonder whales started rocking up.

The minke is a baleen whale, meaning it has bristly baleen plates instead of teeth. This species filter feeds, devouring tonnes of tiny fish a day, in contrast to toothed whales, which can potentially take on prey such as the colossal squid. Most baleen whales migrate thousands of miles per year to feeding and breeding grounds, covering vast distances and depths.

I admire these mammals both for their extreme intelligence and for their immense importance to the planet. On average, 1 whale will sequester around 33 tonnes of carbon in its lifetime – to put that into perspective, the average tree will sequester 1 tonne. When it dies, the whale descends to the bottom of the sea floor, creating a literal carbon sink for thousands of years. These giants are equivalent to swimming forests! Like the seabirds' guano, whale poo also encourages the growth of phytoplankton, which in turn produces oxygen. To me, one of the most important ocean conservation efforts of our time is to restore our great whale populations. If the whales can't thrive, neither can we. I loved explaining all of this to as many visitors as I could. Having listened to me for hours, Meta paid me the great compliment of saying that I was good with words. I was thrilled.

Later on, I asked Meta and Kevin to come over to the cottage for a meal. The three of us had been so enthralled by the day's encounters that we hadn't wanted to leave the platform for lunch, just in case something happened while we were away. We didn't realise just how hungry we'd become until the topic of food came up. I radioed Rachele and invited her along too.

I was delighted to be cooking for everyone – it was going to be so nice to sit down together for the first time since meeting. I'd packed enough food for a month … but it needed to last a month, so I asked everyone to bring some sort of side dish. With everyone's donations I made a delicious mishmash of Indian, Chinese and English pub grub-influenced dishes. It was random but very tasty. I love cooking, especially for people who may not have eaten much plant-based food before. I love proving that you can create just as much deliciousness on a plate.

When we'd finished eating, we decided to make the long journey of twenty seconds to Rathlin's perfect little pub, McCuaig's Bar. Maybe it was the island air or because we skipped lunch but after a few pints of perfectly poured Guinness, we were soon all hungry again. The smell of chips wafted through from the Hungry Seal, a new food van that had just opened that night beside the pub. I rushed over and ordered four portions of chips with lashings of salt and vinegar.

It was wonderful getting to know Meta, Kevin and Rachele better. It was a brilliant night – I chatted to so many people in the pub. Every table had a dog lying beneath it, waiting for peanuts to be drunkenly dropped, and in the distance I could hear music coming from Charlie's boat. I sat watching the sun go down when an overwhelming sense of peace washed over me. 'Rathlin, you beauty,' I called loudly, toasting this magical place with my friends and another pint.

Walking home, I came across a drunken man running his hands over the leaves of a plant on the road outside my cottage. It was rather strange. 'Can I help you?' I asked, intrigued by his peculiar behaviour. 'Poor boys' blanket,' he muttered. 'Not seen one of these beauties in years so I haven't – just feel how soft it is.' So I stroked the leaves of this towering plant too. It was a section of overgrown vegetation I hadn't paid much attention to before now. Towering a hefty two metres tall and with yellow flowers, the plant looked as if it belonged in a tropical rainforest. I asked why the plant had that name and the man told me that the long, soft leaves could act as a blanket for someone who has nowhere to sleep. Apparently the real name is common mullein, but I preferred his.

The next morning, I slept later than usual, feeling a bit tender after last night's pints. I only had twenty minutes to get myself sorted for a day on the cliffs but I managed to get ready with time to spare, so sat on the sofa, ready for a mindless social media scroll. Suddenly Francis barged through the door. He explained an earlier ferry had come and that he needed us to open up earlier than usual.

Woken by Francis, Rachele came running down the stairs, pulling on clothes and panicking. We raced to the centre and when we got there, Francis attempted to apologise for barging in on us with an extra bag of treats before he left promptly.

On cue, as the last of the visitors walked through the doors, Meta and Kevin turned up – calm, collected and fresh faced. They could see that Rachele and I were a little flustered, but we didn't have time to explain.

I guided the visitors down to the platform and gave them a tour of the cliffs. The reserve was alive with life and the Big Five tick sheets were full of sightings. A little girl at the corner of the platform started to get upset that she couldn't see a puffin. I offered her my binoculars and positioned them carefully so that they pointed towards Puff and Finn, who were playing with sticks by their burrow. 'Look now,' I said. 'This pair are my favourite – the one on the left with the scratch is Puff – she comes up here sometimes to say hi.' As Puff tripped and tumbled over rocks, the little girl laughed with delight.

I then panned across the colony, introducing her to every pair of birds that I'd picked out and named. Fulmars, kittiwakes and guillemots covered, as I reached James and Michelle, I could see some unusual activity in their nest. 'Wow, we've got lucky here,' I said. The raucous razorbill pair were turning their singular egg and we got a glimpse. 'I can see it,' the girl said with excitement. I was pretty excited too – every turn of an egg or return of a partner created so much anticipation for the moment I'd finally see a chick.

Next an elderly lady called me over to ask me what was happening at the ledge under the platform. There was movement in Freddie and Fanny's nest – their chick was finally hatching! 'The first baby of the nesting season,' I shouted. I couldn't believe it.

The young life was weighed down by the gunky membrane of the egg. So much had happened inside five minutes. It was an honour to watch these changes unfold hour by hour, day by day. After the gunk gradually fell from the young bird's face, one of the parents regurgitated mushy fish from its stomach, providing the chick with its first meal. Before I could talk to anyone else, my radio rang. It was Eleanor. 'Ruby, we have a bit of a bird problem in the centre.' She assured me that she wasn't joking so I rushed up the many steps to see what on earth was the problem.

When I arrived, there were pigeons everywhere – in the toilets, behind the till, in the office, on top of the vending machine. There must have been about twenty inside the building, with a further fifty outside. Half the visitors were laughing, half were flustered. I asked Eleanor how it had happened. 'It happened so quickly!' she said. She'd gone to clean the toilets only to find this chaos when she arrived back.

I could see that the pigeons had rings on their legs and when I looked closer at one, there was a telephone number. It turned out that they had come from a town in Devon, over three hundred miles away. The person who picked up confirmed that they were racing pigeons and said that they'd either find their way home or try to survive where they'd ended up. I was all too aware that if they stayed here, a peregrine would probably eat them.

The first course of action was to try to shoo the pigeons out of the centre using mops. Most left but there was a young female that used the automatic doors to her advantage and kept coming back in as if she owned the place. She perched on top of the vending machine, refusing to leave.

Eleanor announced that her name would be Barbara. 'Yes, I love it so …' I started to reply but before I could finish my sentence, Eleanor added a funny twist. 'Look at the name of the vending machine that she's sitting on. It's called "snacky" – why don't we call her Snackybabs?' 'Even better,' I said, nearly crying with laughter. For the remainder of the day, #Snackybabs was the star of the show. Everybody had to get a selfie with her.

* * *

When I got home that afternoon I felt ready for another solo adventure. I unpacked my work things and filled my bag with paints, pencils and paper before setting off to cycle to the East Lighthouse. In a summer dress with my muddy hiking boots on and with the wind through my hair, I felt free. My eyes were tired from squinting at work and my muscles felt tight from the constant stair climbing, but there was nowhere else I wanted to be.

As I sped up Rathlin's gorse-lined lanes, a view of the north-east side of the island opened up and before long, I could see the lighthouse's revolving light. I discarded my bike in a mossy layby before navigating my way through a path of spiky heather. Gazing out towards Scotland, trying to identify the neighbouring islands of Islay, Mull of Kintyre and Jura, I wondered who could be looking back at me from across the sea.

I wandered over to the cliff edge which sat above a noisy sea cave. There was a weathered information board beside the lighthouse that told an infamous tale. The Scottish king, Robert the Bruce, had just suffered a devastating defeat. He had been forced into exile and found himself stuck on Rathlin, hiding in a cave. Robert realised he had two options – to flee or to attempt another battle for Scottish freedom. As he weighed up his choices, he spotted a spider hanging from a thread on the roof of the cave.

Six times Robert counted the spider trying to swing to a nearby rock to affix a line and create a web. Six times the spider failed but it did not give up. Robert the Bruce had attempted to achieve Scottish independence six times and, like the spider, he had failed all six times. On its seventh try, the spider succeeded and Robert took it as a sign that he should try again. He returned to Scotland and the rest is history. This story is also why lots of people think spiders are lucky.

As I lifted my head from reading, a dart of orange shot across the sky – a puffin. It came in, circling the lighthouse like Puff did when I was at work. I kept watching as the bird fished in the craziest tidal surge, pushing deeper and deeper down into the cold sea. It looked hard; the prize must have been worth it. Like Robert, who tried, failed and tried again, this puffin kept bobbing up and down in the crashing waves, time and time again. Eventually it surfaced and flapped vigorously out of the sea to a grassy spot on the cliff. It sat for a well-deserved rest with half a dozen sand eels in its beak. I focused in with my binoculars, watching the puffin gobble them up.

I laid out my blanket, together with a flask of tea, biscuits and my notepad. 'I found Robert's cave,' I noted. 'Every day this place feels more magical.'

There was a particular anxiety that I used to feel that didn't exist here, a heaviness that lifted. Trying to find a word that would explain this feeling, I dotted down 'detached' but I didn't mean it in any negative sense. For the first time in my life, I was detached and the only thing that mattered was me, the slowing beat of my heart and the pounding crash of the waves. As a child of divorced parents, the only well and able-bodied sister of three girls, and a lifelong, hyperactive ADHDer, guilt consumed me on a daily basis. But not here.

In England, for example, I felt pressure to see my dad more. I'd spent the majority of my life living with my mum and the guilt I felt about it ate me up. My ADHD energy, compulsivity and eccentricity have often felt a burden. I might have been labelled as 'loud, outspoken and distracted' in school but here on Rathlin, or doing the jobs that had brought me joy in recent years, my employers, reviews and references described me as 'confident, knowledgeable, attentive and alert'. As someone who's senses are heightened, every flicker of light, noise and movement can throw me off track but in the presence of nature, I feel 'normal,' my head is clear, yet full of wonder and fascination for everything around me. My neurodiversity is a positive force for good, I know that now, years later, from seeing the world in a different angle.

My sisters Lydia and Mabel have had to deal with what no children should – illness and disability. Lydia has Ehlers-Danlos syndrome, a connective tissue disorder, which causes dislocations and aching joints. In my younger years, I felt guilty that I was physically healthy, but mentally, I was struggling. Being surrounded by ill health led me to a state of hypochondria and anxiety.

I didn't have the life of a normal fifteen-year-old. On the day my peers at school were taking their first GCSEs, I was cycling to Poole Hospital A&E to give my mum fresh pants and socks, and a few snacks. She was at Mabel's bedside, watching her deteriorate from anorexia. I then left my mum and sister to head home and look after Lydia, making sure she had taken all her medication and eaten a decent lunch. The next day, as I got to school for exams with no small effort, I was taken aside by a SENCO (Special Educational Needs

Coordinator) and told off for not attending the day prior. Despite knowing my situation, she went on to say, 'You are so silly. You need good grades. What if, say, your mum died – how would you get a decent job to look after your sisters?'

For all the years since, I'd carried all this anxiety with me. But there, by the lighthouse, like a wound gradually healing into a lightened scar, I could feel tension from the past starting to leave my body. I thought again about the puffins on the lower ledge, and wondered what Puff had been through to get her scar.

The distance between this tiny island and the mainland had psychologically detached me from problems I used to face – if only I'd known it was this easy. I slurped my tea, feeling lucky to have met some wonderful friends and discovered new places. I was grateful to have found a new, bird-resembling family. The landscape and its wildlife had welcomed me with open arms.

With a coastline like my beloved South West Coast path, this place was full of familiarity yet so far away. In discovering this new place to call home, I'd found a landscape where the earth met the sky – a land where I could begin again and learn to fly.

Chapter 5

Flourishing

When the world sprang to life.

The morning of the summer solstice I woke incredibly early. This longest day of the year, when there is usually glorious weather and the loudest birdsong, always lifts me.

In 2018, I went to Stonehenge. It was incredible to experience an ancient pagan tradition that celebrates nature. After a night of music under the stars, the sun rose through the gap of the henge and the summer officially peaked, shining a renewed hope over the land. Rathlin gave me the same sort of feeling as I headed outside that morning.

I sat, enjoying the sunshine. It was my first day off in a week, but the only place I wanted to be was the Seabird Centre. It had been a whole twelve hours since I was last there … With absolutely no logical reason there was something in my head telling me that something would happen at the Seabird Centre on the summer solstice and I was worried I'd miss it. I decided to distract myself by going for a swim. As I did several lengths of the harbour I came across the adorable line of eider duck chicks hiding behind a boat.

I climbed out and dozed on the wooden pontoon. Before long I'd turned into a sun-dried tomato, with salt crystallising on my skin. I heard the ferry coming in and saw that someone was waving to me from the deck. When I sat up, slightly dehydrated, colourful speckles littered my view. I rubbed my eyes and soon realised it was Craig. He hadn't told me he was coming to Rathlin. Overjoyed to see him, I waved back and jumped off the pontoon to swim home and greet him.

Craig told me that his granny had improved a bit and she'd told him to come and see me. This was amazing news – I was delighted that Lizzie was doing well – but, with no warning, I hadn't been able to prepare for Craig's stay at all. Even though it was just me in the house, there was a super-strict RSPB rule that friends and partners could not stay in property owned by the organisation. I was so tempted to smuggle him in, but this was an island, and word would spread. Then I remembered Jane mentioning her spare cottage on the outskirts of the Craigmacagan Trail to the east of the island. I told Craig we could either camp or stay in an old cottage a mile away. At first he didn't think I was being serious and laughed, but then realised by the look on my face that I was.

Craig knew I was running low on supplies so had brought me some fresh food, including four cartons of orange juice, fresh croissants and chocolate. We sat at the kitchen counter, enjoying breakfast together. As we demolished the last few crumbs, Craig asked if we could explore my favourite places. 'Well, the westerly cliffs where I work are my favourite,' I said. I'd been longing to go to the centre that morning but I'd had no excuse until now. 'Let's go see if any chicks have hatched,' I suggested. So we hopped in the car and left for West Light.

Eleanor was at the welcome desk and looked very confused as we arrived. 'And what are you doing here on your day off, Mrs?' I explained that Craig needed to see all my favourite nesting pairs and that I had a feeling some different species eggs were going to hatch today. Looking surprised that I couldn't stick even one day away, Eleanor told us to enjoy and waved us goodbye as we ran down to the platform.

I showed Craig Puff and Finn, and Busy Lizzie, who – as usual – was on her patch of grass. He loved watching them. I really hoped to see some activity in the nests. Most of the auk parents had been turning their eggs for days and it felt as if a gorgeous chick would hatch any minute. I asked Craig if we could stay a while before I showed him another part of the island. It was approaching midday and the sun was blazing. Craig was getting impatient asking how long we'd have to wait. Just five minutes later it happened – a ball of fluff emerged from its shell. The razorbills tilted their angular necks back, almost laughing with joy. The parents nudged the young hatchling to encourage a sign of life and then created a winged wall across the nest, ensuring nobody would come anywhere near their brand-new chick.

Even though I wasn't working I ran up and down the seabird platform to tell all the visitors what had happened. 'I knew it!' I said smugly to Craig, 'It's the solstice – nature always provides today.'

After snapping a few photos, Craig asked me where we could head next. I suggested the Kinramer South trail. 'The skuas that nest there have made me too scared to walk the path alone,' I admitted. 'Since you're here, perhaps we can do it together?' 'Essentially you're saying, please come, I don't want to get dive-bombed alone?' he replied. 'Precisely,' I said.

We left my car at the Seabird Centre and set off to walk the trail. It felt like old times in Cornwall – just us, a hiking bag and the ocean. We stumbled upon an incredible section of steps that led to the sea. We were tempted to descend but it looked too overgrown so we moved on.

Halfway around the trail I could see West Lighthouse from an entirely different angle. As I looked towards the seabird reserve from a neighbouring section of headland, I admired how proudly the white lighthouse stood over the bellowing waves and contemplated what a feat of engineering it was. We sat for a moment and drank some water before Craig opened up, telling me how hard it had been on the farm. He explained how seeing his granny get better then worse then better again was like a cruel rollercoaster. 'Thanks for moving to Northern Ireland, for leaving Cornwall – I know you live here, but it feels as though you're close by.' We stood up and carried on hiking, squeezing each other's hands.

At the path's summit we could hear the skuas call, all while we were talking about Craig's granny, about death. The eerie atmosphere created by the sound made me want to leave. We headed back to the car, leaving behind the skuas' territory and the heavy subject at hand. 'There you go, you weren't eaten alive by skuas, so maybe now you can hike here without me?' Craig said. 'Do you want to be left here alone?' I joked.

We spent the afternoon touring round the rest of the island before visiting Rathlin's highest hill at Ballygill. Blissfully unaware it was so late, we came through a thick layer of cloud to find an amazing sunset. Knocklayde Mountain on the mainland, the Paps of Jura in Scotland and where we stood were the few peaks clear of this blanket, which stretched across the neighbouring isles. The landscape felt otherworldly. Craig had needed this break and I could see that Rathlin was already improving his state of mind.

When we arrived back, I darted over to Jane's to collect the keys for her cottage. I put something warmer on and we left for a few pints of Guinness at McCuaig's Bar. Liam and Sean, the RSPB island wardens, were there having a drink so we joined them. I introduced Craig and they all got on well instantly, talking about tractors and soil. There must be something that clicks when farmers meet farmers.

Liam asked what our plans were for the night and suggested that we should keep an ear out for corncrakes. Corncrakes are surprisingly small birds, little bigger than blackbirds, and very secretive, spending most of their time concealed by tall vegetation like nettles. Their presence is given away by their unusual rasping call. Corncrakes are summer visitors that migrate to Africa for the winter – their bright chestnut wings and trailing legs are unmistakable, but sightings are rare.

Intensification of farming practices had caused the corncrake to become locally extinct on Rathlin and throughout most of the Northern Ireland mainland. Now the island farmed in harmony with nature, and after a few years of volunteer-led conservation work replanting vital nettle habitat that had been lost, Liam was hoping for signs of recovery. He'd already heard two birds call towards his side of the island, but hoped for more. 'Tonight it's mild, the day is long and the vegetation is high – I'd expect to hear the corncrakes call loud and clear,' he told us. Excited by the prospect of hearing a sound that had been restored to this ancient landscape, I asked what time would be most likely. '2 a.m.,' Liam replied. 'The best wildlife encounters require dedication.' We were up for the challenge. I bought a bottle of wine from the bar for our late-night adventure and we left the pub with a skip in our step.

Bags on backs, we headed off on the coastal road towards the Craigmacagan Trail. This route rose high above the northeast side of the island and would give us the best chance of hearing the birds, I thought. Not long after we set off Craig shouted, 'Look!' and pointed to the shore. He was absolutely certain he'd seen a flicker of light. I walked closer to the glowing tideline and realised it was bioluminescent plankton – tiny marine organisms that can emit light when disturbed by a predator or motion. A real rarity to encounter – Mill Bay became a whole lot more magical that night. The stars were bright and the sky was clear and the world felt alive around us.

I couldn't resist a dip in the sea so I stripped off and ran in. Craig sat on the sand and skimmed smooth pebbles onto the sea's shimmering surface. Every bounce lit up the patch of water surrounding it. It was beautiful, it was midsummer, but it was still the north Atlantic Ocean … I soon got dried and dressed and we carried on hiking.

We opened the gate leading to the Craigmacagan Trail, and after half a mile of climbing, found a bench and sat patiently waiting for the corncrake's call. I cracked open the red wine for a late-night tipple. Craig held my warm hand and I rested my head on his shoulder, pulling in tighter to his side. We sat in silence, happy to hold each other. It was that perfect sort of silence where neither of us had to speak, because we knew exactly how the other felt. In that echoing silence, came a rasp like no other. It was the corncrake. We were delighted. 'What time is it?' Craig asked. It was 2 a.m. on the dot – Liam had been right.

We drank up and clambered back down the trail, getting closer to the calling corncrake, which seemed to be nestled beside Jane's cottage. Like a stick being run over a metal school fence or a comb being flicked quickly, the call was like nothing I'd heard before. At first, it sounded like there were four or five, but it was in fact just one corncrake's mighty call bouncing off abandoned old buildings nearby. I wanted to bottle the crakeing sound, the flickering stars and vibrancy of the sea so that I could revisit this wild night whenever I wanted. To me, the corncrake's cry was a reminder of the importance of using our voices for species whose sounds are disappearing from our world. The corncrake's call became louder through the night. We didn't sleep one bit, but it was so worth it.

The next morning, we left Jane's cottage to get some breakfast at mine. Once we felt a little more alive, I oiled my bike while Craig fetched a skateboard and some rope from my car. Recreating one of our favourite Cornish pastimes, Craig tied one end of the rope to the bike while I held the other, so that the bicycle was pulling me along on my board. We got weird looks going through the harbour.

At East Lighthouse boats travelled past us, away from Islay, most likely carrying litres of whiskey from its distillery. 'Islay' means island in Scottish and though is spelt differently to the English version, is still pronounced the way you would say 'Isla.' We realised how much we both loved that name. 'Would you call your future kid Isla?' asked Craig. 'Definitely,' I replied.

We could see the next Rathlin Ferry coming in from the other side of the island – Craig had to head home as he had a shift at work in a local garden centre that afternoon. We raced back to the ferry and I waved goodbye.

During the remainder of my afternoon off work, I sunbathed in Church Bay. A painted lady butterfly landed on my face, inspecting a daisy I'd put behind my ear. Kevin and Meta walked past me as I snoozed in the midday sun. 'Hello there,' Meta called. 'What're you doing tonight?' 'Nothing, I think,' I replied, in a bit of a daze. Kevin suggested we go for a hike together since he and Meta were leaving in a week. June had disappeared before my eyes, and I was upset they were leaving so soon. We arranged to meet by Rue Point at six.

When we met up later Kevin was attempting to pull a sheep out a sheugh, and did so pretty successfully. Five seconds later, it ran away. 'Right, let's go,' Kevin called, acting as if nothing had happened. We walked towards Rue lighthouse, reminiscing about our wildlife encounters over the last few weeks. Meta looked teary. 'I'll never forget this month here with you!' she told me. Neither of them wanted to leave. It was beautiful to walk along the coastal path in the evening sun. But after a while, a cluster of common gulls became irritated by our presence and started to dive-bomb our heads. We ran back along the lighthouse path towards our cars with twenty-something birds swooping down on us from every angle, squawking.

Back home, I didn't feel like staying inside, so wandered

around the tideline, collecting shells to decorate my room. I met Ali on the beach – she was looking for a volunteer to help her to move her donkeys across Rathlin to a new field. It was late but I didn't mind – I was delighted to help.

We walked up to collect her three beautiful donkeys. I realised I'd met one of them on my first full day. She shook a bucket of food to entice them over and then put a halter on each of them so that we could safely lead them. 'I've never walked a donkey before – any tips?' I asked. 'Just don't stand behind them,' Ali replied. 'Got it,' I said, way more confidently that I felt, now worried that one of them would kick me. Ali gave me the lead rope of one of them and we set off. Meandering down the roads from Kebble Cottage, we headed towards a field beside Church Bay. Every so often the donkeys would get a rush of excitement and start trotting ahead, pulling us forwards, but for the most part, it was relaxing. The donkey I was leading insisted on eating every bit of greenery that lined the road. 'We don't have time for this,' I told it.

Ali asked me how June had been as we walked, so I told her all about our incredible encounters at West Light – the whales, skuas, seals, and so much more. She then told me all about her time working at the Seabird Centre and what it had meant to her. It felt so wonderful to be part of the history of this place, at least for a little while. Ali told me about how many more puffins used to line Rathlin's cliffs. I felt both sad and hopeful that one day these numbers could return.

'Right, here we are,' Ali called. We released the donkeys and they galloped around the new, fresh patch of meadow that had a better view than most people's bedrooms. The donkey I'd been leading found a nice dusty patch of mud and started rolling exuberantly. I headed home to bed, excited to return to work the next day.

All too aware that Kevin and Meta would leave in a few days, I soaked up every minute on the cliffs with them and did my best to exhaust their rich collection of travelling stories. Rachele's partner Eddie had made them a thank you present – a hand carved puffin with their names engraved on it. We said goodbye, and Kevin and Meta promised to come back soon.

The morning after their departure everything felt slow; we could all feel Meta and Kevin's absence. By contrast, the seabirds were racing back and forth to their nests and the cove was louder than I'd ever heard before. The horizon was a dotted black line of auks, flying across the summery sky in their thousands. Hunting for fish was their top priority.

A few minutes after I positioned myself down beside the lower ledge, Puff emerged from her burrow, wide-eyed and seeming a little confused. She went back in, and then came out and then waddled back. What was going on?

As I looked around me, scanning the cliffs for predators, I came across lots of new auk chicks on every ledge including

Ottie and Orla's guillemot hatchling, which was still half in its shell. I started to wonder if Puff and Finn's chick had hatched. Finn landed with a huge catch of eels and took them down into the burrow – was this the couple's first baby feed? Watching a species' breeding behaviours when everything takes place underground can be confusing and involves a lot of guesswork.

Busy Lizzie came tumbling in way too fast, rolling over before flapping herself back up the cliff. She preened herself in the morning sun before closing her eyes. I'd never seen a puffin sleep before, mainly because this usually happens in burrows, tucked away. She looked so peaceful. I often wondered how old she was, especially because she reminded me so much of Craig's granny. I had some clues that she was a good age. The exact proportions of a puffin's beak change with the passing years. With time, the bill deepens, the upper edge curves and a kink develops at its base. One or more grooves may also form on the red portion of the beak as the bird ages. Busy Lizzie's beak ticked all these boxes, so I knew she was an old bird. By contrast, Puff and Finn were sprightly. There was no certain way to determine their age but I figured they must have been young as their bills weren't as broad as those of adults.

My morning visit complete, I headed to the platform and opened the binocular hut, ready for the first bus of people. When the visitors arrived, the new chicks were the main attraction. As I looked up to the razorbill couple, James and Michelle, I could see their little hatchling stretching its wings on a rather dangerous ledge. It was so tiny and vulnerable; it blew my mind how well its parents protected it from the prevailing winds. Slightly below the guillemots, the kittiwakes looked busy. Focusing in on Kitti and Kat, I could see their fuzzy chicks in the warm glow of the sun. The kittiwake ledges had already seemed full, but now they were completely crowded with newly hatched, hungry chicks, and there wasn't a single space left.

This week, with so many young birds hatching, the predators were out in full force, just as they were at the start of egg laying. My heart broke when I saw a young kittiwake, just hours old, being carried away by a great black-backed gull. It was also pretty cool.

That night, the harbour was busier than usual and tourists had come from miles around to stay a night in one of Rathlin's B&Bs. As I ate my dinner, a very large yacht came looming through the tiny entrance. I couldn't help but feel it looked so out of place here. I wandered outside to take a look when another, much smaller boat arrived with a Cornish flag on the bow. It was full of young people. I ran over to ask where they'd come from. 'Falmouth,' a girl replied. 'I've come from Cornwall, too. It's great to see you,' I called. 'Right on!' the crew shouted back.

Later that evening, Rachele called me to explain she would be bringing the new residential volunteers over the next day, and asked if, in the morning, I could focus on getting the centre really tidy, to make sure they had a great first impression.

The next day the weather had turned and the sea was so rough that the ferries couldn't sail. With nobody here, I was left with no other choice but to enjoy the adorable seabird chicks. The downy fluff of the guillemots and razorbills, the pom-pom-like appearance of the fulmars and teeny tiny bodies of the kittiwakes kept my smile wide all day. Although I loved it when the platform was full of visitors, it was nice to have the place to myself for once.

Just as I arrived home, Liam called out to me from a field behind my cottage. He was holding a small, brown food-waste bin. He opened it slowly and showed me a corncrake chick. 'A cat caught it,' he told me, 'but we managed to get it back and it seems to be okay.' It was hard to believe that – with all the threats facing vulnerable species like corncrakes – pet cats could be so dangerous too. I stared at the chick's long legs, which were twice the length of its body. It was strange but gorgeous. I was incredibly lucky to see a corncrake – never mind a chick – since they are notoriously hard to spot. Liam had located the nest where he thought this corncrake had come from and set off to return it to where it belonged.

The next morning Rachele arrived bright and early with the new volunteers, Paddy and Beth. Knocking on my door after catching the earliest ferry, Rachele introduced me to the excited couple. They were clearly ecstatic to be here and I knew we'd get on. 'Right, shall we show them around the reserve then?' Rachele suggested, which was definitely a nudge to me to get dressed. Hearing Paddy and Beth's oohs and aahs on the way up to the centre reminded me of how blown away I'd felt when I'd first arrived too.

We stopped at Kinramer Cottage so that Paddy and Beth could drop their luggage off at their new home. Rachele went in with the couple to give them a short tour while I stayed in the car. While I waited, a sound déjà vu struck a chord inside me – a high-pitched 'pew, pew' whistle, sung again and again. I felt like I knew this sound but couldn't quite put my finger on what it was. It clicked after a few more bursts. Stepping out of the car, I looked up to see two skylarks bouncing around above my head. Seeing them took me back to being a child on the South West Coast Path, trudging along sticky muddy trails, looking up to see the skylarks. 'Pew, pew,' my sisters and I would shout, mimicking their calls as we pretended to fly along the narrow cliff-top path. I hadn't heard this call since I was little.

This melody had reignited memories I'd completely forgotten. During a quick Google, I read about the skylark's steep decline across the south of England in recent years, which is probably why I forgot who they were. I thanked Rathlin for reminding me. In that moment, I wondered what other species and the memories woven into my encounters with them I might have forgotten.

Rachele, Paddy and Beth were soon back in the car and after a few minutes of driving, we arrived at the centre. I gave Paddy and Beth the grand tour and a rundown on the previous weeks' sightings. Then I took them both downstairs and their faces lit up. Our eyes were all instantly glued to the tiny chicks. They were doing such a great job of clinging on to their patches on the rocky shelves. They instinctively knew to hold on.

Paddy and Beth asked if they were allowed down to the lighthouse platform, 'I just saw some puffins down there, you know,' Beth said. I knew ... I wanted to take them down there but felt worried we'd get caught away from the viewing platform. I also didn't know if more than one person's presence would scare the puffins away, plus, I enjoyed gatekeeping my little relationship with them. That platform was my secret spot.

'It's just that … it's too messy down at the bottom, we may get in trouble, but we can go into the lighthouse itself,' I said. 'I could show you the lighthouse keeper's bedroom?'

'Let's do it,' Paddy replied.

The bedroom had been left untouched, as it had been years ago, as if a keeper still lived here. A yellow lighthouse keeper's coat hung up above the bed – I was tempted to try it on, but didn't want to move anything. The job meant being alone for months on the side of this exposed cliff in all weathers – it wasn't for the faint hearted. A radio sat beneath a leaking window, ready to hear news of an incoming storm. I could only imagine how hard a life this would have been, but the view probably made the hardships worth it.

Rachele radioed down, asking if we were ready for the first bus of people, 'Yeah, all good down here,' I replied, before running up to the top floor to open everything. 'I'm not too sure I know enough,' Paddy said. So I pointed everything out to him – the Big Five birds, the binocular hut, the guides and the spot where people could have their lunch. A crowd of tourists arrived at the platform. 'If you guys want, you can listen to what I say to the visitors for this first bus?' I said. 'You can steal my facts and have a go when the next lot come down. Does that sound okay?'

Proudly pointing out my seabird friends to the visitors, I could see Beth and Paddy's inspired faces out of the corner of my eye. I talked about the puffins' courtship and pointed down to Puff and Finn, before describing the fulmars' puke defence mechanism, highlighting Freddie and Fanny's tiny chick below our noses. I then turned everyone's attention to the sea, emphasising the depths guillemots can dive to, while encouraging people to look for Ottie's and Orla's poo-marked heads on the sea stack ahead. We also looked above to see

James, Michelle and their chick snuggled on their ledge, and turned to Kitti and Kat, whose precarious nest was overflowing with chicks – three to be exact.

A second busload of people clambered down as the first lot left, so Paddy and Beth had their turn talking to visitors. They did so well. I walked up and down the platform with binoculars, making sure everyone could see the wildlife they wanted to and took my camera out to capture the chicks with their parents.

As the afternoon wore on, a 'fuck, it's warm' flew from Paddy's mouth as he wiped the sweat from his forehead. The concrete platform really locked the heat in. 'It's due to get warmer,' Beth said. 'In fact, I think we could be approaching a national heat record tomorrow.' She'd heard that the temperature might reach 32 degrees. We were definitely in for a hard shift tomorrow but I wasn't concerned about us – it was the seabirds I was worried about. These north-dwelling birds weren't used to such high temperatures and there wasn't any proper shade for their chicks.

After a few hours' more work, everyone hopped in my car and we raced down to the bay, dripping with sweat. Before Caitlin left the island that night, she joined us, desperate to cool down. We watched staff from the Hungry Seal dive off the pontoon – fully clothed. If we'd been hot working outside all day, I couldn't imagine how they'd coped in a chip van. 'Our turn,' Beth shouted. We pulled on our swimsuits and leapt in, letting out screams of joy. The cold sea's embrace had never felt so good. We swam back to the shore and lay on the warm sand before heading in for dip number two, then three. An old windsurfing board lay marooned on the rocks beside the beach, just waiting to be pulled out. Caitlin, Beth and I attempted to pop up like surfers on a wave while Paddy pulled us along the shoreline. My chest was sore from laughing. We all left to go home high on sea, sun and seabirds.

Driving up to the harbour to collect some staff the following morning, I rolled down the windows for some cool relief, but was met with a draught of heat. I like warm weather but this … this was far too hot. That morning seemed like any other. We opened up and I walked down to get the platform ready and greet my beloved puffins before the staff and visitors came down for another day.

But something just didn't feel right. It was quieter than usual, the heat was sickening and the light was blinding. Most mornings I'd stand and scan the colony upon arrival, but that day I felt too exhausted by the heat so sat inside the lighthouse for some respite. The first bus of people arrived – I walked out onto the platform and began pointing out seabirds. But when we turned to look at the cliffs, we were met with scenes of suffering. A few days of this warm, dry weather and a line of fulmar chicks lay dead. They had been parched in the sun, literally baked to death. I was horrified. After everything – all the energy and time to get here, the mating, nest-building, egg-laying and hours of protecting – it was awful that the chicks had died because of the weather. I explained this to the visitors, who were equally upset by what they saw.

Seabirds spend long periods at sea, searching for enough food for their growing chicks. Although the parents take it in turns to fish, allowing one to stay near the chick, if considerable time passes, the parent protecting the chick will also flee the nest in search of sustenance – hungry, weak and in need of some food itself. Climate change is increasing the sea temperature, driving seabirds' food sources into deeper, cooler waters, and forcing parents away from their nests for longer stints in search of food. This longer time away from the nest can leave the chick vulnerable. Freddie and Fanny's chick was tucked away in a sheltered crevice – it was lucky to have escaped the sun's lethal heat.

I couldn't help but feel guilty. Yesterday afternoon, while we were playing in the sea and sunbathing, these tiny, helpless chicks were starving as they baked in the rays of sun that killed them. It was hard spending the rest of the day on that platform, with the consequences of my species' actions staring me in the face. I dropped Eleanor back at the ferry before embarking on a solo dip at Mill Bay, a lot less cheerful than I had been than the day before.

The next morning news stories filled my social media feed, showing pictures of Brits basking in the sun, with articles talking about the extreme warmth in such a positive light. As I drove into work, in pain for the fulmar parents who'd appeared visibly distressed by their chick's death, a man was being interviewed on the radio, talking about the benefits of a warming climate. He discussed how growing tropical fruits and British wine will greatly benefit the 'economy'. A gift shop owner was talking about how her local business was thriving due to the influx of tourists in the heatwave, again talking about the climate crisis as if it was some sort of lovely benefit to society. How was the media turning this global emergency into an interesting talking point? I wanted to call in and tell everyone why this was wrong but ended up just screaming into my steering wheel with rage.

I felt trapped between two conflicting worlds. I was about to enter a reserve that had been ravaged by the effects of climate change, yet all the bulletins could talk about was a thriving economy. But what they failed to see is how there would be no economy, business, tourism or life without a healthy functioning planet. I felt so disconnected to the human realm and went to sit with my seabird friends, grieving the needless loss of at least twenty young birds. Before long the female peregrine came swooping in to take the dead chicks' bodies. At least they're feeding something else, I thought, trying to find something positive to cling onto. I ran down to check on Puff, Finn and Busy Lizzie and was comforted by their presence.

On the Friday, after days of rocketing temperatures and mounting high pressure, the darkest clouds I'd ever seen rolled in, soaking the sea pinks and sun-baked soil. The rain was so refreshing and much needed. There was a smoky hue rising from the smacking rain over the top of the warm sea.

Paddy, Beth, Rachele and I stood outside, getting soaked in the lashing rain as if we were plants in need of a watering. We contemplated the weeks' losses while looking ahead to a new day. The clouds departed quickly – it had been like a sporadic, tropical downpour, fleeting yet flooding. The sun peeked through once again and we felt renewed.

Rachele asked us if anyone fancied a barbecue that night. She suddenly remembered we were all plant-eating youngsters and added, 'a veggie barbecue of course…' We were all keen, and I promised to collect some rolls from the Co-op shop on my way home. It felt as though we covered every topic – including religion, climate change and, of course, birds – as we ate together.

As we shared our stories and experiences, we found one that bound us together. We were all going through or were very close to someone experiencing loss. Paddy's mum had just passed, something I couldn't even begin to comprehend; Rachele's granny was slowly slipping away after the death of her grandad last year. Beth and I were partners of people experiencing deep waves of grief. What I soon realised from our conversations was that we had all ended up here a little lost, in search of some healing – and here we'd also found each other to bounce off. After the most spectacular sunset, we packed everything away and I dropped Paddy and Beth home.

After such deep conversations, I found it hard to switch off so parked beside the path that led to an old coastguard hut and sat alone looking towards Scotland. The moonlight shimmered bright and a lone sailing boat slowly crept past. It was the Cornish sailors I'd met before. I jumped onto the bonnet of my car, wrapped a blanket around me, opened my sketchbook and drew the faint, dark outline of Islay, mesmerised by the bright summer sky. Nearly a quarter of a mile away from land, the people on the boat started shouting, waving at me. Once they'd caught my attention, they pointed to the left side of their boat.

When the fluke of a great whale rose high above the saline sea I wondered if I was dreaming. It sank back into the black beneath and I never saw it again. Was that real? I thought. I rubbed my eyes, unable to keep up with Rathlin's dreamy offerings.

Chapter 6

Gliding

Be brave like a jumpling.

As the summer went on, Craig and his dad managed to get lots of work done to our caravan. It was now nearly finished. Craig sent me pictures on a daily basis and before long he was moving furniture in through tiny caravan doors. Our empty oblong now resembled a home.

I couldn't sleep much in late July. Rathlin is much further north than I was used to and it was light almost all the time, with just a sliver of dark blue sky in the early hours of the morning. One night was so bright that – even with my blinds fully drawn and sheets covering my face – it felt as though the sky's light could still find its way into my eyes. I was off work the next day, so I accepted that I wasn't going to sleep and walked out to the harbour.

I lay at the edge of the harbour's rickety pontoon watching shooting stars, feeling tiny on this island, perched on a rock called earth, floating in space. I could see every inch of detail in the rocks at Rue Point. It reminded me that I hadn't been to that area in a while, so I set off on my bike to the south-east tip in the early hours of the morning.

The seals were sleeping: there were a few on the rocks and a few in the water, bottling. Bottling is a description of the way seals sleep in the sea, tilting their heads upright so the water doesn't splash over their faces. While they sleep, like most marine mammals, seals use unihemispheric slow-wave sleep to keep one half of their brain active while the other rests to ensure they can regulate the opening and closing of their nostrils. They look just like bobbing bottles in the water. I love whoever coined that term.

I got off my bike and I walked the final stretch to Rue Lighthouse. The water had been so calm by the harbour – it wasn't until I reached the outer edges of the island that the tidal currents seemed so wild and fierce. I spent hours rambling along the coast, clambering up and down to find tidal pools, often with fish in them that had been stranded when the tide went out. The world around me slept as I wandered alone, in awe, with the moon and stars as company. My senses were heightened, tuning in to the nocturnal world around me. The torch on my phone lit up the bright eyes of hares in the distance and the corncrake's call cut through even the loudest of crashing waves.

The sun met me at 5.30 a.m. Cormorants and shags circled Rue, out looking for an early morning catch, as the sun's glow crept up behind the headland. I cycled back to the harbour flying down hills, accidentally eating flies.

Halfway home, I stopped for a cold swim. The island felt

serene. No ferries, shops or people were operating yet. I sat on the shore, trailing my fingertips through the surface of wet, sticky sand and accidentally flicked a periwinkle from a sandy hole. Its tiny body emerged from its bright, orange shell and it made its way across the beach. Then I came across the teeniest tiniest sea star I'd ever seen. Beside the sea star was a collection of cockles, and slightly beyond the cockles came a flock of turnstones, doing exactly what their name describes – flipping over stones to find food. At low tide, purple sandpipers started to forage, searching for worms. It was all very peaceful, but then a freak wave flooded the tideline, disturbing all of the species from their morning routines. It was my cue to finally go home.

All of these encounters with tiny beings reminded me that there is always a little life going about their day somewhere nearby. They do not care about you, or me, or any of our worries, and I find comfort in that.

On the final part of my journey home I glided along the coastal road and spotted some storm petrels tightly skimming the sea's surface. Northern storm petrels are among the most challenging birds to study, yet we know they too are declining. Feeling far too confident, I took my hands off the handlebars and held them out, mimicking wings. Gaining more momentum as I headed downhill, for a moment I felt as if I too were flying.

I decided to use this Sunday off work for some overdue self-care. I scrubbed the sand from my body, sleep from my eyes and mud from my feet, plastering up scratches from barnacles and moisturising my weather-beaten cheeks. That night I slept so well, catching up on two nights of sleep in one.

Monday came quickly and I was excited to be back with my birds and their babies. Arriving to work after a flurry of heavy rain the night before, I noticed that the seascape was a shade or two darker and petrichor scented the air. A couple of rock pipits that nest in the lighthouse's guttering piped up, singing sweetly for visitors. With every flap of this bird's wings came a shooting high-pitched noise. They flew around the colony looking for invertebrates among the rocky terrain. I wondered if they got on the seabirds' nerves, like a bee hovering a little too close to a picnic.

I'd not been away from the reserve for long, yet I could see such a difference in the chicks. I could just about make out the kittiwakes' distinctive black collars beneath their downy fluff. The guillemots' and razorbills' frizzy fur coats seemed smoother and more grown up. Only the fulmar babies seemed exactly the same, like white pom-poms with beaks.

I scurried down to the lighthouse to see my three puffins as soon as I could. It had been a whole twenty-four hours and a catch-up was long overdue.

Busy Lizzie was being nosy, peeking her head into Puff and Finn's burrow. Within a matter of seconds, she was shooed away by Finn, who was being highly protective. That was most likely because they had a chick tucked inside. I wondered what it looked like.

On the day it hatches, a baby puffling would neatly fit into the palm of someone's hand but it grows quickly into a larger ball of fluffy down feathers. When the young bird is between thirty-eight and forty-four days old – when it fledges the nest – it resembles an adult but has a greyer face and bill. Puffins only gain their distinctive colourful bill when they nest during the breeding season, so a chick doesn't look like the iconic puffin we all know and love until it's between three and five years old. I didn't know when Puff and Finn's chick might have hatched but I hoped to see it appear sometime in August.

I sipped my coffee behind the grassy mound and Puff came

to investigate. Stomping her big, orange feet, she tilted her head and looked up at me. I wondered what she thought. Finn then called and Puff wandered back off. As I hurried back up to work I met Beth, who told me she'd seen me at the bottom platform and asked why I was there. Caught red-handed I had no excuse so I came clean. 'I was seeing those puffins you spotted the other day, but please don't tell anyone else I was there. I'll take you down, but for now let's forget about it.' 'That's amazing,' Beth replied. 'So, every day, when you disappear, you're seeing them?' I promised to take Beth down the next morning before our shift.

The next day, we headed down to the lower platform and I showed the puffins to Paddy and Beth as we peered over the vertical drop. It felt nice to share what was such a big part of my Rathlin experience with two like-minded, lovely people.

Over the next days, nests turned into preening stations as the seabirds' young lost much of their downy fluff. Some kittiwake chicks grew faster than others and I wondered if this was down to certain birds being favoured by their parents. By the weekend, a handful of kittiwakes were receiving flying lessons, hesitantly launching from their nests to feel the air flow under their feathers for the very first time. Circling the cove, always with at least one parent nearby, they wobbled in the wind, sometimes crashing into rocks before trying again.

I met Jane as I drove home. 'I'm having a meal at mine tonight – would you like to come?' she called from her car window. 'Absolutely,' I said and promised to bring crisps – Jane's and my favourite.

I got changed out of my uniform into some nicer clothes and headed to Jane's. Liam and Ali, and Jane and her husband were sitting around the table with a selection of photo albums and an array of food. No one else around that table followed a plant-based diet, yet everyone had made something that I could eat. Ali had even made a vegan tiramisu just for me. I was so touched. Everyone had brought their photo albums to show me Rathlin as it had been for the last hundred or so years.

I saw Liam as a young boy, exploring the island, Ali when she first arrived and Jane with her parents hiking around Rue Point as a child. I felt so lucky to be shown such precious memories. Most pictures had been taken before phones existed and every single shot felt like a slice of treasure. Witnessing how much I was falling in love with the place, Ali and Jane wanted me to see what Rathlin meant to them and the history that connected them to this special island. I was touched that they showed me. I noticed the time – it was after midnight, so I thanked Jane for such a lovely evening and headed home by the light of the moon, thinking about all the people who had lived here before me.

Back at work, many of the young auks were nearly fully grown. Every nook and cranny was full of chicks eager to fatten up and fledge. All the auk parents were so close to achieving their aim: one summer, one healthy chick – a chance to create a new generation of seabirds. But there were predators everywhere – the ravens, peregrines and great black-backed gulls were circling, keeping a watchful eye out for fledglings.

Nearly all 150,000 guillemots were expected to leave the sea stacks over the next week, encouraging their three-week-old fledglings to follow. These part-grown and flightless guillemot chicks are famously known as jumplings because they jump when they fledge the nest, sometimes falling hundreds of feet to the water below. Their small bodies are surprisingly tough and can take quite a hard impact. Usually the female guillemot parent helps their chick down slowly to lower ledges while the male parent waits in the sea to greet it after its turbulent journey. Once in the water, the jumpling will feed at sea with its parents for a few weeks before flying further out.

Since auks rely on safety in numbers, when one chick decides to fledge, many more follow. Because of this, there's usually a certain evening when the jumplings start jumping and then within a few days, nearly all the guillemots are gone. This jumping phenomenon usually begins on a mild July evening as the sun goes down and it starts to get dark. The darkness means the chicks are less likely to be targeted by a predator.

There was an aching in my chest as I observed the chicks' behaviour over the following days. I felt like a mother whose children were about to leave home. I'd watched Ottie and Orla for so long and couldn't believe that they would only be here for a short while longer. After just a few weeks of being alive, their chick was about to take the biggest leap of its life – a jump into the unknown.

Every day Paddy, Beth and I watched them even more closely than before. The strong bond between the seabird parents and their chick was quite remarkable. Occasionally, we saw both parents and their young nestled up beside one another and momentarily – in their little nook – the hectic nature of the seabird colony seemed peaceful.

One sunny Tuesday afternoon at work during late July the anticipation was rapidly growing. The guillemot chicks' wings were flapping, and their parents started nudging them, preparing the tiny creatures for their massive leap. All the guides that I'd read told me that these behaviours indicated that the departure was getting closer. I checked the forecast for the next evening and it was good. This was surely going to be my best chance to witness this magic.

The next morning at the cottage, I packed breakfast, lunch and dinner. I wanted to spend the entire day on the cliffs to see as much as I could of the guillemots and razorbills before they left. At lunch I hid in the binocular hut, not wanting to spend one second away from the nests. I peered out of the window and had a clear view of Ottie and Orla. It seemed there was a lot more movement today – their chick was waddling into other couples' nests, exploring further than it should ... almost looking as though it was itching to fly the nest.

I closed up that evening, waving goodbye to all the staff, volunteers and visitors and told Rachele I would lock up and head home, but instead I stayed. Once I'd locked the doors I ran back down the stairs to set up my camera and eat my dinner.

The sun was setting behind the furthest sea stack and there was a sound like nothing I'd ever heard – the guillemots' moans were long and loud, like sirens echoing up the cove. I could no longer make out Ottie and Orla's nest as the cliff became shadowed in darkness but I was able to see some nests slightly closer to me. The male parents left first, flying gracefully down to the water, from where they then started calling at the top of their lungs to their chick, still on the sea cliffs high above. Though the parents all sounded the same to my human ears, I knew that each chick could recognise their parents' call and would be drawn down to it.

I took my time over my dinner and went to the platform edge as the sun finally began to sink into the horizon. I was there to see the first guillemot jump. Hopelessly flapping its tiny wings, it had no chance of flying, but did the best it could to get to the sea. My heart was in my mouth – it felt like this brave chick, the first to go, was falling for over a minute. Eventually the tiny jumpling fell into the sea beside its calling parent and was embraced by an incoming wave. It was okay.

Before long, night fell and under cover of relative darkness, many more chicks fledged to the sea. I watched for a few hours more – the visibility was surprisingly good because the skies remained so bright. A cacophony of noise echoed around the lighthouse and I wondered what the other seabird species made of the guillemots' loud departure from their nests. I wanted to stay all night, but I was working early the next day so reluctantly headed home.

The next morning I told Paddy and Beth all about the jumplings, and asked them to stay and watch with me that evening. There were still quite a few chicks that needed to fledge. Paddy and Beth agreed immediately.

So that night, I did it all over again. After heading back to my cottage for a quick dinner, I collected Paddy and Beth from Kinramer. They were incredibly excited. As we arrived at the centre, we could see that the clouds out to sea were doing something incredible. The high pressure had led to a natural – but very rare – phenomenon called Kelvin-Helmholtz clouds, caused by two separate layers of air in the atmosphere moving at different speeds. The clouds looked like rolling waves. It was an incredible start to our evening.

Paddy, Beth and I sat together on the cliffs, like children with front-row seats at the cinema. Dusk brought the calls of tens of thousands of guillemot parents echoing up the cliff walls, and we watched the chicks jump.

These wild nights are engraved on my brain forever.

Over the coming days, the cliffs emptied and only a few slower, older guillemots remained, including a bridled pair that I'd not seen before. Bridled guillemots have a white line around their eyes and are quite rare – I was so glad to have seen them.

While some razorbill chicks had already fledged alongside the guillemots, the majority of them still had to leave. Within each seabird species, the birds can arrive and therefore leave at different times. This means that each stage of nesting can be slightly staggered, with some birds leaving later or earlier than others.

One afternoon towards the end of July most of the razorbills were starting to make their journey to the sea, except they did so in a slightly more cautious style than the guillemots. Not all of them jumped; in fact most razorbill parents slowly guided their young over the sheer rocky faces – a kind of reverse rock climbing. They hopped down from their nests ever so slowly – some chicks took hours to arrive at the sea. While the razorbills usually wait for the cover of darkness, that day most didn't and many were taken by gulls and skuas. It was sad to see ... I hoped that, at least, James and Michelle's young would make it to the sea alive. Paddy, Beth and I spent our shifts watching them intently, feeling so nervous for the chicks that it was hard to narrate what was happening to visitors.

That Friday, my dad surprised me with a last-minute visit to Northern Ireland. He had some annual leave to take, so turned up with a few days' notice. A keen cyclist, he planned to visit me on Rathlin and cycle to all three of the island lighthouses in one day. First Rue, then the East and finally to see me at West Light. I was working on the platform and received a radio call from Eleanor, who was upstairs, to tell me that my dad had arrived. She jokingly said he was very handsome and laughed. 'No one needs to hear that about their dad!' I replied – and looked up to see him running down the steps.

We ran to meet each other at the bottom of the steps and hugged. I'd not seen my dad in months and I'd really missed him.

Almost immediately I handed him some binoculars and pointed out James and Michelle. 'These razorbills should start leaving any minute. I've watched them all summer long! Their nest building, egg laying, the chick hatching and growing has all led up to this moment.' We waited patiently together and eventually James and Michelle's family set off. 'Look, now, now, now!' I told Dad, before running up and down the platform urging every visitor to bear witness to this departure.

The young chick was moving off its ledge. A flap, flap, flap indicated that it was ready to go. It steadied itself before taking a big leap away from the only comfort it had ever known. It wobbled everywhere, then fell, then tumbled, then picked itself up again. The parents, never too far behind, gave it encouraging nudges every so often. An hour later it had arrived at the sea, but if I hadn't had a watch, I would have sworn that it had taken days. Now camouflaged in the sea with its counter-shaded body, it was safer from predators. Countershading is an adaptation that I find so clever – most auks have this outwitting outfit of a white belly and a black back. The white blends into the sky above and the black blends into the dark sea below. At sea – with nowhere to hide on the waves, no burrow, no ledge – these birds rely on their evolutionary excellence. The tiny razorbill chick would now float at sea for a few days before venturing out deeper into the vast Atlantic Ocean.

There was something so nice about having one of my own parents beside me to watch seabird parents help their young fledge. After sharing this experience, we arranged to meet in the harbour for a drink after I finished work, and Dad headed off to get some food.

Hour by hour, the cliffs were emptying of auks and as the thousands of birds gradually left, the noise started to soften. I hadn't prepared myself for how this would make me feel. The puffins were due to leave throughout the first few weeks of August – just around the corner – and then there would be no auks left – just me, the kittiwakes and the fulmars.

The platform was lined with captivated faces that day. Although there were also some tears for the chicks that did not make it, it was overwhelmingly positive for us and the visitors. Paddy and Beth looked emotional: it was already their last day, and like the razorbills, they too had to leave Rathlin that night.

I was dropping Paddy and Beth to the ferry for 5 p.m. so we had to lock up the centre quickly. I parked outside Kinramer Cottage so they could collect their bags and then we made a trip to the harbour. 'It has been so wonderful meeting you. I'll miss you guys,' I said, giving them both a huge hug. Every time any residential volunteers left, I felt emotional, partly because it felt as though they were taking a chunk of the seabird season that we had shared with them. Each couple experiences a unique month that would never play out quite the same way again. There are few people who can understand how living on Rathlin made me feel – but these volunteers, though only here for a month, they got it.

I parked up and walked round to the pub to see my dad. Already two pints in, he looked merry. He bought me a drink and told me all about his cycling adventures, and about his new home. Since I'd moved to Rathlin, my dad had taken a leap into the wild himself. Giving up his house and half his belongings, he had moved into a Mercedes 609D in the heart of the New Forest, one of few places in the southern counties of England still offering some wildness. He had worked in the NHS for nearly his whole life and was mentally fatigued so decided to make a radical lifestyle change, not only to save money, but also to connect with nature on a much deeper level. He'd struck a great deal with the landowner – looking after her vegetable patch and chickens in return for a rent-free field.

He congratulated me on finding my own slice of heaven. 'This is the life, Rubster.' 'Well, cheers to that,' I said, as we clinked glasses.

I didn't drink that often before Rathlin, but somehow since living here I'd turned into a regular at the pub. It was the friendships – neighbours and locals all wanting to meet over 'a few' all too often.

Because of the house rule, my dad had to stay in Jane's hut, so we drunkenly wandered the mile-long road to his home for the night. 'Right, some things you have to know before I leave,' I said. 'If you hear a very loud bird, that's the corncrake; if you're not too hungover in the morning, watch the sunrise at Craigmacagan; if you're needing a pee … well, you'll have to go outside; if I'm already at work when you are up and about, return the keys to Jane.' I was working the next morning so we wouldn't see each other until I returned to the mainland for the weekend, so I gave him a hug goodnight before walking back down the road to my place.

Saturday was different: no volunteers, no guillemots and barely any razorbills. I had a rare shift upstairs, helping in the visitor centre to sign people in, but all I wanted to do was sit with the birds. Rachele asked me to restock some items in the gift shop and refill the vending machine; then after lunch I could go downstairs for a final shift with the razorbills before heading off the island. I got into a rhythm, just trying to get these mindless tasks done quickly. Before too long I had nearly completed my jobs and was all set to go for some lunch, when suddenly someone inappropriately touched me from behind. My stomach turned. I quickly turned to see who it had been, but whoever it was had fled. There were two or three men that it could have been, who had been standing behind me. I felt powerless and angry. It made feelings I had buried deep down re-emerge. I had been sexually harassed before, quite a few times, it's something many women have to deal with in life, but I didn't expect it to happen here, where I had felt so safe.

I ran down the steep steps to the seabirds with an hour left before the centre was due to shut. I told them what happened, like they would understand. I wished I could fly off with them out to the Atlantic too. The incident made me look at every man on the platform differently. Was it you? Or could it have been you, or you? I thought, tormenting myself.

I was initially stunned but after a while I told Rachele what had happened. By that time, that busload of visitors had left the reserve, so Rachele wasn't able to find out who it had been but I filled out an incident report form and we talked about ways to make the reserve as safe as it could be for all of us who worked there.

I left work and drove to the ferry, determined not to let this man's actions ruin my day. Stepping onto the *Spirit of Rathlin*, I clung to the safety bars as the incoming swell rocked the boat. Feeling vulnerable – like a ship to the elements – but strong enough to keep pushing through, I walked up the creaking metal steps, to a perch on the middle floor. Ever since I'd seen the dolphins bow-ride this boat, I'd always had my camera with me, on and ready during every crossing.

The ferry set sail and left the safety of the harbour. We swayed back and forth, and some passengers became seasick. I closed my eyes and took myself to the place I always went when bad things happen. I daydreamed about a scene under the sea, where I was at one with auks, diving to depths of hundreds of metres. Suddenly a wave at least five feet high crashed over the side of the ferry. I frantically wiped the salty water off my camera and laughed at the shock. Thirty minutes into the passage and we were only halfway – the push and pull from the swell had slowed the journey so I stared out towards the horizon, trying not to feel sick. Standing still, attempting to balance on the slippery, ocean-washed floor, I glanced over to what I thought looked like a breaking wave, a rather large breaking wave ... That's unusual, I thought. Maybe a dolphin? Again, a minute later, a spray of seawater came shooting upwards. I focused on what I at first thought was a solitary dolphin leaping out of the water, but from my years of wildlife guiding I soon realised that the magnitude of the spray was too great, too perpendicular ... This was no dolphin!

I managed to focus my camera in on the section of water this creature had just disturbed, hoping it would breach again in that exact location, and sure enough it did. When I watched it leap for the third time, my mouth flew open in disbelief. I instantly knew what this was – a thresher shark! I screamed very loudly, so loudly that an islander called Ruari asked me if I was okay.

Within a matter of seconds, the entire floor of that ferry knew about the thresher shark. Some people could see the shark when I pointed it out, but most missed it. I couldn't help but cry – I knew that this was a once-in-a-lifetime moment. I captured the shark's penultimate leap, which was incredibly graceful and metres high out of the water. I caught three pictures of it breaching, flying and falling – which transformed my day in three simple seconds.

What makes this encounter all the more extraordinary is that the thresher shark spends most of its time in the deep, cold waters of the open sea and rarely strays into coastal areas. To survive in these colder waters, they have evolved to be endothermic. This means that they can keep their body temperature higher than the temperature of the surrounding water. They do this through a specialised heat exchange system, which allows them to conserve heat produced through internal body mechanisms such as muscle shivering. Thresher sharks use their extremely long tail to hunt. They herd smaller fish into tight shoals, swim at them and thrash their tail like a whip, stunning some of the fish and making them easy to catch.

I couldn't help but feel this was another massive nod from nature, like my first puffin encounter; it felt like a thank you of some sort, a much-needed boost. This shark, a symbol of struggle but also strength, came out of nowhere when I needed it most.

I called Craig, then my dad, then some old colleagues from the wildlife boats to tell them all about it. I don't think anyone really believed me until I sent them the pictures. That's when everything changed. I showed the ferrymen my images – they were blown away. I uploaded them to social media and within half an hour I was contacted by BBC News Northern Ireland.

Dad had borrowed Craig's car and picked me up in Ballycastle. I felt bad being so distracted but he understood the craziness of the situation and waited proudly as I was interviewed by journalists. Before I knew it, what had happened that day at the Seabird Centre was completely behind me. I was swamped with media enquiries and the local Irish Whale and Dolphin Group reached out to me, telling me that I had sighted and photographed the first recorded thresher shark in Northern Irish waters. 'Well, you can add that to your CV,' Craig said when we arrived at the farm. Craig showed my dad and me around the newly renovated caravan – it was beautiful and so modern that it felt like we were standing in a very stylish apartment. My dad turned to me, looking impressed by all Craig had achieved and said, 'You've got a keeper there.'

A momentous day called for celebration – Craig's parents, Alan and Rhonda, bought us all pizza, and we enjoyed some quality time together. Alan asked me all about Rathlin life, so I started telling him about the corncrakes. Looking bewildered, he interrupted my story. 'I know corncrakes,' he said. 'They used to arrive here on the farm every summer when I was a young boy. I haven't heard their call for twenty-odd years.' I could suddenly see the child in Alan, a middle aged- North Antrim farmer, come back to life. He reminisced about the excitement he and his siblings would feel when the corncrake calls echoed across the fields, and described the way they would try to find these elusive birds nestled in the nettles at night. He went on to explain how he could never sleep due to their constant calls, and I understood. But I could only understand because I got to live somewhere special, somewhere unusually biodiverse, where conservation was at the heart of life.

Unless we have the privilege of living somewhere relatively wild or knowing people who tell us stories of how nature was before it drastically declined, we don't understand how the world should feel, smell and sound. In fact, most people of my generation are experiencing something called shifting baseline syndrome.

Shifting baseline syndrome is a phenomenon that describes

the way in which each generation gets accustomed to the state of nature as it is. My generation wasn't alive fifty or so years ago so doesn't truly understand what we've now lost. It's a dangerous thing, because if I hadn't listened to the stories Alan tells of corncrakes on the farm, I wouldn't be able to visualise what that was like, and I would accept today's standard as it is. We simply cannot do that. With worldwide nature declining by 67 per cent and UK nature by 41 per cent, most of us alive today, especially young people, have never got to experience what our environmental landscapes should actually feel like.

My farmor, which is Danish for father's mother, instilled a love of bird watching in me. Every year she would buy the new RSPB calendar and test my knowledge of the birds that were pictured each month. While I memorised species well, I realised I hadn't seen many of the birds in real life, although my farmor had. My grandma Margret, my mum's mum, would get me to help in her garden, and often pointed out the birds on her roof, also acknowledging ones that she didn't see any more. My grandad Lester, who passed down his beautiful binoculars to me, would often tell me about encounters he'd had with species that I had never even seen. This generation all lived in a world with much more nature than the one I grew up in and I envy them for that.

Did you grow up to the sound of skylarks? Did you hear corncrakes in the countryside local to you? Have you ever seen a large starling murmuration with the backdrop of a crisp winter sunset? The more time that elapses after species loss, the more likely we are to experience widespread shifting baseline syndrome, meaning that we're running the risk of creating younger generations that are completely detached from nature, unaware of what's missing.

That weekend I kept the assault at work to myself and

revelled in the aftermath of my thresher shark sighting. On the Sunday, Dad, Craig and I decided to explore County Donegal together. Even as we travelled along the most remote roads of Redcastle and inlets of Inishowen, press were trying to contact me about the shark encounter. I had three telephone interviews and, much as I loved that the story was being shared so widely, I craved some peace. Soon the signal dropped, the calls stopped and we arrived at Malin Head, the most northerly point of Ireland. The horizon was so clear that we could see for miles, the water was so blue and still that it mirrored our reflections. Without hesitation Craig and my dad jumped from the high rocks but I stood back for a second or two, gathering my courage. My stomach scrambled looking down at the ten-foot drop. If the guillemots can do it, why can't you? I said to myself and leapt in. It was instantly refreshing.

Once submerged, I could see the sun's rays piercing through the photic zone, highlighting the moon jellyfish and planktonic species surrounding me.

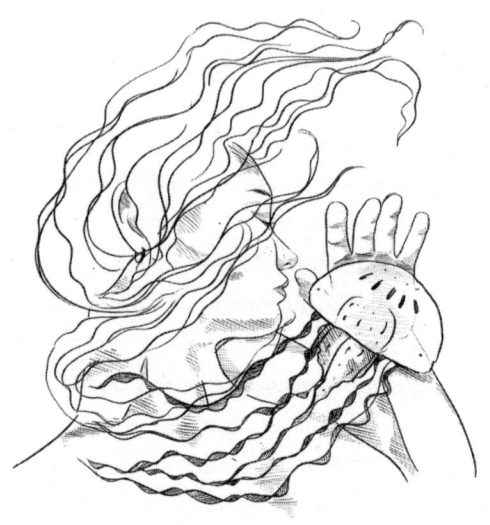

I floated, mesmerised by these sparkling organisms. My sea-bleached legs wafted around, stirring the marine soup. The crystal-clear waters of Malin made us believe, just for a few minutes, that we were in the Caribbean.

We swam then sunbathed then swam then sunbathed all day long. I could have repeated this cycle forever but it was time to go. As we drove home via the coast road I could see the West Lighthouse shining on the west side of Rathlin. It looked like a tiny white blob, all those miles out at sea. It was the first time I'd seen the Seabird Centre from the mainland. I can't quite describe the pull I felt back to it – that tiny segment of Earth meant a lot.

We took a turn inland, to a dual carriageway that took us nearly the whole way back to the farm. The mass of concrete, the lights, the car fumes, the noise was all too much. This human world felt more brutal than ever before. During the drive Craig told me that it was Lizzie's birthday the next day, and not just any birthday – she'd be turning the grand old age of ninety. I couldn't believe he hadn't told me before now. Luckily I had my sketchpad with me and spent the rest of the journey drawing a bunch of flowers for her. When we arrived back, I gave it to Lizzie and she was delighted.

I was sitting with her, talking about the flowers flourishing in her garden and all the life I saw it supporting, when Rhonda and Craig came rushing in to tell Lizzie she was runner-up in the Antrim in Bloom gardening awards. Craig had secretly entered her for the competition a few months earlier. We were all overcome with excitement, happy to have Lizzie's gardening efforts recognised. Craig shed a happy tear at the brilliant news.

The next day my dad had to leave, and I was back to work on Rathlin. After an emotional farewell, Craig left me to the slipway in Ballycastle before taking my dad to the airport.

When I arrived at the slipway, three fishermen came running up to me with questions about the shark. They were still stunned that I'd seen it and managed to get pictures. When I boarded the ferry, the captain took me aside to congratulate me and shook my hand. I blushed. The islanders boarding the ferry that morning were treating me like a celebrity. It was surreal, because I could still hardly believe what had happened myself, let alone make sense of it to anyone else.

Travelling back over, I had a newfound love for Rathlin Sound. I couldn't help but think of all the unseen life below the waves. I got back to the house, made lunch and drove up to the centre. I needed a quiet Monday on the cliffs to decompress. I couldn't wait to wander down to the reserve and see how many razorbills were left. Every step down to the cliffs was a step closer to what felt like home. I turned the last corner before the view opened up.

It felt so empty.

There were still plenty of puffins, fulmars and kittiwakes but the lack of guillemots and razorbills felt substantial. The guillemots' and razorbills' departure reminded me how fleeting life is – one minute I was longing to see their eggs hatch, the next they were all gone. No Ottie or Orla, James or Michelle or the other hundred and fifty thousand plus of them. All that remained of them was their poo, changing the colour of the cliffs from black to a whitewashed mess. There was something quite funny about seeing so much guano – with so few birds, August visitors would soon think it was all the puffins doing. The reality of two species leaving focused my attention on the remaining hubbub at the reserve.

I was glad to be back on wilder land.

Chapter 7

Surviving

When the predator took its prey.

Our new residential volunteers Kirsten and Tom arrived during the first week of August. They had travelled from St Andrews in Scotland, ready for an adventure. I took them from the harbour to the reserve and showed them around, worried that they would be expecting to see more birds. I felt sad that they weren't going to witness the auks in all their glory, but soon realised that you can't miss what you haven't known, and that the place still seemed incredible to newcomers. One glance down to the grass and Kirsten yelped with excitement. 'I'm so glad that the puffins are still here,' she said. Not for long, I thought to myself.

I took Kirsten and Tom down to the bottom of the lighthouse, my secret section of the reserve, to introduce them to my favourite puffin personalities. Now Paddy and Beth had been there I didn't mind showing others the lower lighthouse ledge so much. When we got there, I couldn't see Puff, Finn or Busy Lizzie on the grassy lower cliffs, but we could see lots of puffins rafting on the water below. Rafting is when puffins sit on the water together in large groups, a behaviour they display just before they leave their nesting sites to migrate. Usually during the last weeks of the puffins' breeding season, as their young become strong enough to take care of themselves, the parents raft on the water's edge, feeding and conserving energy in preparation for their long journey back out to the ocean.

So many puffins were bobbing up and down, floating in the breaking zone of the waves. Kirsten and Tom thought it was incredible. Although I hadn't seen this cool phenomenon before, all I felt was a terrible knot inside. I worried that Puff, Finn and Busy Lizzie had left for good and that I wouldn't see them again.

We had to leave the lighthouse platform and start work but I didn't want to go until I saw at least one of my three puffins. I'd never come down here and not seen any of them. Rachele radioed down. 'Will you three be okay on the platform for the morning? Ruby, can you show Kirsten and Tom the ropes?' So that was it. I had to leave. Once we got back to the viewing platform, I pointed out my selected pairs to Kirsten and Tom, but this time, instead of the Big Five, it was the Big Three… soon to be Two.

Kirsten and Tom flung their bags inside the binocular hut and made themselves at home. I was really enjoying their company after a few days alone. Visitors arrived so we stopped nattering to each other, and Kirsten and Tom watched as I narrated the antics of the cliff's residents to visitors for twenty minutes. After my spiel, they had their turn talking to visitors. They were both great with the public – so much so that I felt I could confidently leave them and take an extremely early lunch break. I was itching to run back down to the puffins' lower ledge.

Minutes felt like hours as I sat with my eyes glued to the burrow … before too long, my break time was nearly up. I'd lost hope but just as I packed my bag, ready to climb back up the lighthouse steps, Puff and Finn emerged with their gorgeous, beautiful, tiny puffling. I couldn't believe it. I'd never seen a puffling before – and the fact that the first one I saw was Puff and Finn's made it all the more special. The little bird looked nothing like a puffin, just a walking ball of grey fluff. Seeing their little chick come out from its burrow to the vast open world made me feel protective and proud. My emotions changed as quick as Irish weather, from loss to pure joy.

Before I had a chance to capture this moment on my camera, the tiny creature was shuffled back inside the burrow by its parents. Moments later, Busy Lizzie finally came flying home to her perch with, as ever, many sand eels in her bill. The record for the most sand eels held in a puffin's bill is eighty-three, but that afternoon, I reckon Busy Lizzie beat it. When she landed back on her grassy ledge, she laid the fish down and then gobbled them up, one by one. Halfway through her feast, a flock of black-backed gulls came swirling in to steal her food. 'Go away!' I shouted, getting caught up in the birds' wild fight for food. Falling backwards, Busy Lizzie started to tumble down the cliff. The gulls had stolen her food and flapped their wings so exuberantly, they'd made her slip. I didn't leave until I knew she was okay. She eventually re-emerged from behind the cliff edge, looking rather lethargic.

That afternoon I regularly peered down to Puff, Finn and Busy Lizzie's ledge from the higher platform, watching every bit of movement that I could. Elsewhere on the reserve, the kittiwake chicks were restless, all growing quickly in such tiny nests. The fulmars were super attentive, with the parents' swapping shifts, coming back every few hours to regurgitate fish to their chicks' desperate mouths.

At the end of the day I told Kirsten and Tom my plans for hiking Craigmacagan and they said they'd join so we set off a bit later from my cottage. We approached the loughs lining the trail and flickers of blue flew around the water – they were Irish damselflies. I approached, lying down on the wet bogland in an attempt to get a shot on my phone. Kirsten and Tom laughed as my body sank into the wet ground, but it felt wonderful to see the damselflies so close, knowing how endangered they'd become. Damselflies are beautiful and fascinating. Their courtship behaviour is elaborate and unique, with males and females linking together to resemble a heart shape. The nymph can be found in different types of freshwater habitats – including rivers, lakes, ponds and bogs – but they are becoming harder to find because they will only inhabit especially clean habitat. The Water Framework Directive found that none of Northern Ireland's 496 rivers, lakes and coastal waters reach 'good' status – no wonder damselflies are declining so fast.

We walked to the trail's peak, discussing our shared love for animals. Kirsten and Tom were setting a good pace, but I found myself lagging behind, distracted by bugs of all kinds. Hoverflies, cinnabar moths and wasps buzzed around hurriedly, like they were all running late for a train. Eventually we made it back to the start of the trail where I handed out coffee and a packet of biscuits for us to share.

The next morning I arrived to work feeling like I'd known Kirsten and Tom for years. They slotted into our centre routine well and were excited about everything – even cleaning the toilets. Scanning the colony day by day, I could see puffins leaving their burrows more frequently. We didn't have much time left with these beautiful birds. I had no clue how I would last a month and a half longer here without them. The pink thrift and oxeye daisies were starting to wilt as the puffins rafted in greater numbers, a sign that their season here was truly coming to an end. Baby puffins fledge after thirty-four to sixty days, and will head out to sea alone.

Just before these iconic birds leave to fly hundreds of miles back out to sea, they display some unusual behaviours – a period known as 'puffin silly season' on Rathlin. As their chicks get stronger, parental responsibilities reduce, so the breeding birds dart around the cove as if they are having fun with their new-found free time. This can also occur in March, before breeding responsibilities begin. Puff and Finn circled the lighthouse together; they would soon be leaving their breeding ground for another year.

Busy Lizzie, however, stayed put on her ledge.

I ate my lunch, watching Puff and Finn's every move, when Craig called. 'She's getting worse, Ruby. Palliative care are here – the nurses are putting in a bed and some weird machines.' I sat, stunned … Lizzie had seemed so well at her birthday. Although I'd known she wasn't going to get better, I hadn't expected her to go downhill so soon. 'Shall I come home now?' I asked, biting my nails as a busload of expectant-looking visitors poured onto the cliffs. 'No, it's okay, I'll let you know when to head back. We'll be okay.'

The next morning I didn't see Kirsten or Tom en route as I usually would and wondered if they'd mistaken today for our day off, but when I arrived at West Light, they were already there – I could hear them shouting joyfully, 'Another pod! Look, another one!' I ran towards the commotion, looked out to sea and saw thousands of common dolphins. I'd never seen a megapod like it.

The common dolphin is the most abundant cetacean in the world, with a total population of six million globally. Most ordinary pods consist of between two and thirty dolphins but on occasion, depending on the species, dolphins form superpods or megapods. These contain hundreds and even thousands of dolphins. Massive pods like this often occur when multiple groups of dolphins converge during hunting or mating. However, these large groups don't stay together for very long, so it's very unusual to see them. We had got so lucky. I asked Kirsten and Tom why they'd arrived to work so early; they explained they'd been out for a sunrise hike and had seen this megapod, so followed it round the cliffs to the Seabird Centre. Liam had been right – the best wildlife sightings really do require dedication.

Beside the dolphins were hundreds of gannets diving – there must have been lots of fish under the water. With its torpedo-shaped body, a gannet can dive into the water at sixty mph from more than sixty feet above the sea, closing up its two-metre-long wingspan and activating internal airbags to protect its organs just seconds before plunging into the water.

Seconds later, a line of small, black speckles appeared

alongside the dolphins, darting through the choppy sea. It was as if someone had sprinkled salt and pepper onto the surface of the waves. I grabbed my camera in attempt see what this could be and as I gained focus on one of the fast-moving blobs, I realised that they were Manx shearwaters! Magnets on the water, they are like hovering black sheets of paper, skimming above even the choppiest sea surface.

Although they are a rarity, this was the time and place to see them. As summer turns to autumn, Manx shearwaters start to return to their wintering grounds, often gathering in large numbers off the west coast of the UK and Ireland to feed before their long journey home. These birds are often seen out at sea and are very rarely spotted on land. They undertake a flight that sees them move down the west coast of Africa before crossing the Atlantic to spend midwinter off the coast of Brazil.

Manx shearwaters are known to be the longest living birds in Britain, the oldest recorded being 50 years, 11 months and 21 days old. It was ringed on 17 May 1957 on Bardsey Island in Wales and was caught by a ringer on the same island on 8 May 2008. These birds are the kings and queens of countershading – they have a chalk white belly and the darkest black back.

As if there wasn't enough marine activity, a minke whale made a cheeky appearance to the back of the dolphins, clearing up any scraps the megapod had left behind. What a feeding frenzy! I tried to film this spectacle on my mobile but no matter how hard I tried, I could not get the scale of it all in one shot. I put my phone away and embraced the incredible encounter which went on for a solid thirty minutes. Just as we thought it was over, a few hundred more dolphins came leaping past.

While we watched the dolphins I asked Kirsten about her

love of hiking. She told me about her many solo adventures around Kinramer. After the assault the previous month, I'd felt a bit afraid to be alone on a reserve, and the tales of haunted forests had got inside my head, but after hearing Kirsten's stories I was determined not to let what had happened stop me.

Rachele, Jean and David arrived. We felt awful that they'd just missed the dolphins. Then the Puffin bus arrived a few minutes later with more than thirty visitors on board, all ready to see some wildlife. Francis wandered in, looking for a coffee in return for a packet of oaties. He asked me how many puffins were left and asked if I was ok. He understood how attached the staff get to the birds. 'It's difficult. I have a few friends that will be hard to say goodbye to,' I said. I felt the heaviness that generations of Seabird Centre staff must have experienced at this time of year.

All this talk of puffins sent me running down to the lighthouse ledge, missing every other step, practically flying. Busy Lizzie was the only puffin there. 'Hey you, where's Puff and Finn?' I asked. On cue, I noticed Puff flying towards the ledge and Finn waddled in from the other side of the headland. The pair were spending less time here now their little puffling was on the edge of independence. Reconvening at the cliff edge, they sat side by side tapping beaks and let out the most beautiful, deep cooing noise. It was as if they were congratulating each other for all their hard work raising a chick. The puffling stood at the opening of the burrow, ready to leave any day. It would fledge just after the adults, following in their wingbeats, hundreds of miles out to sea.

Many people think we are the only sentient beings that can experience complex emotions such as love but I completely disagree. These puffins that I'd had the pleasure of watching every day over the course of months had showed more love and trust towards each other than I'd seen in many human relationships. An attachment so strong, it endures long distances, crosses oceans and lasts decades. A love like a puffin's is a love to cherish.

It was as I sat, marvelling at their companionship, that Puff clocked me. She seemed to purposefully walk in my direction before flapping up onto the grassy mound that I was hiding behind. I lay on my stomach, looking up at Puff standing over me.

There was a lengthy moment of silence as I studied Puff's orange-outlined eyes. Then she bowed and nudged two pebbles down the mound. They came rolling towards me; I picked them up and clutched them tightly. I couldn't believe it. Puff tilted her head endearingly and flew away, back to Finn. The feeling that Puff had given me these pebbles made me believe that everything was going to be okay. More valuable than gold to me, these little pebbles would be my memento of Puff and Finn.

The radio abruptly crackled to life. 'Rachele to Ruby, over. I'm down on the platform but I can't see you. I just need to talk to you about the binoculars.' This was bad – I should have been up there getting everything ready. 'Oh yes, I'll be right with you, won't be a moment,' I replied, racing up the stairs. Rachele told me we'd been losing binoculars – ten had been stolen in just a month, all costing a few hundred pounds each. She asked all of us staff to tie some blue ribbon to each pair, so that if anyone ever deliberately or accidentally walked out with them around their neck, we'd be able to spot them easily. As I tied bows to binoculars, I daydreamed of Puff and those pebbles.

The last Puffin bus approached and I was excited to end

my day on a high. There was just an hour left before I could leave work and get ready for a sunset hike but a group of men on a stag do had other ideas. One of them tumbled down at least half a dozen steps, hurting his hands and his knees. He didn't look great – parts of his chin, knees and palms had been skinned by the gravel path below. Using a first aid kit, I cleaned his wounds and applied pressure to the cuts on his knees, which were especially bad, before bandaging everything up. The man was mortified. 'Right, let's get you back up to the bus,' I said, helping him to his feet, but he said that he really wanted to see the puffins first. I found this quite sweet, given his battered state, but slightly annoying given it was closing time. 'Aren't you too injured to walk the rest of the way?' I asked. 'No, I'm grand. Thanks for all your help! I'll buy you a pint at McCuaig's,' he promised. Rachele told me I'd done more than enough for one day and said I could go home.

After dinner I drove to a vantage point just beside Kinramer cottage, parked my little car and set off.

I navigated my way to the starting point, a small lough beside Ballygill, where I came across my first ever twite – its call reminded me of a stadium full of people cheering. I walked closer to the lough and a coastal path began to make itself known. I felt like I'd stepped back in time as I wove in and out of fairy trees. As I climbed higher, the entirety of Rathlin and the surrounding sea came into view. I hiked this path for a few miles before entering a dark wood – this section scared me slightly and I felt relieved to leave it. As the sun started to set, it lit up the lavish lines of spider silk that were woven throughout the heather, and its warm tones poured into the navy sea. A snipe sat on the edge of some fencing, watching my every move and I stopped to take it all in. I was halfway through the walk, and things were going well, but it would be dark soon and I was still a few miles away from the road that would lead me back to the car so I followed the path down into some fields.

In the distance I could see a herd of inquisitive cows slowly walking towards me. Among them was a big bull. I knew I shouldn't run, but it's all I wanted to do. I resisted the temptation but I sidestepped as quickly as I could across to the next field only to find that there were many more. I knew that the cows belonged to Liam so I called him. 'I need your help,' I said. 'I'm eating me dinner, what's the matter?' he replied. 'How friendly are your cows? I'm currently finishing the North Kinramer trail and, well, I'm a little stuck.' I thought he'd hop in his pick-up and come to the rescue, but no … 'Right, well, they're friendly enough, just give us a call back if one starts chasing you, okay?'

Not very reassuring but it was the best I was going to get. As I walked through the field, contemplating how unglamorous a death via cow crush sounded, the cows that had been following me at a distance began to run. 'Oh crap,' I screamed. I ran towards a section of stone wall that led into another field and jumped over it without thinking. I'd escaped the cows but was now covered in heather and gorse pricks, bleeding with a hundred-odd tiny cuts. I decided cuts were better than cows, so tiptoed through the field of spikey heather and gorse to reach the tarmac road that led to my car. This hike had been as rewarding as it had painful. As I checked my cuts and looked for ticks I got a call from Liam. 'Well, ye escape the cows?' he said, ever so casually. 'Just about,' I replied.

As I reached my car, the island plunged into darkness and a shooting star shot across the horizon, so I made a wish. When I finally got home I put Puff's pebbles on my windowsill, next to my other treasures – my shells, flowers, and seabird eggs and feathers. I cleaned all my scrapes and scratches, and felt so excited to tell Kirsten about my solo adventure – proud that I'd got out there again.

* * *

Wandering down to Puff and Finn over the following mornings, I watched as they gradually spent far less time on land. Their little puffling began to appear more often, peeking out of the burrow. On rare occasions, I'd catch a glimpse of Puff returning to the burrow, checking her chick was ready to leave. The puffling's naive curiosity made me feel anxious – there were so many obstacles ahead, so many miles to cover, so much to learn. The thought of climate change or some other human action threatening this chick's future brought me great sorrow. But that was out of my control, and in that moment I had to remind myself of a line I'd written in my diary earlier that summer: 'You can't change the world, but you can change your world.' I did not hold the power to ensure that this puffling would go on to live a safe, healthy life, but I could be part of the puzzle to get it there. After seeing the energy, resources and love Puff and Finn had invested to raise just one healthy chick, I was more determined than ever to make a positive impact on this world, for future generations of human and puffin alike.

The next week, more puffins left, thousands every day, making me feel lonely on the cliffs. Nearly all the auks had abandoned us and each new morning brought less time with my beloved trio. Scared to see them go, I sat with Puff, Finn and Busy Lizzie, thanking them for being such great friends; for being there every morning; for putting on magical flying displays for visitors; for being beautiful and unique.

Later that working week, there were very few visitors at the centre. It poured sporadically and all remnants of the guillemots and razorbills were being quickly washed away. The quieter rainy days were always cleaning days. Scrubbing pungent, fish-smelling guano from the safety bars, I peered down to the lower ledge hoping to catch Puff and Finn checking on their little chick, just one last time. Straining my eyes to see the hundred-odd puffins rafting on the sea below, I saw a seal floating on its back just metres along from the stacks. At first I thought it was dead, but when I focused my binoculars I could see it wriggling ever so slightly. It had a dark teal fishing net tangled around its neck. It must have been in great pain. I had to do something, at least try, so called up all the local wildlife charities, hoping one could come to its rescue. I also called the warden, Liam, but nobody could help. I even considered climbing down there myself but this seal – in the waters of a protected, busy, seabird reserve, being tossed and turned by waves, sixty feet below – was too hard for anyone to reach. It suffered for hours before it died. Devastated, I broke down crying, startling every visitor in the reserve that hadn't been aware of what was going on.

Every needless loss felt so huge, so unfair. Why was an animal so beautiful dying so that someone, somewhere could have fish and chips. I felt angry. Rachele asked me if I'd like to run the visitor centre upstairs since the seal incident had distressed me so much.

A few hours later I remained watery-eyed, unable to move on like everyone else had. Living on Rathlin, spending most evenings or mornings swimming with seals, it was like I had just watched one of my own die. This passing felt personal, like an attack on nature, an attack to my wild friends. That night I swam at Mill Bay and apologised to the seals on the rocks for what the humans had done to their friend at West Light. On the walk home, I picked up every inch of discarded rope I could find and filled a wheelie bin with mackerel line, monofilament and rope as well as crab crates. Turning my anger into action, removing heaps of pollution from the seals' haul-out spot, I felt a bit better.

The next morning was Friday 13 August and although I am not usually superstitious, things weren't going my way. I'd woken up to a punctured tyre so had to change it before driving to the reserve. When I got there, Kirsten and Tom had pale faces, as if they'd seen a ghost. 'What's the matter, guys? Are you okay?' I asked. They didn't answer. Whatever it was, they couldn't find the words to tell me.

Feeling sick with anxiety, I looked over the edge to see that the last raft of puffins had gone.

I raced down to the lower ledge shouting for Puff, calling for Finn, as if they'd hear me and come back. Arriving at the bottom of the reserve, probably completing a personal best for the time it took to get there, I couldn't see anyone but Busy Lizzie, who was on the grassy patch. 'Have they gone, for certain?' I asked the wise old puffin. Of course she couldn't reply, but it felt like a no.

I stayed on that ledge as long as I could, waiting just in case they'd come back like they had a thousand times before, but I soon came to realise this really was the end of their season. Just as I thought that was it, Puff and Finn's puffling emerged fully from its burrow for the first time. It had grown so much since I last saw it but it still looked small. I radioed Kirsten and Tom. 'Look down here now, the puffling – I wonder if it's going to try to fly away.' They came down to watch.

With our eyes glued to the minuscule bird, we watched it gain in confidence, flapping its wings vigorously on the edge, before leaping off the grass and flying out to the open sea – the last time it would feel grass between its flippers for several years. I couldn't wait to see Puff and Finn again, to tell them how well their little one did, how brave it was when it jumped, but I'd have to wait another eight months to deliver that news, if they were to make it back, that was.

All that should have remained now were kittiwakes and fulmars, but Busy Lizzie was still there, alone. The only puffin left on Rathlin Island it was almost as if she accepted that this would be her last season. She sat still on the open patch of grass; it was the slowest I had seen this little bird. She had no strength in numbers, no burrow and no partner. I was scared for her.

Saturday 14th came. The cliffs were quiet, it was raining and just a few hardcore birders, battling the elements, had made it to the reserve that day. Through the mist and rain a man called me over, worried that Busy Lizzie was being tormented by ravens. I could see the man staring with fright through his binoculars, like he was watching the crescendo of a horror film. It was hard to watch her grow weaker every day. With the puffins and their chicks fledging through the start of August, a predatory presence again hugged the cove. Skuas and falcons scanned the air and there was an underlying feeling of anxiety.

While I ate my sandwich beside Busy Lizzie's lower ledge, Craig rang. 'Has it … happened?' I asked worriedly. 'No, not yet, but the nurse thinks it could be anytime from now. Will you come home, even just for tonight?' Craig replied. 'Of course, I'll be right there,' I said.

I spent a moment more with Busy Lizzie, uncertain whether or not she'd still be there when I got back. She was wobbling all over the place, and her last five meals had been stolen from her by bird bullies. She just didn't look well – her crimson beak was fading, almost as if her life was draining away.

This puffin wasn't just a puffin – by this stage Busy Lizzie felt like family, as did Puff and Finn. She had listened to all my worries and been a constant comforting presence – all without saying a single word. I wished that I could spend longer with her, but no amount of time would ever be enough. Craig needed me and I had to go.

I sat up, dusted the mud from my backside and psyched myself up for leaving, unsure what my evening would hold, Busy Lizzie cocooned herself in the burrow that Puff, Finn and the chick had once nested in. She must have felt cosy, finally sheltered away from the elements after all that time outside. I could feel it, she really didn't have another journey left in her. 'Right you, I'm off, behave!' I said jokingly, trying to evade a sad situation. I thanked her, blew a kiss and before my eyes could water any more, started my journey up the hundreds of steps to the top.

Once I was back up at the visitor centre, I explained to Rachele why I needed to dart off. She understood and told me to go. I drove my Nissan faster than I ever had before and headed to my cottage. I packed so quickly that pieces of underwear were falling from my unzipped bag as I ran to the ferry. I looked out towards the harbour from the top deck as we neared Ballycastle and saw Craig waiting in the car with his head in his hands. I was upset to see him so upset and disembarked as quickly as I could before getting in the car. I kissed Craig's cheek, put a song on and held his hand. He couldn't speak, numb with pain. We sat in silence the whole journey home but occasionally I squeezed his hand to remind him that I was there.

When we arrived at the farm, the wet weather circling above the hills reflected all of our moods. Everyone close to Lizzie was at the house: a few cousins, brothers and sisters of Lizzie, Craig's auntie Linda, Alan, Rhonda, me and Craig. There were so many of us but so little conversation. To break the silence, I offered everyone a cup of tea. There was a universal yes, so I made eleven cups of tea, all with different amounts of milk and sugar as Craig went to say goodbye to his granny. I watched him walk down the hallway to her room; I'd never seen him so empty looking.

I handed out drinks, passed around some biscuits and then walked down the corridor into Lizzie's bedroom. Craig was sitting by her side. I came in behind, hugging him, again not sure what to say. Although Lizzie was unable to open her eyes, Craig told me she could hear him, he was sure of it. So, while we sat, he told her our plans for the future of Ballyconnelly Farm. He whispered stories of swallows, birds, bees, flowers and foxes; everything we'd experienced here this summer, everything we planned to conserve. To my surprise, Craig then went on to tell her about all the grandkids that would be running around her fields one day. I pinched him and we giggled as the built-up tears poured down both of our faces.

Craig leaned into me and held Lizzie's hand as we sat with her. I took off the rainbow scarf Lizzie had knitted for me and wrapped it around Craig before telling Lizzie we'd do her proud.

There was nothing more to say or do except wait.

It was getting dark and Craig had been at Lizzie's bedside for over four hours so I made us some food. I asked him if he'd leave her so that he could eat, and he did, but it took everything to pull himself away from her. Walking into the caravan – our perfect, tiny home – Craig melted into his favourite armchair and I brought his dinner to his lap on a pillow. He was too fatigued by sadness too sit upright at a table.

That night in bed, I thought about Rathlin, the farm and both Lizzies so completely in the hands of nature. There was nothing I could do right now for them, so I turned to face Craig and wiped his wet face. Eventually the night hushed us to sleep and we drifted off.

When I woke up, the sinking feeling of reality hit. I stared at the ceiling, waiting for Craig to wake. Then there was a knock at the door. I wandered over to find Craig's dad stood looking desperately sad, it was the first time I'd seen him cry. 'She's away,' he muttered, a very Northern Irish way to tell us that Lizzie had passed. Craig overheard from the bedroom and rushed in to see his granny. Not wanting to intrude, I peeked around the door to see him kiss Lizzie's cheek and say goodnight.

Hours passed, tears fell, the carpet was soaked. I had been blissfully unaware of the reality of death until now. Craig asked if we could go for a walk to the highest point on the farm. So we left Lizzie's bedside and took some comfort from the expansive views surrounding us.

'Let's get out of here, honey, and clear our heads. I'll get us some food,' I said. Craig needed some distraction. We drove for thirty minutes to Portrush and inhaled a dozen pastries and a few lattes. We felt horrendously sad yet, weirdly, desperately relieved. We had been waiting in limbo for months – it had been hard to know that Lizzie was living in such pain. Now she was free.

It was nice to get away with Craig, to comfort him, to hold him, to just finally be there for him. All I wanted to do was stay and look after everyone but there was a clock ticking away in the back of my mind. I knew I had to be on Rathlin that night. After much food, a beach walk and a thousand hugs, Craig dropped me back to the ferry. 'I'll be back soon, okay? I'll book a few days off this week, I promise. I'm so sorry for leaving,' I said.

Rathlin had been a place of sanctuary for me, but today it felt quite the opposite.

During the crossing, I thought about death intensely for probably the first time in my life. The infiniteness of it scared me. I felt alone. All I wanted was some familiar company so I migrated to the cliffs of West Light faster than Craig could skim a stone. All I could think about was Busy Lizzie. When I had left these cliffs yesterday afternoon, she was alone, but would she still be there today? I hoped so.

I raced down the lighthouse steps but I couldn't see any sign of her. I reassured myself that perhaps she was just sleeping in Puff and Finn's old burrow – but when I looked at the ledge where she usually was, a peregrine was in her place and black feathers covered the grass. The answer to my question lay scattered before me. The peregrine stood on Lizzie's old patch of grass, picking puffin feathers from its pristine yellow beak. There and then it seemed to me like death itself – the grim reaper in bird form.

I had lost both Lizzies in the space of one day. Alone on the ledge, I sobbed for them both. The island felt cruel and life intense. My crying scared the peregrine away from the platform edge, revealing something that had been beneath its talons – a puffin skull and bill, all that was left of Busy Lizzie. The peregrine must have been hungry. I picked it up and walked back to the grassy mound. I lay there, a place where I'd spent countless hours watching the puffins. The familiarity of the crater my body had created cushioned me.

The first night alone after both Lizzies' deaths was long and tricky; it was hard to not let my thoughts overwhelm me. Over the coming days, anytime the noise of grief got too loud, I put on my flippers and dived into the water at Mill Bay. The sea pulled me in any direction it wanted, reminding me it was in control. A higher power – in charge of seasons, death and new life – the swell I felt wash over my body was part of the same cycles responsible for the food that I ate, flowers I smelt and ground I trod. Here I was, a living, breathing, floating part of it.

At work I'd walk to the lower lighthouse platform, hoping for a miracle, hoping to see the puffins again, to remind me of better times ... but it never happened. That weekend away from Craig was hard and, now that there were no puffins, I felt as though I had lost my Rathlin family. Through the loss, a host of activity happened around me, and life had the audacity to carry on.

I was comforted by the wildlife. The kittiwake chicks were the next to leave, now polishing their flying skills every day. Wobbling in the wind, young kittiwakes occasionally landed on the puffins' old perch for a rest and I missed them.

When I arrived back on the mainland for Granny Lizzie's funeral a day or two later, I squeezed Craig like never before. It was the end of an era. Ballyconnelly Farm has been in the Holmes family for over two hundred years and for a massive chunk of that time, Lizzie had ruled the roost. The place didn't feel the same. We were driven to the local church – generations of Craig's family had been buried in the graveyard there. I didn't know what to expect – it was my first time attending a religious service, and I had never been to a funeral before. When we arrived at the tall building, we found ourselves a seat and tried to make ourselves comfortable. I hadn't met many members of Craig's family and he wasn't overjoyed at the prospect of me meeting a certain few that he didn't like.

I grew up in an agnostic family and to me the whole ceremony seemed cold and austere. Craig and I thought Lizzie deserved so much more. I felt confused throughout the entire service, with the minister reciting fear-inducing bible verses. Weren't we meant to be celebrating someone's life? The one personal touch came from Craig. He read the most heartfelt, detailed and emotion-packed eulogy that brought everyone to tears.

Every time the minister asked everyone to close their eyes and pray, I sat up tall in my seat, looking at the arched ceiling of the church, imagining instead we were standing in a forest of ancient trees that formed a high canopy with an interconnected network of roots beneath our feet. That, to me, was a lot more comforting, a lot more natural and a lot more real.

The whole experience made me think about the way humans perceive and process death. I realised that religion is so often the comfort blanket people use to soften the blow of loss, but worried that it had further disconnected us from nature.

Like every blip of biodiversity that lives for its short time, all biomass must at some point die, including us, making us in turn equal to all life. To avoid this reality, we've added a layer of importance to ourselves – a heaven, a secret club only our species can join. I have a strong memory of being in RE at school. We were learning about Christianity and I asked if the frogs, butterflies and cats come to heaven too.' 'Well, not quite,' the teacher replied. 'Only humans.' This separation felt wrong, even when I was young. I adored learning about Buddhist beliefs, which incorporate the natural world into their death and life systems and hold animals as equal to humans.

The ecological truth is that we are born, we live, we die. This blunt sentiment feels scary to many, like it's not enough. But I can't help but feel it's because we're framing it the wrong way. I find it beautiful that we come from the darkness, and return back to it, to the deep womb of our beginnings, to the organisms that made us, to the mycelium that connects us, to the earth that grounds us. Imagine if humans viewed our afterlife as the longevity of the natural world – I feel we would protect it a whole lot more.

Like the many Indigenous Amazonian tribes that protect the forest because they see their ancestors living through its trees, in that moment I accepted the darkness and my eventual fate too. Maybe the reason that many are so disconnected from, and fear this reality, is because we do not see the beauty and power of the natural world, nor do we worship it. If we did, I feel this magic called nature would be enough.

I let the Christian rituals wash over me and left the ceremony behind, holding onto Craig's wonderful speech.

As Craig's family gathered at the farm for some food, a sombre silence filled the air. I was excited to escape and all Craig wanted was for everyone – especially the long-lost relatives who pretended care – to leave.

'Come back to Rathlin with me?' I asked Craig. There was nothing holding him at the farm, no granny to be there for. 'It'll do you good, it'll do us good – what do you think?' He agreed, so the moment everyone left, we headed back to Rathlin, back to the cliffs, back to Jane's hut for the night. Having experienced one of the best nights of our lives here a few months back, when everything was a little easier, a little warmer, this second stay couldn't have felt more different. We drifted off to sleep, hugging as if we were clam shells fusing together.

That next day, we gave Lizzie our own send-off on the cliffs of West Light. Hiking around south Kebble to find some pretty flowers, we played like kids and Craig told me more of his childhood memories of his beloved granny pottering in the garden. We made our way to the highest point of the path, where skuas had previously scared me away, and stumbled across the prettiest patch of buttercups, which we added to our collection. Heading back to the Seabird Centre, we walked down to the lower lighthouse platform together and both talked about how thankful we were to have had Lizzie in our lives. I said how honoured I was that I'd got to meet her, talk to her about nature, about the farm, about Rathlin; Craig thanked her for being the best granny he could ever have wished for. Taking the opportunity to celebrate Busy Lizzie also, I thanked her for all she'd taught me and placed one of Puff's pebbles on the grassy ledge as some sort of farewell gift. Craig then placed the bouquet of wildflowers down for his granny and we felt satisfied that they'd finally had a fitting ceremony. I sobbed again. 'Pint?' Craig asked.

As we walked into McCuaig's, a member of bar staff came up to me, telling me a very drunk man had come in and given them money for two pints for me as a thank you for helping him out after his fall at West Light. It was a welcome surprise.

We didn't drown our sorrows, but instead sat celebrating someone's life.

When Craig left a few days later, the reality of heading back to work hit me. I was excited but nervous – things would now feel so different. In fact, I'd arrived back just in time because there was no more than a week left with the kittiwakes – they were so close to fledging. Their flying had greatly improved since I last watched them. Proud of Kitti and Kat and their three chicks, I watched the entire family fly together, dominating the cove. I'd miss their loud, unapologetic calls: the cliffs would not be the same without the never-ending 'kitti-wakee, kitti-wakee'.

That week Kirsten and Tom gave me so much time and sympathy, and they really helped me process what had happened, always providing shoulders to cry on during lunchbreaks. My emotions were all over the place, but they made the end of August much easier for me.

There's something so unsettling about saying goodbye,

whether it's forever or just for a while. I've never been good at it and by this stage in the season I was tired of it. My heart yearned for the time when the reserve had been packed with promise.

Symbolically, as the last kittiwakes left the cliffs for good, Kirsten and Tom had to go. 'You guys are good eggs and I hope to see you soon,' I said, hugging them both. Rachele thanked them for being so brilliant and drove us all down to the harbour so that she and I could help Kirsten and Tom load their many bags onto the ferry.

Young kittiwakes danced among the wild waves outside, watching older, wiser birds hunt as the *Spirit of Rathlin* left. Waving and jumping as the ferry headed out to sea, I cried. But the wind carried sweet melodies from the harbour. Charlie's boat had some musicians playing on it – there was a handful of people, beers, dogs. I was in. Taking off my work jacket as I walked over, Charlie handed me a beer and told me to hop on, and I began to feel slightly better. Children danced, their feet clunking around the old, wooden boat, and some of Charlie's friends cooked up a feast inside. The youngest child there, a girl with beautiful auburn hair and a freckly face, took my hand and asked me to dangle my head off the side of the boat. She did the same – apparently we were pretending to be mermaids. 'Sing,' she said. 'Sing like you're a mermaid.' I looked down at our swaying hair reflected in the still sea, and gave it a go.

I rhymed off something about crabs and starfish being our best friends; she giggled and laughed, then made up her own verses about her favourite seabirds and how she loves to swim with them.

The easy laughter of this lovely girl helped remind me how to feel carefree again.

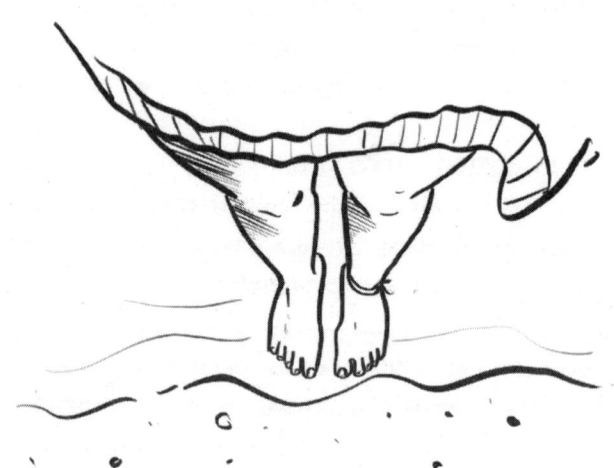

Chapter 8

Thriving

The sea of silk.

As I grieved, I clung to nature like a limpet to a stone. My little puffin friend, the second-busiest Lizzie, was ingrained in my psyche and I missed her on the cliffs, but a host of activity was happening around me and life carried on.

September: a new month, a new start. Only a few fulmars and their chicks were left at the reserve. I'd mistakenly thought that the cove seemed lifeless, but when I looked a little closer, I was met by a lovely surprise. A few pregnant seals were hiding in a cave under the cliffs and the arrival of fluffy grey seal pups was suddenly on the horizon. Another surprise also caught my eye – it was the chick of my fulmar pair, Freddie and Fanny, furiously flapping its wings! At some point over the next few weeks it would attempt to fledge. These discoveries kept me motivated on the now colder, lonelier, rain-drenched cliffs.

The season had ended for most other staff – visitor numbers would soon drop off so Rachele just kept on Eleanor and me. Rathlin felt quiet. On the first Monday of the month, just one person came through the doors. Desperate to keep busy, I entertained the visitor for half an hour before sitting again in quietness, alone. The last pair of residential volunteers were expected to arrive any minute – I couldn't wait for some company.

September welcomed Gordon and Carol, who are parents from Bristol, although Gordon is originally from Scotland.

They burst through the Seabird Centre doors, bringing an energy with them, an energy this place had started to lack. Their excitement was contagious – it was as if they were arriving in peak summer with sunshine and full cliffs.

When I showed them around, the rain was lashing down and the wind almost blew us off the cliff but they didn't care one bit. I discovered that they were a well travelled couple, and that Gordon was really funny: he reminded me of Billy Connolly in his prime. I pointed out the pregnant seals and fulmar chicks in an attempt to show the couple how unique and exciting this month could be, despite the lack of seabirds. I felt relieved that Gordon and Carol were so easy-going given the last few weeks I'd had.

That night I felt tired. The wind blew through the harbour outside, clattering ropes and bells off sails. The emotional fatigue from August had taken its toll and I needed to be kind to myself so I ran a bath, trying to bring more Hygge into my routine. I lay in the liquid warmth for an hour or more and got out only when the water became cold. For dinner, I made a leek soup and decorated it with the bright orange petals of nasturtiums from Granny Lizzie's garden before drifting off by the log burner.

The next morning I woke to the sounds of a stormy harbour and hastily dressed myself for the working day ahead. Cutting it quite fine, I darted up to Kinramer Cottage to collect Gordon and Carol. 'How was your first night's sleep here?' I asked, hoping they hadn't been too cold or disturbed by the wind. 'Transcendent,' Carol replied. Wow, nothing would make this couple unhappy.

The storm had got worse by the time we arrived at the centre. Rachele handed us cups of tea, apologising for the fact that we were about to spend the day on an exposed, very cold platform. None of us minded. Gordon, Carol and I ran down to the binocular hut as quickly as we could but we still got drenched. We hid away, lingering over our tea until the weather had cleared. A rainbow appeared over the cliffs and finally the sun came out. A beam shone down from the dark clouds, illuminating Freddie and Fanny like they were some kind of seabird angels. Their chick was teasing us: every twitch and turn looked like an attempt to fledge, but it stayed put. Meanwhile, down in the cave, one of the two pregnant grey seals hauled out looking very large – like she was about to burst. Seals have markings as unique as our human fingerprints. This particular seal had beautiful markings, so I snapped away, taking pictures from every angle I could.

Like the residential volunteers that came before them, Gordon and Carol were naturals. They didn't know too much about seabirds, but after twenty minutes of listening to me talk to visitors on the platform, they could remember and recite everything I'd said.

Rachele brought down some snacks Francis had left off and we all ate lunch together in the lighthouse. Gordon and Carol asked about the season so far. I explained how amazing it had been, but also told them about Craig's granny and Busy Lizzie. The couple were very understanding, and Rachele gave me a hug.

After lunch we monitored the movements of every fulmar chick. They didn't have much time left here, and it made me think about the day that I too would have to fledge Rathlin, which was now just a few weeks away.

The first few days of working with Gordon and Carol lifted me. Their strong Scottish and Bristol accents made me happy and were something a little different to tune into. I connected to nature more than I ever had before and it helped me. For the first time in my life, I was thinking deeply about death and dying. In doing so, a layer of madness consumed me, yet a clarity and calmness that I had never felt before. The feeling it gave me reminded me of how I felt when I first properly looked at the night sky and comprehended both its terrifying enormity but also its beauty and constancy. I looked at everything in finer detail, the veins of leaves, the lines of my fingerprint, the movements of birds, the magic of mushrooms. My soul would one day lie among these neighbouring jewels of life; it felt only right to get to know these organisms a little better.

Discovering more about the microscopic world gave me a newfound fascination and appreciation for life. I learnt about the wood-wide web and all the mycelium networks that interconnect it, distributing nutrients from decomposing matter across the forest. What's more, plants – from the tiniest microscopic level to trees – communicate with each other. These unseen networks ensure nature's ability to thrive, supporting a surplus of other species. I found nature's underground, nutrient-cycling eternity captivating. This was a kind of heaven that I could get with. There was a world beneath my feet that I was only beginning to uncover, a place that brought me so much comfort.

I thought too about our souls – if they are just gone forever. But I don't believe that. I believe that someone remains with us in the legacy they left behind. Like a droplet that creates a ripple in a still pond, one human existence will touch many others, often unknowingly, creating waves across the world. I thought about the quote that's engraved on my grandad's grave that says: 'To live in the hearts of those we love is never to die.' That was it.

I started to watch the cove obsessively all day. I'd think

about life, climate change and these cliffs so deeply that other people's words just didn't enter my head. I watched orange sea spaghetti drift under the clear turquoise waves for hours, waiting for something to happen, a chick to fledge, a seal pup to appear, a dolphin in the bay, something.

I missed my morning routine of greeting the puffins at the lighthouse's lower ledge, so halfway through September, I decided to start visiting it again. I sat with my daily coffee in hand and stared out for hours. One morning I spotted the cove's peregrine resting on a sea stack that the guillemots had nested on before they left. In contrast to the negative way I'd been thinking about this bird, it suddenly looked majestic framed by the sea. It was too stunning to represent something as dark as death. Watching this falcon as it fanned its wings, an easing epiphany came to me; a reminder that there was something beautiful about nature taking Busy Lizzie, because although it was devastating, such deaths form part of the system to which we all belong. Instead of only despising the cruelty of the situation, I was now able to find some beauty in it. This new mindset helped me to further process the loss of Craig's granny. I felt honoured to have experienced a person and a wild animal, both so unique and special. They were so hard to let go – but so is anything good.

When I arrived back to the platform from the lighthouse, I saw Gordon and Carol leaning over the safety bars, looking incredibly excited. Their eyes were glued to Freddie and Fanny – what was going on? 'Oh Ruby, thank goodness you're here, I thought you'd miss it. Look, I think this pair's chick is about to head off,' Carol said. She was right – the chick was flapping harder than before, picking up height with every wingbeat, lifting further off the ledge each time.

The chick launched off and circled the cove a few times,

using the wind to keep its body in flight. Its parents watched on, looking, I thought, genuinely proud of the miracle baby they'd raised. The chance of this bird leaving Rathlin alive was low, which is why every fledgling's flight felt like a step towards victory. After a few minutes in the skies, the chicks returned to their ledges for a break before practising again.

I had watched this bird hatch; I'd watched as its parents fed and nurtured it, returning to it every few hours; I'd watched as it got flying lessons on its precarious ledge. Now I got to see it taking a massive leap into the big wide world – a world drastically changing but one we can fix if we decide to. This cove didn't only breed chicks, it bred hope. Another three chicks joined in, and then four, and then before I knew it, it seemed as though the entire fulmar chick population was fledging, collectively figuring out how to navigate the wild, westerly winds. The parents left the ledge they'd raised their young on and joined the fledglings circling the cove.

Although the chicks now resembled adults – with no downy fluff – I could still distinguish the adults from the young due to their larger size and darker grey colouration on the wings. Looking out for their young, the adult fulmars would fly close, ensuring predators stayed away while their young undertook vital practice. The fulmar parents occasionally hovered over the platform, keeping a beady eye on the staff below. We heard an occasional squawking call, which we translated as parental praise for the newly fledged fulmars that were trying their best to stay high and dry.

We were all delighted to finally see the chicks fledge. Fulmars are among some of the longest-living seabirds – they have an average lifespan of forty years, but some can live to the ripe old age of fifty – and they return to where they are born year on year. So these fulmar chicks could call Rathlin home for the next half a century, if the climate and all the conditions that come with it allow it.

A lot can change in fifty years; a lot has changed over the last fifty. Over the course of one of these fulmar chick's lifespans two starkly different situations may play out. We will, as a species, either recognise our impact and make drastic societal changes before things get worse, or we will allow further delay, climate catastrophe and yet more nature loss.

In the UK, since 1970, we've seen a 41 per cent decline in overall biodiversity. Fifty years before that, when war was rife and pesticides came into widespread use, nature began to decline drastically and insect numbers halved. Then, in the fifty-year period between 1870 and 1920, in Ireland we lost the mountain ringlet butterfly and last spiral chalk-moss. Between 1820 and 1870, iconic animals such as the great auk – a large seabird that once called Rathlin home – became extinct, as well as insects such as the large copper butterfly and hornet moth. In the fifty years preceding that, the UK's remaining wild wolf population was decimated. So much can develop, be lost be or be gained in a short-long fifty years.

Over the last five hundred million years, there have been five mass extinctions. Each of these was caused by natural phenomena, such as volcanic eruptions or an asteroid colliding with earth. We are now in the midst of the earth's sixth mass extinction – but this one is caused entirely by humans. Alongside multiple other species, the seabirds I'd grown to love were now also at risk of disappearing.

One problem we face is that the way we measure these losses and how we feel about them is often in relation to our own human lifespan. The two hundred years that I've outlined above feels drastically long, but, in fact, it is only the length of one bowhead whale's lifespan. Its life spans four to six of our generations – imagine the species loss its lifetime bears witness to. And in the earth's timeline, two hundred years is just a speck of time. It's vital that, going forward, we consider a time lens greater than our own when we are working to safeguard species. It's the only way to bring nature back and avoid more animals going extinct.

So as I looked towards these fulmar fledglings, whose lives were brimming with potential, I hoped with every fibre of my being that every one of them, and their children's children, and those children's children, and so on, could go on to live how they should. I truly hoped then that the generations to come can look back and be proud of what we decide to do.

Emphasising how big a moment this was to visitors was hard. 'Look, there's some fulmars flying' doesn't quite cut the mustard. I wished I could play back a trailer of their summer to every person I spoke to: show the eggs that had fallen from nests, the chicks that fell victim to the extreme heat or those that were carried off by predators. 'These fulmars made it; at least half of them didn't!' I said. 'This is the result of the most persistent parenting I've ever seen.' The fulmars dominated West Light for the first time all season. It was their time to shine.

Over the coming days, Freddie and Fanny gradually spent less and less time with their chicks and each other. Landing on ledges that weren't their own for short rests, the small family slowly separated. Towards the end of the second week of September, the chicks had become so good at flying that they decided to come check us out. They pooed on us constantly and Gordon, Carol and I spent a lot of time in the lighthouse toilet, cleaning guano from our uniforms.

After much anticipation, the next few days saw the start of grey seal pupping season. A gorgeous little pup was born in the cove below. Gordon and Carol looked down at the pup with their binoculars, ecstatic to see it. The tiny pup's pink umbilical cord was still attached so I could tell it was just hours old. The mum licked yellowy, sticky mucus from the pup's fluffy coat as it suckled – seals' milk is nearly three times richer in fat than humans. I'd worried that the storms battering the island might have injured the mum but all seemed well. The bond between mum and pup was unbreakable – it was beautiful to watch.

During the third week of September, the fledglings were leaving Rathlin fast. I couldn't make out who was who any more, but said a collective goodbye to all of them, hoping Freddie, Fanny and their little (now big) baby could somehow see me frantically waving them off. A massive dollop of poo landed on my shoulders as the last few birds left the cliffs. 'It's good luck, right?' Carol said, laughing. I wasn't even annoyed, because I knew it probably wouldn't happen again this season.

With no seabirds, the cove became silent – it was weird.

Rachele asked us to start cleaning the centre as end of the month drew in. The act of clearing away felt strange: it seemed like yesterday that I was pulling the place together, ready for reopening.

I was aware Rachele hadn't had many chances to discover the island and everything it offered. Conscious that we were running out of time, I asked over lunch if she wanted to borrow a kayak from one of the islanders and go on a sunset adventure that evening along the south coast. At first it was a no, but as the weather improved, she warmed to the idea. Eventually I convinced her so we set off not long after work. I made out that I was more confident than I actually was. Even though I'd worked as a water sports instructor as a teenager, the strong currents of Rathlin scared me. Luckily, the lady we borrowed the kayak from also worked in the coastguard team and pointed us to a safe route.

A line of harbour seals followed us, perhaps intrigued by our paddles. 'This is great,' Rachele called. I was happy she could enjoy the island from a different angle. I pointed out to her many of the spots where I'd spent much of time over the summer and shared some of my best memories. As we paddled back to shore, Rachele thanked me for being a good friend and employee, telling me she could not have done that season without me. I was so touched. When we got back to the cottage she gave me a card and a beautiful, handmade wooden puffin with the words 'Rathlin 2021' and my name engraved on it. Her partner Eddie had made it. It meant the world to have such lovely recognition, and a handmade keepsake of the place I so adored.

The next day we welcomed the last flurry of visitors, all eager to come before we shut for the winter. Edith, who had enjoyed the centre so much earlier in the season, arrived with her mum. I was delighted to see them again but felt worried about the lack of birds. 'There's no seabirds left but a very cute pup has appeared this week that I think you'll like!' I said to them. 'You don't need to worry about that. The main reason

Edith wanted to return was to chat about nature with you again!' her mum explained. Edith asked me if she could use my binoculars again and of course, I let her. Smiling down at the seal pup, she looked so happy. I told her an array of seal facts, ending with 'and at just three weeks old, the pup will be left to fend for itself!'

Edith gave me a huge hug. 'This is what the seal does to her pup, isn't it?' I was taken aback by her lack of reserve and returned her hug. She told me she missed the puffins and I told her I knew how she felt, but reassured her they'd return next year, as long as we look after their home.

'How can I look after their home, Ruby?' she asked. I thought about what to say for a long moment. I didn't want to give her a list of lifestyle changes she could make or politicians to contact – she was far too young – so I replied, 'Care deeply, like you do now, and get others to do the same.' 'Easy!' she responded.

Edith stayed on the platform all afternoon with her mum, looking out for seals and the occasional passing cormorant. In between the many packed lunches they put away, I watched as she wrote notes in her little pink journal about all the wildlife she could see.

The end of the day neared. The little girl and her mum stayed as late as they could. Putting the binoculars back inside my case, Edith's mum thanked me and wished me all the best for the future. I admired their sweet relationship and thought of my own parents. They had also taken me into wild places when I was a young girl and I was forever grateful for those experiences.

Gordon and Carol commented on my ability to chat with kids, telling me I had a special ability to turn into one myself. I took that as a compliment.

As we closed up, Carol said, 'Right, last day tomorrow. Let's go for a pint and a swim.' 'I'm in,' I replied, 'and I'll watch,' said Gordon. It was a cold night so we managed about five seconds in the sea before wrapping ourselves in towels to drink hot chocolate. Then we headed to the bar and I spoke to every friendly island face I saw, knowing it might be the last chance I would have to do so for a while. I thanked everyone I could for the time they'd spent with me and told them how welcoming they'd been. A local man who'd visited the centre a handful of times overheard that it was to be our last day at work the next day, so decided to buy Gordon, Carol, Rachele and me a line of shots. They were a bright green, very questionable colour. To this day I'm still not sure what the drink was but it put me to sleep.

* * *

It was my last day on the cliffs; I could not believe it.

Before I'd arrived on Rathlin, I'd tried to prepare myself for the lifestyle change of moving to an island, but I'd not thought much about the day I'd have to say goodbye. I wasn't ready to go. I never could be. The ocean reflected my emotions: calm and smooth but grey. Just like my tear ducts, the clouds overhead looked like they were going to burst at any moment. There ended up being no visitors that day so our only job was to tidy the place up, ready for whoever would be taking over the Seabird Centre next summer. Would I be back here? Would I still be working for the RSPB? So many unanswered questions came into my mind, but I had to push them aside and enjoy my final few hours. I collected the keys from the office, grabbed a few out-of-date oaties, and headed down the steps for the penultimate time. It was the first morning I saw no wildlife, which felt apt given it was the day that I would also migrate off the cliffs.

I did a deep clean of the binocular hut, lighthouse, platform and picnic areas, remembering the liveliness that I'd experienced over the summer. I remembered all the kids who'd loved discovering close-ups of the seabirds using the binoculars; I thought about Puff hopping up to me by the railings on my very first day; I felt sad again about Busy Lizzie but also thought about how much I'd loved watching her antics.

Every inch of this centre was packed with memories.

We all gathered on the platform to have lunch together. Gordon and Carol were emotional. They had only been here for a month, but that's all the time they needed to fall head over heels in love with this place. Eleanor and Rachele were wobbly too. The ending of this season was ending a significant chapter of our lives. We enjoyed one last lunch, we demolished one more bag of buns, we had one last very British/Irish conversation about the weather and it was nearly time to bid farewell.

I watched the cliffs like a hawk for any wildlife whatsoever. Even a lone herring gull would have sufficed. How one habitat could go from being so full of life to completely empty blew me away.

After we'd cleaned the platform for a few hours more, Rachele asked me if I could do a final sweep of the lighthouse, while everyone else started wandering back up the steps to the visitor centre. My heart started racing. I was not ready to leave this season behind me. I felt as if the steps back up would be my last goodbye.

So, I headed down to my lighthouse ledge one last time. The lights were off, the displays were covered, everything was ready for the winter. I tripped on the second flight of stairs and my yelp of fright echoed through the winding, upside-down building.

Down at the lower platform I walked to the grassy verge where Puff, Finn and Lizzie had lived. The patch on which Busy Lizzie had gobbled up all her thousands of sand eels was a vibrant green – the nutrients from the fish must have helped the grass grow.

I pulled out my grandad's binoculars from their worn leather case. Tracing the horizon back and forth, I could see that the crossing to Islay was velvety smooth. So much sea and so little me. I shut my eyes, took a deep breath of clean Rathlin air and exhaled so heavily that I could have pushed a sailing boat across.

Enjoying a few moments longer, I clung to the grassy floor, attempting, in some way, to root myself here. My mind was telling me to leave, but my body stopped my legs from working. A gust of wind came rushing in from the left and my curls became tangled around my face. As I turned into the breeze to remove the veil of salty knots from my eyes, I could suddenly see a great disturbance in the water half a mile out. I stared intently through my binoculars, hoping for some miracle goodbye encounter. Then nothing.

But I felt certain that there was something there so I stayed and waited. All the lifelessness in the cove today had put me on high alert for even the tiniest movement. Five minutes went by. I began to give up hope and knew I should have started to make my way up the steps when it happened. A fiercely sharp blade of black pierced through the silky water. A lone male orca swam past in plain sight. Here I was, prolonging my last few moments at West Light and I was treated to a spectacle only few in the UK have ever seen. Gobsmacked, hands numb, endorphins overloading – it was a dream come true.

The UK's resident orca community consists of just eight individuals, four males and four females, which means spotting them is extremely rare – yet here I was, seeing one. Known as the 'west coast community' this group of orcas arrives in northern Scotland in early summer to feast on fish. Unfortunately, no calf has been born to this group in over twenty years.

I radioed up to the centre. 'Guys, look out the doors now! An orca, killer whale, whatever you call it – it's swimming past!' Nobody answered. It popped up again. Like a wind turbine blade or the rudder on a ship, the orca had an almighty presence. The dorsal fin of a male orca is up to 6 feet tall. I was honoured to see a species that was facing so many threats, from pollution to overfishing and boat strikes.

Three seems to be my magic number, because for a third and final time this incredible giant broke the surface before it disappeared for good, sinking back down like a knife into butter. The sea of silk enveloped the whale once more.

'Sorry, Ruby, did you call up?' asked Rachele two minutes later. I told her about the orca encounter, feeling bad everyone had missed it. 'Come up and tell us about it,' Rachele replied, nudging me to hurry up.

I wondered where the orca was going, then realised the same questions applied to my situation. The orca was migrating somewhere, like the seabirds. Everything had a time, place and purpose and so did I. These six months on Rathlin with nature had shown me how capable I was, and now I knew I must leave.

When I went to put my binoculars back in their case, I found a scrunched-up, jam-stained, Big Five tick sheet inside. I could see that all five seabirds were marked off on it and an additional list of 'seals, seaweed, jellyfish and jam sandwiches' was written in bright pink crayon. And 'I love seabirds, thanks Roobie, from Edith x' was scribbled on the back. This line of text made me so emotional that I nearly dropped the page.

Thinking about the thousands of Big Five tick sheets used over the summer, I felt great hope that many other visitors had fallen in love with seabirds and all sorts of wildlife.

It can take just one moment in nature to feel a connection to it, which can lead us to become advocates for it. I knew from Rachele that sixteen thousand people had visited the reserve over the summer. The visitors I had spoken with – the old, young, knowledgeable, or just curious – might perhaps have experienced that one moment here with me. I hoped so.

The stains of poo on the cliffs were washed away as heavy rain came. The birds might have been long gone but their impact wasn't. This tiny slip of paper reminded me of the mark I had left.

Many could look at one tiny seabird and ask what impact it has but wouldn't think the same of a quarter of a million seabirds flocking together to act as ecosystem engineers. I realised then that we can't individually change the world, but together? Sure as hell we can try. Like seabirds, we have strength in numbers.

I carefully folded Edith's sheet up and slipped it into my pocket before turning to the lower lighthouse door. I slammed it shut behind me, and the reverberating sound marked the end of an era. I made my way through the lighthouse and locked the top floor's door, before walking to the visitor centre. Every step felt like a step away from home. Keep going, I told myself. Change is good. The further away from the cliffs I got, the more I felt like the young girl in Narnia, wandering out of the wardrobe, back to her normal existence. I arrived through the automatic doors of the visitor centre to find the others sitting snug, laughing, with warm cups of coffee. I was soaking and sobbing. Carol came and hugged me. 'Oh, Ruby my love,' she said, in the best Bristol accent.

Rachele tried to cheer me up by reminding me that I'd just seen an orca. I tried to wipe away my tears, explaining that it had all meant so much to me, and that really these were happy tears. Everything was empty, sterile and quiet – so different to the preceding months. We should have wiped the sightings chart but we didn't have the heart. Instead, I wrote 'orca' alongside 'seal pup' and 'fulmar chicks' for the next season's staff to marvel at.

We did one last check to make sure none of us had left anything behind. 'Right, well, that's that then!' Rachele said. That was that. She was right. The months had come and gone, time had flown, and just like that, we were closing up the centre for the winter. We lobbed our bags into the boot of the blue van, closed the doors and set off for the harbour with Rachele driving. Carol, Gordon and I gazed back at the centre until it disappeared from view.

We stopped off at Kinramer Cottage so that Gordon and Carol could collect their things and then drove down the winding Rathlin roads to the harbour, where we dropped them off. Not a single other soul was boarding the ferry by foot and the couple looked lonely once they'd boarded. 'Stay in touch,' I shouted as I waved goodbye to another pair of incredible volunteers.

Rachele stayed overnight to pack up her things. We sat by the log burner with some chocolate and contemplated the changes that were about to unfold. While I was extremely sad to leave such a magical place behind, I felt excited to see my family and friends and to start my new life on the farm with Craig. Rachele was excited to leave – managing two reserves at once had been quite the challenge and she was ready for a well deserved rest.

The day that followed was full of packing. After I'd crammed all my things into my tiny car, the cottage was empty once again. I unpinned the sketches I'd drawn nightly from the walls, cleared the bathroom cabinets and hoovered the sandy floorboards. I had just one night left on this tiny island.

Distracting me from the final bits of cleaning I needed to do, a golden glow filled the sky outside, calling me to the sea for a final swim. The water was so cold but the warm colours of the sky fooled me into thinking I was somewhere tropical, at least for a few minutes.

I sat on the cornerstone off Mill Bay that I'd used as my clothes line every day and reflected on my time here.

Late spring, summer, and now early autumn, life had come and gone. The seabirds had travelled huge distances with an innate purpose. They must go, and so would I. Now with hours left on this beautiful island where I'd spent six months, I stood still, staring at the bay I'd bathed in every day. I looked out across the water and saw a bustling Ballycastle, the people, the busyness of civilisation. Living here had made me realise I could live a perfectly happy existence without all that. I could hear nothing but wildlife and wind; the pebbles pulled at my feet, and the sea urged me back in, almost telling me not to go. The sun set lower on the infinite horizon, the land became soaked in gold, and I felt nothing but love for this place.

* * *

The next morning, I left the cottage early and saw the most spectacular final sunrise at the East Lighthouse. A bright orange orb rose from the Paps of Jura behind Scotland's wild coast. A tear drop fell into my flask of coffee as the sun blinded me. I thought of Puff and Finn and wondered where they were now, all these weeks later. On my way back, I paid Knockans Viewpoint a visit, marvelling at the land mass I'd explored during my time here. I thanked Rathlin, blowing it a kiss, before I arrived back at the harbour.

Rachele was standing with her things by the door. We hugged goodbye before getting into our cars.

On the drive to the ferry port the island beamed. It was a glorious day, which made it even harder for me to leave. Staff from the Co-op shop waved as I got closer to the ferry. I reversed onto the metal ramp of the boat and said an emotional goodbye to the crew as the ferry left the harbour.

I left my car and ran to the top deck to watch the island shrink in size. I felt the distance grow – it was bitterly hard.

There were no dazzling sightings for me on this crossing but looking back at the island and down at the sea below brought to mind all the remarkable things I'd seen.

I could just about make out the white speck of my cottage as we approached the mainland, and I resolved then to take everything that I learned on Rathlin with me to whatever new life awaited.

Chapter 9

Reflecting

The decade to diversify.

I left Rathlin and headed straight to the farm to be with Craig after weeks apart. As I parked my muddy car in the drive, he came running out of the caravan to welcome me home, hugging me through the partially open car door. We caught up – it turned out that while I had been spending my time looking out to sea, his way of coping with losing his granny had been to tend her garden. By this point in the year, everything in nature was on the cusp of winding down. I made dinner and told Craig about my remaining weeks on Rathlin, highlighting the wondrous orca encounter. Over the coming days, I got stuck into lots of jobs around the garden, which included cleaning the greenhouse glass and organising seeds into a neat folder. This time spent together, pottering about in the garden after so long apart, was incredibly precious – but parting again was on the horizon. I hadn't seen my mum and sisters for six months and I needed to go home.

The journey back felt surreal. I arrived at the oppressive concrete docks of Belfast, sailed across the Irish Sea to Liverpool and travelled down from the north to the southwest of England by train, feeling a newfound horror at the nature-depleted landscape.

Travelling for hours on the train through acres of bare, baked fields, I struggled to spot a single bird. There was so little to see that I understood why other passengers spent the time on their phones, waiting for the uninspiring journey to pass. Going through large towns and cities felt like venturing through an alien landscape. I spent an hour in London, waiting for a connecting train, feeling suffocated by its buildings and business, firmly understanding that my natural habitat is one with grass beneath my feet and vast views. I felt angry that the opportunity to experience this kind of living wasn't accessible to everyone, and guilty about the joyful time I'd just had. Why should we have to travel to the far-flung corners of the archipelago to find such bountiful nature? If it benefited and taught me so much, I could only imagine the good it would bring to others – especially young people.

The journey home solidified in my mind that it should be our inherent human right to access wild spaces, full of nature. And that nature has every right to exist and spread itself across the land as we do.

When I got to Parkstone station, I bolted off the train and ran down the straight road that led home. My mum and sisters were waiting for me at the door of our house with a drink and some cake, having perfectly worked out the time I'd arrive. I hugged them closer than I ever had before. Everyone told me I looked incredibly different; although I'd worn suncream, my face collected many freckles and my skin had become more tanned than ever before. My mum and sisters asked me what living on Rathlin was like. It was hard to begin to explain. I started by telling them about the sea eagle and golden hare encounters and then got on to Puff, Finn and Busy Lizzie. Once I started, I couldn't stop – so we sat at the kitchen table talking for hours.

My mum kept prodding me for more stories, proud to hear what her daughter had been up to, but I wanted to know how her summer had been. Unbeknownst to me, my mum had been struggling with fatigue, aches and a whole host of other symptoms. She didn't want to worry me so hadn't let me know that she'd been diagnosed with psoriatic arthritis, an auto-immune disease. I felt awful. She told me that the medication she'd started to take had helped, but taking it meant that she had to shield from people, especially with COVID-19 cases still so high. It turned out that my sisters and mum had all had their own health battles this summer while I was having the time of my life. In that moment, the guilt I'd left at the cliffs of East Lighthouse found its way back to me and I became teary. Lydia, as she always does, reassured me and insisted that I shouldn't feel bad.

Being away from Rathlin was beginning to feel incredibly weird. I no longer felt so relaxed about walking out of the house with a completely bare face and wrinkled clothes, so the next morning, when we all went for a coffee at our favourite local café, I put on a nice frock and applied some make-up.

Most mornings that week we sat outside in the sunshine, absorbing both the sun's rays and each other's company. We are a creative family: I showed everyone my sketchpad of nature doodles, Lydia modelled the latest clothes she'd made and Mabel played us her newest song while mum listened delightedly, like the proud mum from *Little Women*.

Not long after I'd got home, my mum drove me over to my dad's. He made us lunch using the vegetables he'd grown. We sat eating in a beautiful wildflower meadow. Instead of being enclosed by the four walls of a house, my dad was surrounded by towering native woodland on four sides. I could relate a lot to him after living on Rathlin – we both knew what it was like to feel isolated but so connected to the natural world. I camped for a few nights on the land before heading back to Poole to stay at my mum's again.

The relatively nature-full place I'd grown up, Dorset, now seemed light on real wildness but if I looked hard enough and went to the right places, I could just about find something that sufficed. The earliest chance I got, I cycled to my wildest local nature reserve, Arne, and met a beautiful array of wintering wildfowl, including my very first spoonbill. It was lovely, but so … predictable. Go to a nature reserve and find wildlife – tick. On Rathlin, the rare wildlife I'd experienced often jumped, skipped or flew past me unexpectedly, not just on the nature reserves but anywhere. I'd become greedy for cool encounters.

Old Harry Rocks shone in the sunlight; the chalky needle cliffs welcomed me back to familiar land. Scaling the Purbeck hills with my mum and her new Romanian rescue pup Laszlo a few days later, the trails were lined with nostalgia, but the path felt different. A place that once felt mountainous now seemed tiny. The towering Paps of Jura I'd faced every day had changed my perspective. My mum and I talked nonstop, making up for the time we'd lost.

We left the South West Coast Path and dropped down

to the white sandy beaches of Studland. Since I'd moved to Ireland, this place had become home to a few large-scale rewilding projects, which felt incredibly exciting. Dorset Wildlife Trust had released beavers here for the first time in over four-hundred years. I couldn't spot any but just knowing they were here was hugely exciting. There was also seagrass habitat conservation taking place in a number of sites along this sandy stretch, with the aim of reversing the decline of the local spiny seahorse population. Seeing the area where I'd grown up becoming home to such ambitious rewilding efforts gave me hope.

The time to bid Dorset farewell came. As ever, it was hard to say goodbye to my family but they all knew I was destined to live in a field, so encouraged me to chase that dream. With Irish family roots, my mum hoped that she'd be able to move closer to me eventually – this very thought helped me leave when it felt too hard to say goodbye. I left during the second week of October. Inspired by the conservation initiatives on Rathlin and in Dorset, all striving to bring nature back, my new mission was to focus on my own patch – the farm – which was a place I could have some real influence.

The journey back to Ireland felt like another step into the unknown. It was just as scary as when I'd left Dorset for Rathlin, because, although I'd spent time on the farm, I'd never lived there. Craig wasn't going to go anywhere else both because the farm wouldn't go to anyone else and because the opportunity there was immense. This journey was to a new life. As a twenty-one-year-old woman, the prospect of forever shook me. Most people my age were still pondering where they want to live and who they need to be, and here I was with my future almost laid out in front of me. I was as excited as I was scared.

When I arrived in Northern Ireland, I had an urge to hop straight back over to Rathlin, but promised myself that I would only return in 2022. That would be my reward for moving somewhere new.

When I got to the farm, home was the caravan that Craig had renovated that summer. The stunning view that looks out to evergreens meant there was always a green glow reflecting up through the windows to the ceiling. Our living space wasn't big but it was all we needed. I loved our life there and felt at home rather fast.

We spent hours outside, figuring out how we could bring more nature back to the farm. Our list of ideas soon grew too big for a page, so we created a large document that idealistically mapped out the next ten years at Ballyconnelly Farm.

Craig's parents farmed the land but the plan was that we would take it on in the future. In the meantime, they were supportive of our ideas, though sceptical at times. While the land wasn't anywhere near as intensively farmed as others, over the years a few ecologically damaging practices had come in, including intensive potato and dairy farming. Craig's parents came from a generation that was incentivised by government departments to intensify agriculture but they were excited to see us steer this farm in a new, nature-friendly direction, and happy to let us use some areas of land for something new. They put a huge amount of trust in us … we just had to work out where to start.

Craig had just started a new role as an ecologist for a consultancy; I was also keen to get a new role in the conservation sector. That meant there would be two incomes to help support our plans, which included growing more vegetables and getting underway with some rewilding projects.

Knowing that I was looking for work, Rachele forwarded

me information on a vacancy at the RSPB, this time in the policy and advocacy team, working on campaigns and communications for the charity. Although the job seemed a little less nature-facing, the ad said that the successful applicant would be mainly working from home.

I applied, went for an interview and got the job.

For my first day, I went to the headquarters in Belfast to meet members of my team. On the front desk lay a new laptop labelled 'Ruby Free, Policy and Advocacy'. It all now felt so real. My manager, Daithí, rang to ask if instead of meeting him at the office, I could come to Stormont to take part in a political briefing. I hadn't even had an induction let alone worked a full day, and I was already off to brief politicians. 'Of course, be right there!' I said, with as much confidence as I could muster.

I arrived at Parliament Buildings, went through security and walked to a fancy meeting room where I met my team for the very first time. Everyone called me 'thresher shark girl,' over Ruby. I was surprised they already knew who I was.

After some brief introductions, politicians began walking through the doors. Daithí reminded me of the campaign at hand and showed me the pledge we wanted them to sign. He encouraged me to traverse between the MLAs, telling them why they must work for nature to revive our world. At first I was nervous but I reminded myself this was nothing I hadn't done before. I pretended the politicians in front of me were visitors on the seabird platform, ready to be inspired. I kept linking our policy points to the seabirds I'd become so close to and pretty soon I realised I knew so much more than I'd previously thought. Signatures flooded in and my team were impressed with how well I'd pitched in, especially since I'd been thrown in at the deep end on my first day.

Now I had a regular stream of income, I could think about our farm projects and ideas more seriously. My life became full of contrasts. I went about surveying the fields of Ballyconnelly Farm every morning and evening, attempting to log everything that called this place home. And during the day, I sat at my office desk at home planning campaigns and events, trying not to get distracted by the rolling hills outside my window. I donned a fancy blouse for Zoom calls and waterproof trousers for the hours spent knee-deep in grass.

Aside from having to face the ever-present reality of species loss, my job felt exciting and renewing. The work on the farm, however, soon felt hard – there were many practical challenges and obstacles to overcome, and a lot to learn. I found myself googling how to bring back certain species of insect and bird, when to plant the right trees and how to make healthy compost.

Occasionally the task ahead of Craig and me seemed mammoth. Sometimes environmental gains we'd made were counteracted just a few feet away. In a field that the fox family used, we left the margins to grow wild and high. I often saw flickers of the foxes' dark red coats through the grass there and felt joyful we could provide such dense habitat for them. Then one Saturday morning, as I toured around the farm picking blackberries from the hedgerows, I saw a man in the field behind ours, behind the foxes' den, walking back and forth with a cylinder of pesticide on his back. He was spraying the ground close to where the foxes lived, polluting the environment we had done so much to protect. But there was nothing we could do; the man wasn't on our land. The foxes were safe on our side of the hedge ... but not his.

During the third week of my new job, COP26 (the 26th Conference of the Parties) took place in Glasgow, Scotland, to discuss action against climate change. It was the first time this event had been held in the UK, and every environmental group across every corner of GB and Ireland was ready to demand better. I was tasked with helping to organise the Global Day of Action protest in Belfast as part of my new role. I needed to come up with ways to engage new audiences in time for the rally, so developed a group called the 'Youth Campaigners,' believing that it is extremely important for young people to have a strong connection to nature in order to campaign for it effectively.

Within a few weeks there were more than thirty members; a small army of passionate nature lovers from a variety of backgrounds.

I soon noticed inequalities within the group – for example, a lucky few had been on safari in Africa but some hadn't ever visited a nature reserve. Seeing this disparity left a sour taste in my mouth, but it wasn't surprising. In the UK, over 80% of young people live in urban areas and in Ireland, this stands at 63%. This and the cost of living, alongside decreasing public green and blue spaces, is a contributing factor to the reality that a third of young people have never watched wildlife. Believing in equal access to nature for everyone, I started to plan a variety of trips for the group, including one to Rathlin the following summer.

As COP26 kicked off, an anticipation grew, a hope that maybe this time something truly ambitious would be agreed. The demand for fast-acting policy change was being amplified by passionate young people and the world was watching. On Saturday 6 November 2021, we rallied to revive our world, resulting in the biggest environmental demonstration Northern Ireland had ever seen, with approximately three thousand people gathering together. It was inspirational to see people from many walks of life speak up about the nature and climate emergency, demanding that political leaders act now to quite literally save the natural world.

The Youth Campaigners and I had the important role of leading the march to Belfast City Hall. Our message was clear: we wanted to see Northern Ireland's first ever Climate Bill passed by the Assembly. We were the only country in Europe without a Climate Act. I was marching with the seabirds in mind, thinking about how severely affected they are by the changing climate.

The Intergovernmental Panel on Climate Change (IPCC) had that autumn declared a 'code red for humanity', which should have served as a wake-up call ... but it did not. At the COP26 discussions in Scotland, the final clauses of the agreements were not ambitious enough because the oil, gas and farming lobby were too loud. The UN chief stated it was time to go 'into emergency mode', ending fossil fuel subsidies, phasing out coal, putting a price on carbon, and delivering a climate finance commitment to the most impacted countries – but the conference didn't achieve any of these goals. In fact, during the final hours of negotiations, wording in the agreement was weakened, for example, from 'phase out fossil fuels' to 'phase down fossil fuels'. While many leaders fought for an agreement far more ambitious, most protected economies over ecology.

In the days following the conference my eco-anxiety was crippling so I escaped into the wilder parts of the farm, letting nature hold me as it once did on Rathlin. Worldly issues washed away when I lay nestled in the grass, looking at the sky.

A few weeks after COP, the world returned to normal, and it felt as if everyone had forgotten that we were still in the midst of a climate and ecological crisis.

After the protest, my next task was to meet a group of agricultural journalists at a range of nature-friendly farms. The idea was that the journalists would write about what they'd seen, showing the farming community the great benefits of this type of agriculture. I walked around the farms with my knowledgeable colleagues Ruairi and Mark, who are both passionate about changing food systems for the benefit of nature.

The passion of the farmers really shone through. When one of the farmers was speaking to the journalists, he said, 'When many people think of our landscapes, the line Emerald Isle usually comes to mind, making it sound lush and positive, when really what we're looking at are green deserts, miles upon miles of deforestation and species loss.' I agreed so strongly and it felt refreshing to hear this from a farmer.

One of the most inspiring schemes was the County Down Farmland Bird Initiative, an RSPB landscape scale restoration project comprised of eighty farms in County Down. The objective of the initiative is to halt and reverse the decline of farmland and wading birds, with a focus on red-listed species. Two farms in Downpatrick stood out to me when we visited. Since its involvement in the scheme, the first was now home to endangered species such as hen harriers. The wildlife corridors and winter barley that the farmer had planted encouraged smaller mammals to come and forage, in turn providing a vital food source for declining predators such as barn owls, of which there are under thirty breeding pairs in the whole of Northern Ireland.

The second had areas that are likend to a giant bird table. Crops of quinoa, beans and barley provide food for pollinators in spring and summer, and seeds for birds in autumn and winter. Although agriculture is currently the leading cause of habitat loss and contributes over a third of Northern Ireland's carbon emissions annually, I started to believe it could also be the solution to so many of the problems we face. By making a few small changes and keeping nature as their focus, both farms have seen flora and fauna thrive, showing how nature can bounce back if we give it a chance. I couldn't wait to try to do the same at Ballyconnelly Farm.

Taking inspiration from our farm's rich heritage, I wandered through the ancient byres and old brick buildings, reflecting on the farm's history while also looking ahead to our vision of a wild future. Over the last two hundred years or more the farm had operated in many different ways, and had at times been worked intensively – but soon nature would be our USP.

More motivated than ever before, Craig and I wanted to bridge the gap between farmers, consumers and the growing number of environmentalists we surrounded ourselves with. With Craig having grown up in a conventional farming family and me coming into the picture as a naturalist and conservationist, we hoped that together we had a unique perspective that allowed us to start understanding the needs and priorities of each group. We started spending time building a social media community, connecting people of different ages and backgrounds with the aim of healing the damage we have caused to local landscapes and spreading awareness about what impact intensive agriculture has today.

We started to develop a handful of conservation aims to protect and preserve priority species on our small patch. In line with the UN's Decade of Ecosystem Restoration, Nature Positive Initiative we started a project called '30 × 30', which aimed to rewild and restore thirty per cent of thirty acres by the year 2030.

Over the following weekends, until the winter became too cold to bear, we planted many native Irish hardwood tree species like oaks and horse chestnuts. Next, we widened the openings of fresh springs to create small wetland areas that would benefit a variety of species. I spent some time over the winter researching wildflowers and their important role in farmland biodiversity and Craig became a nerd about all things to do with regenerative agriculture, buying books about agroforestry, permaculture, soil science and organic farming. Regenerative is a conservation term and approach to farming systems that aims to enhance ecosystems, often reversing the damage that might have been previously caused by intensive farming. It focuses on improving soil health, restoring biodiversity and increasing food systems' resilience to climate change.

While Craig's dad still needed to make a living from the areas he farmed, he made positive changes to improve soil health and increase biodiversity. Steering away from destructive practices like deep tilling and applying masses of inorganic fertiliser to the land, Craig's dad planted perennial grasslands with the aim of enriching the soil and storing carbon. We need soil, not dirt; after all, soil is like a layer of planetary skin. It's a living, breathing layer of land that is home to millions of organisms that go on to support the wider food web. Working to make the field margins more diverse, we planted corridors of blackthorn to form dense hedgerows and on areas that weren't being used at all, we planted more trees.

I often looked out across the surrounding landscape, which was devoid of wild habitat, and felt shocked at the extent of the land area being used for meat and dairy production. It was a stark reminder of how land-hungry this industry is.

I have been eating a mostly plant-based diet for nearly ten years – a choice I made so that I could live more healthily, ethically and sustainably. Aside from the odd egg from my neighbour's pet chickens, I made this decision when I was fourteen after finding out about the cruelty involved in meat, dairy, fish and egg production. Subsequently, I discovered the impact most animal farming was having on the environment, so decided I didn't want anything to do with it. A few months after changing diet, my cystic acne cleared and my energy levels increased greatly, bringing me added health outcomes. Knowing all the benefits this way of eating brings, I would love more people to do the same, but I know there are great complexities and that this is not a straightforward topic. Tradition, taste and money can all be obstacles to change. I am often asked about why I eat this way, and there are many points I could make, but recently I've begun to answer simply: 'If I have the option to eat in a way that doesn't harm animals, I'll take it.'

Staring out into the vast fields of monoculture grass which would be used to feed cows, to feed … people, I knew this system was incredibly inefficient. Livestock uses a hundred times as much land to produce a kilocalorie of meat as it would to produce a kilocalorie of plant-based alternative. The solution seemed obvious to me – grow more plants and rewild the remaining land for nature, but currently, governments do not incentivise this sort of activity.

It's not that I believe animals don't belong in farming systems. In fact, I'd advocate for mega-herbivores, such as cows and ponies, but I don't personally value these ecosystem engineers as a product. Brushing up against flora, disbursing seeds across land, tilling their turds into the ground, promoting soil health and boosting organisms, as they are ecologically known, mega-herbivores are vital for a wild and healthy landscape. However farming these species to feed a growing human population isn't sustainable or necessary.

And necessary is the key word for me. I can get everything I need in my diet – beans, seaweed, fruit, vegetables, nuts and legumes – from local plant-based sources and the environment benefits. If I was in a situation where I had to catch a fish, I would. But the point is I don't need to in order to live a healthy, happy life, especially when many fish species in UK and Irish waters are facing such huge declines in numbers. I don't judge those who live differently to me and have a deep respect for indigenous cultures still clinging onto their wildness who may still have a need for hunting, which is why I'm passionate about bringing the world of plant-based/vegan cuisine to a wider audience, via a compassionate, non-judgemental lens.

I sat dreaming of a day where I could rescue my own small array of animals that could be ecosystem engineers, and friends. I imagined vast forests stretching to the distant horizon and I longed to see humans adopt a kinder existence.

As I thought more about the food we eat, I began to

grow my own. Lizzie had always grown vegetables and we were determined to do the same. We enjoyed the last of her leafy greens and thanked her for the bounty she'd left, before planting hardier varieties of onion and squash.

Our social media page gained traction quickly and a local teacher got in touch to ask if we would run a nature and food-inspired workshop for local primary-aged pupils. After running one, we ran another and before too long, multiple schools were asking us to come in and deliver talks. Whether bird watching, participating in nature-inspired drawing classes or tree planting, the children loved getting involved, and we always took the chance to tell them about how food production and nature conservation can co-exist.

As we got closer to the deepest, darkest days of winter, Craig told me that he felt like a piece of the farm was missing. Less daylight meant less time doing the activities that kept us busy. He missed Lizzie so much.

I checked on our young whips in the snow, hoping they'd survive the frozen ground, while Craig prepared veg beds and mulched ground ready to plant wildflower seeds in the spring. This was my first winter in Ireland. At first it felt incredibly romantic, with our view of snow-capped mountains. I enjoyed being nestled in our little caravan but after a while it felt a bit tedious. The grey weather seemed to stay for weeks, the cold was often bitter and the rain, sleet or snow never-ending. The afternoons became darker and the season seduced us into sleeping earlier and earlier.

One day in December, we'd finally had enough of sitting around so packed the car and set off to the north coast for a day of surfing. The sky may have been white and snowy, but our hearts felt warm. Using the longboard I'd had custom-made in Cornwall, named 'The Salty Cetacean', I paddled out to catch the glassiest of waves. It was the first time I'd ever worn a wetsuit hood.

I travelled home to Dorset for Christmas while Craig stayed on the farm. I was so excited to get home and see everyone and tell them all about what we'd been doing on the farm. A week into my stay, Craig phoned me up and told me that he had come down with COVID-19. I felt awful for him. Not only was this his first Christmas without his granny, but now he had to be away from everyone. Isolating away in our caravan, Craig attempted to recover as his mum brought him an excess of Christmas dinners and all the leftovers.

I travelled back in early January. The new year felt weird – 2022 seemed to creep up out of nowhere. The further away we got from Craig's granny's passing the more pressure we felt to make her proud, most of all through taking good care of the garden.

With the lengthening afternoons, snowdrops pushed through the soil and we started riding the fast trajectory towards spring, spending many days on the coast in and around Ballycastle. We collected buckets of seaweed to use as natural fertiliser for the veg beds. One stormy morning when we arrived at the beach, the geography of the tideline had changed because the seas had been so wild over the preceding days. Warm island days felt far away.

The sound of sand dunes rustling rung in my ears all the while starlings danced above our heads. I loved looking across the water to see Rathlin's comforting silhouette lining the horizon and Islay standing proudly just beyond, in the midst of the crashing elements. Craig and I swept the beach for kelp while other people played with their dogs. A few days earlier, I'd seen a post online about a litter of collie-lab-lurcher puppies that needed rehoming, just five miles from

Ballycastle. I was waiting for the right moment to show Craig the cuteness I'd stumbled upon. 'Coffee?' Craig asked. 'Puppy?' I replied. I showed him the page of pups and suggested we should rescue one. Without hesitation, Craig said yes. Both Craig and I had always wanted a dog, and now that we were living on the farm and almost always at home, we knew that the time was right. While Craig had grown up with dogs, I'd never had one myself, and felt as if I'd waited my entire life for this moment.

We both felt so nervous as we got closer to the moment when we'd meet the pups – this had to be right. When we arrived, the mum – a gorgeous collie called Daisy – and four excitable puppies ran straight towards us, jumping up and nibbling at our clothes. At the back of the room a lone puppy, black with a gorgeous white chest, looked chilled and unphased. I was instantly drawn to it. Craig sat down on the sofa beside Daisy and this gentle, subdued pup came snuggling in on his lap. She had picked us. 'This is the one Craig,' I whispered. He agreed, falling immediately in love with the puppy in his arms.

'What shall we call her?' he asked – but there was no real question. We both knew straight away that her name should be Isla.

Isla, a calm yet beautiful bouncing ball of energy, arrived and revived the farm. We missed Lizzie every day, but Isla was certainly a distraction, and we knew that Lizzie would have loved her. Falling so deeply in love with my dog, and the farm that housed us, a new sense of belonging washed over me. The days got slightly warmer and the early perennials that Lizzie had planted years ago bloomed. The crocuses grew higher than ever, and we felt her presence living on through this tiny patch she had so lovingly tended for so many years. Eternally alive through nature, she had left behind an imprint that was as certain as the seasons, a legacy of patience, love and nurture. Craig found comfort watching the frost recede from the farm. He felt it was his mission – that summer and forever more – to look after her garden and the farm. As soon as the weather allowed it, we started propagating seedlings ready for the veg patch and garden, with the aim of keeping this place vibrant, just as Lizzie would have liked it.

Walking around the garden, we tried to remember exactly how it looked last summer. We planted and planned meticulously, ensuring the garden would be as lovely this year as it was before. The start of February was warmer than usual, and fooled us into thinking we could sow more than we could. We primed the soil and planted wild primroses; Isla bounded around, often making a mess of our work, but we didn't care. Having her with us in the garden added so much joy.

Occasionally I had to leave Isla for a few hours, which always felt hard. During the first few weeks of February, I helped organise events on local farms with the Youth Campaigners. Our task was to dig up nettle rhizomes for the RSPB to plant on Rathlin just before spring. These rhizomes would grow into a dense nettle habitat for the corncrakes, providing them with space to nest. It felt good to give back to the island after all it had done for me.

Away from my farm idyll, the clock was ticking for the Northern Ireland Assembly to pass a Climate Bill into law. As it stood, there were two climate bills progressing through the Northern Ireland Assembly. The first (Climate Change Bill) had been put forward by Green Party MLA, Clare Bailey, and the second (Climate Change No. 2 Bill) was proposed by the DUP Agriculture and Environment Minister, Edwin Poots, in opposition to Clare's more ambitious bill. The Climate Change No. 2 Bill was progressing through the political stages more quickly and had more chance to be made an Act, so the focus was on adding ambitious amendments to it.

Soon it would be the end of the Assembly term and the climate legislation was only at the consideration stage, so it needed to progress quickly. On the day of the Bill's consideration, when MLAs voted on adding amendments, it was my job to organise a rally outside Stormont. We protested together, calling for a 'net zero' amendment to be added to the Minister's Bill. As the MLAs debated amendments inside Stormont, a collective of passionate environmentalists grew outside. Ten of us held massive letters reading 'NET ZERO NOW' before some of us engaged in a tug-of-war PR stunt. On one side were scientists and young people; on the other were regressive politicians, holding briefcases, money and their parties' rosettes.

To my surprise, a large counter-protest took place, organised by the Ulster Farmers' Union, which opposed the original Climate Change Bill, on the basis that they thought it threatened livelihoods and farming. This couldn't have been further from the truth. In fact, by acting on the nature and climate crisis, we would be directly aiding the protection of livelihoods and farming, safeguarding future jobs. Yes, the bill would set out an array of carbon targets, and this would require farming systems to become less environmentally destructive, but through a well-planned, just transition – from damaging practices to nature-friendly farming – there would be an opportunity to future-proof food systems and nature, benefitting everyone.

After twenty-five hours of debate, days later, the results came through. The net zero amendment had passed, and the Bill would now be made into legislation. This was incredible news, but there was one caveat – the Assembly had also passed an amendment from Edwin Poots, limiting the reduction of methane to just 46 per cent by 2050, which wasn't as ambitious as we'd hoped.

But it was something. While news of climate progress made headline news, I watched as farms surrounding ours burned peat, slashed hedgerows, spread fields with slurry and sprayed pesticides. On the other side of the water, the UK government had just granted licences to new oil and gas projects in the North Sea, undermining the net zero targets that they had committed to at COP26.

While we celebrated small steps in the right direction – pieces of paper in fancy buildings – the natural world struggled. Trying to create an uplifting blog for the RSPB to celebrate the new Bill, I found myself reading new research about the prospect of puffins disappearing from our shores before the end of the century. We weren't acting fast enough.

I did all I could, both in and outside work, to campaign for nature, taking part in protests, panel discussions and more. I found solace in tending to the farm, ensuring that at least the nature on my doorstep could thrive.

In the name of 'work,' I carried on organising the trip back to Rathlin for the Youth Campaigners, excited to show them the place that had done so much good for me. To ensure they'd get to see lots of wildlife, and in the hope of good weather, I planned our trip for July. The upcoming seabird season was so close that I thought about it every day. Were Puff and Finn on their way home to Rathlin across the expansive Atlantic? What difficulties would they encounter?

Just as I'd once checked on Puff, Finn and Busy Lizzie every morning with a coffee, Isla, Craig and I now walked the perimeter of our favourite field early each day, excited to see what wildlife we could find. One day, as the sun peeked out from behind Slemish in the distance, Isla accidentally flushed a snipe from an area of field we'd rewetted earlier that winter. The simple changes we'd made to help nature were starting to work. We looked up to see five buzzards above our heads and I told Craig that his granny would be immensely proud if she could see all the work we'd done.

Everything felt so right, so peaceful until the roar of chainsaws crashed into our delicate soundscape. I realised the plume of buzzards we'd seen were there as the result of some great disturbance in a field not far away. Closing Isla into the house, Craig and I ran towards the noise. We came to an ancient stretch of trees that were being torn down. These trees didn't belong to us but they adjoined our land. We had spent hours sitting in this shaded stretch, tucked in below the branches as if we belonged under the trees' ancient canopy.

This practice of clearing trees and levelling the ground is ironically called 'improved grassland' when it's nothing more than degradation. It's done when a landowner wants to have more land for intensive farming practices, such as housing more livestock, or growing potatoes, grass or wheat – it's how we've ended up with the patchwork landscape we have now.

We watched helplessly as an acre of established, mixed woodland was ripped up, destroying a mosaic of habitat. A ball of rage built inside my chest. I began to ask myself why I bothered – what was the point in caring, breaking my back to plant trees while others tore them down. I started screaming at this dinosaur of a farmer, who was ripping away the only good thing about his land – although I realised he couldn't hear me over the chainsaw. Craig held me back, trying to calm me down, but it made me all the more upset.

I didn't set foot in the field that neighboured the decimated trees for the rest of that month; I couldn't face seeing the aftermath of the massacre. The farm, the place to which I had retreated to get away from the atrocities of the outside world, was now an uncomfortable place to be.

Craig could see how much I was struggling and tried to persuade me what we were doing on the farm was making a difference, and that things would get better. He handed me his granny's gardening award from the year before and asked me if we should enter this year, in an attempt to distract me from my sadness over the trees. 'Let's make this garden and the farm so colourful, messy, nature-rich and beautiful that it annoys the hell out that old fuddy-duddy across the road. Let's get this place full of insects, birds, butterflies and flowers and win the award for nature, and for Lizzie.'

'Let's do it,' I replied.

We were a team, both so emotionally invested in making this place everything we wanted it to be.

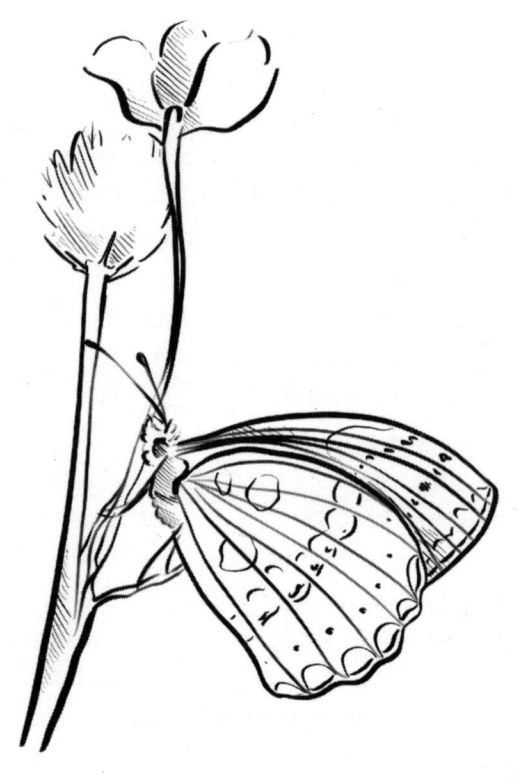

Chapter 10

Returning

A note to nature.

Much as I loved living on the farm and working on Lizzie's garden with Craig and Isla, Rathlin was always on my mind.

Having dug the nettles with the youth campaigners earlier in the month, an opportunity arose in late February to clear some habitat on Rathlin so that the rhizomes could be planted along the borders of the corncrake fields. That weekend, earlier than planned, I decided to go back to Rathlin for the first time since leaving. I couldn't contain my excitement.

On the ferry I met Anne Guinchard, the Conservation Officer for the Causeway Coast who was leading the day's activities, and we discussed how much we both loved the island. The heavy sound of the engine brought back memories. As we travelled out of the harbour and across Rathlin Sound, I soon remembered how rough the crossing could be, but Anne's calming French accent distracted me from the waves of sea sickness I began to feel. I felt great joy, but also piercing pain, knowing Rathlin was no longer my home, knowing I would have to leave so soon.

During the crossing, I asked Anne when the puffins would arrive. She said she really couldn't be sure – last year their migration had been off schedule, and they arrived much later than normal. But she did tell me that the usual time is between 25 and 28 March. I asked what might have cause the delay last year and she told me that the months of extreme winter weather and easterly winds could have thrown the birds off track. Harsh weather caused by climate change is having increasing negative impacts on seabird populations. The intensity and persistence of extreme weather events are forecast to increase with a warming planet, which could have far-reaching consequences. These include causing a delayed breeding season – affecting the birds' mating, egg-laying, incubation and hatching times, and also affecting the ecosystems that rely on a puffin's multifaceted, well-timed cycle.

They would get here eventually, but for now, I'd just have to guess and hope for the best during my next trip. As I set foot back on the island, I waved across the way to a few familiar faces, who I was so happy to see again. Arriving on Rathlin felt like coming home.

We walked to a field just behind the harbour, where we ate a few biscuits and drank cups of tea from the back of Liam's car before getting stuck into the conservation work. Working around the perimeter of a very thorny field, a few other RSPB colleagues and I cleared overgrown bushes, creating space where nettles could be planted. These nettles would provide the perfect habitat for the corncrakes. I felt excited to be part of their journey to recovery. After the bramble clearing had finished, we hiked back down to Church Bay for a well-earned lunch. Being back was so surreal. We walked past my old cottage, with its lights off and the bedroom bare. I wanted to rewind to a time when I was standing inside it, looking out.

We ate our sandwiches together, and some colleagues started talking about the soon-to-be-published seabird census. Last year I'd seen Liam spend hours at sea in his little boat counting the birds and was intrigued to know the results.

They weren't looking good.

Although the count of common guillemots showed Rathlin as the species' largest colony in UK and Ireland, the overall number of seabirds nesting had declined – rather a lot. A reported total of 407 Atlantic puffins was the lowest population ever counted, indicating a substantial decline over recent decades. This was unsurprising but deeply saddening. That afternoon as I left the island, I couldn't help but feel worried for all the birds I'd fallen for on Rathlin.

Throughout the next weeks Craig and I spent more time in the garden than ever before, in the hope of winning the gardening prize in Lizzie's honour. I longed for the day I could journey back and actually see the seabirds again, so my never-ending list of farm jobs was a good distraction. Swallows arrived at the farm and made themselves at home in our potting shed, evicting us to the greenhouse. They had arrived unusually early, and I felt immensely happy to watch one of my favourite birds frolic around the fields brimming with insects. Isla loved chasing the forked-tailed beauties through the hazy, humming fields, although she was never able to catch up with them.

Slightly warmer days arrived; hawthorn buds emerged and the sweet smell of coconut from gorse came drifting through the air. I planted more seeds of all varieties, imagining a colourful rainbow of fruit and veg in the bare soil beds before me.

I tried to prove my farmer-girl worth to Craig, getting up early to water the seedlings.

Rathlin's puffins are usually punctual visitors, but I couldn't be sure if they would be late again this year. The weather had been really calm so I hoped this meant they'd arrive slightly early, or at least, on time. I wanted to be the first to welcome the puffins back to Rathlin, so I left home early on 25 March. This would be my treat after a long winter on the farm.

Halfway through the ferry crossing I looked out, expecting to see handfuls of floating clowns on the islands' surrounding waves, but there weren't any. I kept my hopes up as I walked along the path to the Seabird Centre from the harbour, but even when I got to the cliffs there was no sign.

As much as I knew they might not be there, I was still so disappointed. My excitement had been building over the past weeks as I planned this trip, hoping to return exactly when my beloved puffins did. Interrupting my growing disappointment, a group of guillemots surfaced from a tremendously cold dive, lifting my sunken spirits. By this stage, I was happy to see any form of seabird.

Trying to put hopes of seeing Puff and Finn – or any puffin for that matter – behind me, I enjoyed watching the guillemots, razorbills, fulmars and kittiwakes from the cliff top and decided not to visit again until my trip back with the Youth Campaigners.

On the journey home, Craig called me to ask how my day had gone, not expecting to hear a deflated voice. He was able to cheer me up by telling me that Isla had found a fallen baby bird, 'I think it's a starling. Isla came and let me know – she was so gentle.'

Almost as soon as I'd driven into the farm, Craig was at the car window with a fuzzy, downy chick who he'd named Sherry. 'You're the wildlife medic, now work your magic!' he said. I took Sherry to the greenhouse and put her in a warm, dry corner before looking under pots, trying to find worms and woodlice for her to eat. She chirped and chirped without stopping; Isla sat looking up at me, bewildered by the little bird's sounds. I'd planned to keep her there, but it was still spring and the nights were cold, so I took her inside our caravan.

Craig and I walked around the farm with Sherry, lifting logs to teach her how to find food. It didn't take long for her to figure out how to hunt for herself. After not too long, she built up her strength again. Looking after Sherry was a good distraction from a lack of puffins the week before. After weeks of rehabilitation, Sherry looked healthy and strong; she was developing a beautiful adult coat of feathers and her appetite was fierce. Before long she didn't actually need us any more but she stayed. Now semi-wild, whenever I left the caravan, I could expect a starling to fly onto my shoulder within seconds.

Now that I wasn't living on an island, I was able to volunteer as a wildlife medic again. I had trained to respond to wildlife in distress just before I moved to Cornwall and had over the years attended a number of wildlife-related incidents, including dolphin strandings and seals stuck in fishing lines. During April of that year, I was called to a poorly seal with a laceration on its hind flipper, a few injured birds and a poorly hedgehog. Every time I returned to the farm, I knew Isla could smell other animals, and Craig joked that she thought I was cheating on her.

After deciding that my next journey back to Rathlin would be with the volunteers, I tried to wait patiently. Putting all my energy into the farm throughout May, growing, planting and sowing, I spent hours in the garden with Craig.

We went to Portglenone Forest every other day to see the carpet of bluebells that had emerged. I loved seeing Isla experience new sights, sounds and smells but I'm not sure she liked the strong scent of wild garlic too much.

During early June, a Scots pine became host to some nesting sparrowhawks. We couldn't see them, but their distinctive and rapid high-pitched call echoed down from the top branches. These hawks nest relatively late compared to most other birds of prey. This is because their breeding season is timed so that their chicks hatch when woods and hedgerows are full of recently fledged songbirds, such as tits and finches, who are young and inexperienced, making easy pickings for the parent hawks.

We decided the best way to enjoy summer solstice 2022 would be to camp on the wildest field on the farm. I found that the Irish word for solstice was Grianstad, translating to 'sun stop' – it's the oldest known word for solstice, and reflects the way that the sun moves so slowly that day that it almost seems to stop. We roasted ourselves silly by the fire and listened to the crickets croaking into the night. At 2 a.m. we reminisced about hearing the corncrakes call at that time the year before on Rathlin. No corncrakes called on the farm but plenty of other species came out to sing, including crickets and a distant owl. As the summer went on, we maintained wet boggy areas for waders; dense native trees for buzzards; shelter for starlings; seed-giving tree varieties for declining farmland birds; churned up ground for song thrushes; ponds for insects and frogs; dells for foxes; and areas of rotting dead wood for bugs and fungi. The more we managed the farm in the right way, the more we saw – surprisingly quickly.

Throughout all this time, Sherry had hopped between our shoulders as we pulled vegetables from the beds, waiting for us to dig deep enough to unearth worms. We'd become so used to this little bird's presence and never thought she'd leave, but she did. Across a period of three days towards mid-July she spent less time at our sides and more in the trees above, until one day we never saw her again.

The true warmth of summer finally shone on my back. At long last, that weekend, I would take the Youth Campaigners to Rathlin. The forecast was for bright weather and calm seas, and best of all, the puffins were definitely back.

I met the volunteers in the harbour, about fifteen of them. Everyone looked thoroughly prepared, with their best outdoor gear on and a camera to hand. We boarded the ferry and I took the group up to the best deck for spotting wildlife, like the kind ferryman had done for me on my first day. Months after my previous, failed expedition, I was more excited to see the puffins than ever before. I held the pebble Puff had given me tightly, hoping it was a good luck charm.

The ferry sliced through Rathlin Sound and a small auk appeared in the distance. I took out my camera to get a closer look at it bobbing up and down in the waves – it was a puffin. I felt relieved to be back. The volunteers were delighted to see the seabirds fishing at sea, and didn't put down their cameras until we arrived. The harbour was so clear and the eider duck chicks had appeared again. Though only for one sunny day, I was sharing a place that had done so much for me with other young people who could benefit from spending time in this paradise.

We met Liam at the harbour – it was so nice to see his face again. Anne and a few other RSPB colleagues were there too, all eager to welcome us to the island. The first item on the agenda for the day was a wildlife walk around the corncrake fields near Church Bay. The rhizomes that had been planted in the fields had grown into tall nettle plants. Anne told us that the four nesting pairs of corncrakes that had been spotted the year before had returned. It was exciting for our volunteers to see the habitat that they'd helped to create supporting such an endangered species.

Next we went on a stunning walk along the Craigmacagan Trail. From an elusive golden hare to a flutter of skylarks and seabirds fishing at sea, wildlife put on an amazing show. The yellow, purple and pink of blooming heather lit up the island and tiny black flies covered every inch of our trousers. Taking selfies with some friendly cows after a spot of birdwatching, the volunteers got to know each other better and new friendships formed. It was lovely to watch. After lunch, we spent the afternoon getting stuck into community activities. Rathlin's location means it's a magnet for washed-up marine debris, so we set about cleaning the southern beaches. We removed six big bags of litter – more than twenty kilos. Later that afternoon, the volunteers helped to clear out a community polytunnel and got stuck into some nature-friendly gardening for some islanders.

With all the volunteers happy and busy I saw my chance to go to the seabird reserve for the first time since I'd left. So, I jumped onto the last Puffin bus of the day and made my way there. Francis and I caught up like old friends and he updated me on all things Rathlin. When we arrived at the car park my heart started racing.

Now managing the centre, Jean greeted me with open arms, happy to see a familiar face. 'Ruby, how're you?' she called as I ran through the centre doors. 'It's all happening down there!' I quickly said hello to the new staff before running down to the reserve. I couldn't bear the suspense. The last time I saw the cliffs they were completely empty, but as soon as I went through the doors I could hear that this was no longer the case. As I turned the first corner, I could see that the sea stacks were covered with hundreds of thousands of seabirds. I cried happy tears – I was so thankful to be back again.

Because I'd arrived at the platform in my RSPB uniform, visitors assumed I worked at the centre, and started asking me lots of questions. At first, I was unsure whether I should start a spiel, not wanting to tread on any new employees' toes but soon gave in, telling people all about the birds and their incredible lives. This wasn't my job any more – but I couldn't help it. Looking down, I saw a young boy and his friends ticking their sightings off on my Big Five sheet. It was so nice to see the sheets still being used a year later. The visitors thanked me, telling me I was very knowledgeable; I was so happy to be back in my element. As much as I adored the policy and advocacy work I was now involved in, I missed the human interaction that came with my old job… and I missed the seabirds.

I leant over the platform, eager to see if I could spot any puffins down at the lower lighthouse ledge, then, as soon as I could, headed down the staircase to see if Puff and Finn had returned for another season. I opened the rusting door of the lower lighthouse platform and stepped out to see a puffin on the lower ledge. I edged forwards, praying to the nature gods that I would see a grey scratch on its bill. The bird tilted its head but I couldn't work out if it was Puff. I decided to get back into old habits so I lay on the grassy mound I'd used as my birdwatching perch last summer. I waited and waited, when suddenly another puffin emerged from its burrow, the same burrow Puff and Finn had used last summer. The puffin turned its head my direction and a grey scratch was highlighted by the sun – it was Puff, and beside her, it must have been Finn, a year on.

Mating for life, a puffin always returns to its partner, to its same burrow on the same ledge for the same months every year. It felt I was being reunited with long-lost family, and I found huge joy in sitting with them. I told the birds about everything that had happened since leaving as if they would somehow understand. Seeing the silhouette of Islay reminded me that I had so much to update them on – the farm, Isla, my new job. I also told them that their puffling had made it, flying out to sea not long after they left … and that their neighbour, Busy Lizzie, hadn't.

I sat for a few minutes longer, humbled again by the vastness of this landscape.

After snapping a few more pictures, I had to leave again to get back to the volunteers in Church Bay before they started to wonder where I'd gone. As I wandered up the steps, feeling the distance grow between me and the seabirds, I realised that this place wasn't just part of my past but also my present and most definitely my future. Even if I wasn't physically here, Rathlin would always be with me, and the wildlife that called it home would surely return as long as we looked after it. I walked over to the new seabird staff on the viewing platform before leaving and told them there were two special puffins they should meet on the lower lighthouse ledge. I waved goodbye to the seabird colony, more motivated than ever to help create the changes I wanted to see in this world.

When I arrived back at the harbour with Francis, I gathered the volunteers together and we made our way to the ferry. I waved goodbye to Rathlin once again. I knew this place would always be a home for me, just like Dorset, Cornwall and the farm, and I found comfort in knowing I could come back whenever I pleased. Home isn't a building or even a collection of people, it's a feeling – that you leave a piece of yourself behind in the places that mean the most.

Heading back on the ferry, the volunteers couldn't stop talking about the different species they'd seen, including some they hadn't ever seen before. I could tell that many of them had formed a deeper connection to the natural world that day and were, like me, already excited about the prospect of returning. The fact that a year on, my new job had allowed me to take a collection of young people from different backgrounds to the island, to experience nature that would set their hearts on fire, felt so perfect. Seeing others fall in love with Rathlin, a place that very much changed my life, was so precious.

That evening I arrived back to the farm feeling ecstatic. Isla came running in from the flowery fields and Craig welcomed me with a glass of homemade squash made from foraged elderflowers. My little life on this planet felt all right. As Craig asked me about my day, I picked up Isla and kissed her sunshine-scented head. 'Wild,' I replied, and told him that I'd finally seen Puff and Finn again. Craig put dinner on while I looked through pictures from my return.

We sat in a field eating a dinner of kitchen-garden veggies, and I contemplated my day. Although I felt sad to not be working another season on Rathlin, I felt glad of my new career ventures and my life on the farm. I missed knowing what was happening at the Seabird Centre every day, but now I had another habitat to look out for. That evening's walk around the farm ended with us lying down beside a horse chestnut, one we'd planted the year before. These trees have an average lifespan of three hundred years so we hoped that the grandchildren we might have would climb and play on this tree, while telling stories of their nature-geek grandparents who'd planted it.

White-tailed bumblebees did the rounds on towering, pink foxgloves beside our heads. The Bombus (bumblebee) genus has over two hundred and fifty species, and is vital to plant pollination. We hoped to see more of them on the farm as the summer continued. Now that we saw ourselves as the wardens of this patch of land, every new species we saw felt like such a win. July found us again: willow shot up in abundance and swallow-tailed moths began to hover around them. One evening we followed the bats that were swooping above our heads and stumbled on a glittering substance beneath our feet. Craig instantly realised it was bat excrement. Possibly the prettiest poo I'd ever seen, it was shimmering. Glittery scales from the moth wings had created this effect – a great sign of ecosystem health on the farm.

Over the coming weeks, oxeye daisies shot into flower as different species of dragonfly, some that we'd never seen before, dotted colour across the boggy landscape. Craig and I worked hard in the garden, trying to replicate Lizzie's diverse palette of colours, and revelled in the abundant feverfew, nasturtiums and calendula that grew. Her favourite flower, the fragrant sweet pea, shot up, reminding us of her again.

Hopping between the farm buildings, fields, bogs, veg patches and garden, tortoiseshell butterflies seemed to follow us wherever we went. Our cabbages became host to some large white butterfly eggs. Thankfully we had planted too many of them and felt quite happy if a few succumbed to the hungry larvae – if anything, we felt content knowing our veg aided the butterflies' existence.

The quietly majestic rowan grew masses of berries for the birds. One of Ireland's finest native trees, the rowan bears vibrant red berries after flowering in May or June. Found in a range of habitats, from the banks of streams to rolling hills, rowans are incredibly beneficial to wildlife, supporting an abundance of biodiversity. Interestingly, the *aucuparia* species name translates from Latin as 'bird catching'. The cow vetch or tufted vetch stood out against the greenery of our hedgerows and beautiful, speckled wood butterflies basked in the sun on the shed doors. We were delighted to be building up a detailed list of species resident here.

The garden resembled a rainbow and the veg patch became a haven for pollinators with its flowery borders. We took note of everything we saw – among others, we spotted a male common blue butterfly resting in a bath of wildflowers and a ringlet butterfly taking shelter in the fields of mixed, multi species grass.

In late July, friends came to the farm and I made some delicious food for everyone with the vegetables we'd grown. It felt so rewarding to create meals for people I loved, using the produce I'd grown myself. The morning after our mini foodie festival, a beautiful elephant hawk-moth landed on some remaining leafy greens and it reminded me of the moth-trapping nights on Rathlin.

Fitting our jobs in and around farm life was often a challenge, but it was all worth it. After one particularly tiring day of digging, creating a nutritious compost that would go on to feed our trees and plants, calls from waders wafted in from the distance. Nature always revealed itself when we needed it most.

By August of 2022 we had seen species such as snipe, Irish damselflies, linnets, lapwings, sparrow hawks and goldcrests on our farm. Trailing through the fields with Isla, we'd seen a huge increase in the farm's general biodiversity with higher and varying populations of native flora and fauna, such as orchids, bees, wasps, and moths as well as invertebrate and amphibian species. To top this off, our camera traps had given us an insight into our increasing mammal populations of badgers and foxes, always wandering between their vast network of corridors.

My time on Rathlin had shown me that if you give nature space, it will take it. Sometimes I wondered when we'd be able to get a break and finish the work but realised that we wouldn't – the seasons don't rest so neither could we. The month progressed, and the swallows that had been nesting in our potting shed started to fledge. They got daily flying lessons from their parents. Each of the four fledglings sat on an old cable and waited patiently for their turn to take to the skies, until one day they took off for good. We hoped they'd return after migrating to Africa, so didn't move an inch of the material that surrounded their nest.

As we approached the anniversary of the day Lizzie passed away, we decided to plant a tree in her memory, and hoped she'd be proud of all we'd achieved. A few moments after digging the hole for the oak, a starling began to sing in the hedgerow beside us. The bird – an adult female – then flew down to the ground and began looking for insects just like Sherry used to do. After a few minutes it flew away and joined a group of starlings in a nearby sycamore tree. We remarked that it could've been her, and felt comfort at the idea that she'd found some other friends to fly with.

That night, as we finished dinner, Rhonda and Alan walked into our living room with proud faces. Craig and I had wondered if Rhonda was holding a secret – she had been smiling down to the floor in an attempt to conceal her excited face all day. She showed us a letter and explained that we had won the Antrim in Bloom gardening competition. On the anniversary of Lizzie's death, Craig and I felt so proud of all the nature-friendly work we'd done so far. Exactly a year on, now armed with an award, we were delighted to gain recognition for the legacy we'd attempted to carry on. Rhonda explained that the mayor would come and congratulate us in the following days, awarding us with a first-place plaque to add to Lizzie's collection of prizes. Craig was overwhelmed with joy.

A spark of biodiversity amongst deserts of grass, the farm acted as a beacon of hope, providing spaces for a vast array of species, including some that are endangered. A few days later, the mayor and the council's head gardener came to present us with the award. The head gardener told us that she loved all the messy, colourful flowers and wild and rare species our garden was home to. This was all we'd dreamed of – we were advocating for a future we wanted to see by showing what it could look like and felt proud to call the farm home.

Shaking the hands of the mayor as we accepted our prestigious prize, the cameras snapped and we both welled up. In the following days our names appeared in local newspapers and gardening magazines and journalists got in touch with us, interested to interview the 'young environmental couple'. The questions that we were asked time and time again were; why we did it, why we cared, what motivated us – all questions I struggled to answer concisely… all questions that stayed on my mind.

Although Craig and I were overjoyed, he couldn't help but feel sad that Lizzie wasn't here to see all we'd achieved. I told him 'Lizzie knew. She knew we would go on to do this.'

We had to celebrate somehow so decided to visit the coast for a walk before going for a fancy meal in a restaurant. As we set off Craig said, 'Let's go to Fairhead, to get a view of where it all started.' 'What do you mean?' I asked. 'From Fairhead, we can see the Isle of Arran, which isn't far from Cumbrae, where we met. We can also see Islay, and Rathlin.' It sounded perfect.

So we drove to Fairhead, to say thank you to these places, all islands, that had inspired such change in our lives. We hiked from the car park to the coast path that would lead us to Fairhead's peak, a view I'd gazed at from Rathlin for countless hours.

Copper beach leaves crinkled, crisp and burnt, glowing in the summer breeze while a salty smell came up from the shore, hundreds of feet below. Standing on top of Fairhead with Isla, looking over hundreds of miles of the west coast of Ireland and Scotland, we wondered what lay ahead of us.

As I traced the lines of the clouds, ocean and land with my fingers, I drew myself in as part of the surrounding world. Holding Isla tightly, we watched as boats crossed the Sea of Moyle, passing the three lighthouses of Rathlin.

I thought again about the journalist's questions – about why we did it and why I cared.

Standing up for the destruction of the natural world and all the life within it was my natural, human reaction when I learnt it was under threat when I was fifteen years old. In response to this question I thought, would you ask a fish why it needs the sea, or a puffin why it needs its burrow? This way of talking about our world, as if it's something we can choose to be interested in, is the very start of our disconnection to it.

Why did I care?

Is it my neurodivergent brain that can't switch off, that feels things so deeply, that wonders often and perhaps thinks a little more freely?

Is it because I'm a young person, not tied to capitalist systems or job roles that could control my voice?

Is it because I've lived so incredibly remotely, alongside nature, that my relationship to it is a given? Standing here, I was reminded why I cared so deeply. I was now so connected to nature and so profoundly concerned about its decline. I lived with it, heard it, saw it, understood it. I think we'd all feel this way if we had the ability to stay in touch with the wilder side of ourselves – it's just easy to get sidetracked along the way. Rathlin changed me because I had the opportunity to immerse myself in true wildness: to feel small and irrelevant in the amphitheatre of seabird life; to sleep and wake alongside the natural world, observant of its changes; to grieve, thrive and grow with the ever-present support of nature by my side.

Having had the privilege of living somewhere so remote

and wild, my anger towards humans and their reckless actions had lessened, perhaps surprisingly. I'd come to understand that we are not an inherently evil, careless and cruel species; it is the structures that govern our lives, that tell us to not care and separate us from biodiversity, that make us this way. No wonder the world is so confused.

While we can all do something to help nature in our daily lives, the people who I hold responsible are the ones who hold the power, who control the narrative, who benefit from the depletion of earthly resources – the top richest one percent who are responsible for over 52 per cent of carbon emissions, billionaires, corrupt governments, that largely rule our world.

It is up to us to hold those with power accountable for the nature loss we face, not for individuals to drown in the burden of it all. This understanding had lifted a huge weight from my shoulders. I, for far too long, had felt guilty for simply being human, being alive, taking resources – I should not have ever felt this way, as part of the earth's ecosystem. Trying to compensate for the actions of those above me is not healthy, nor is it right. Eating, living and consuming in an ethical and sustainable way helps me day to day and does make a positive difference, but the blame now has to lie with those responsible.

Having lived a wild life, so close to wildlife, I've come to know my place in the world and I now do not blame myself for being here. I belong here.

It's not a coincidence that I find myself on a planet where I can drink its water and consume its fruits. I have evolved mutualistically in this world, alongside countless other wild species of plants and animal, so of course being in the wildest of natural environments made me feel more human.

After spending hours watching the strong tides recede we left the awe-inspiring views of Fairhead to go for dinner. 'Back to work?' Craig smiled across the table. 'We've won an award, but a lot needs doing. You in?'

'Yes,' I replied with determination.

And from that day on, we spent every moment that we could working on our farm, attempting to campaign for nature, build back biodiversity, and reconnect other people to it. Using the lessons that summer taught me, I found peace in restoring my forever nest. This time there was no leaving in sight, only growth.

Weaving our wildness together, Craig and I had caught the everlasting bug that is nature restoration, always thinking about what we could do next.

Like nature, nothing was ever certain but one thing was for sure – we were creating our own truly wild life.

Act Now for Nature

As I finish writing this book, world leaders have agreed on a new Global Deal for Nature. The last global targets, set in 2010, were not ambitious enough, and we have seen the consequences in nature's sharp decline. The Protected Planet Report (2020) shows that only 17 per cent of the world's lands and 8 per cent of the world's oceans are currently formally protected. However, the landmark 'Nature Positive Framework' that has been passed – and that every country on the planet must adhere to – could now change this. At the heart of the new deal is the core goal of protecting and restoring 30 per cent of land and sea by 2030, with the aim of halting and beginning to reverse the decline of nature by the end of this decade. While this is amazing news, the difficulty will be getting governments to act on these promises because they are not legally binding.

This can be hard considering species loss is not always noticeable. The threat of extinction doesn't feel so immediate – doesn't seem so loud – when wildlife declines so silently. Floods and forest fires appear as clear evidence of climate change on the news. But species loss, to a human population with increasingly little access to nature, is less spectacular and tangible. Although we've seen a 67 per cent decline in global biodiversity in just fifty years, most people do not notice this loss in their own lives. It's clear that physically being apart from nature – not being able to see, hear, smell or feel what we're losing – is half the problem.

Biodiversity loss is a complex issue, with different species and habitats facing various threats in lots of places: from the way we manage our land and seas to urbanisation, pollution and non-native species. The solutions are not black and white – there is no single, easy fix, but like the tiny pieces that make up a larger puzzle, we must solve the whole by connecting the parts. Conservation initiatives are often framed as regional rather than global solutions, when in fact the species in question intrinsically form part of a wider natural system. The puffins that migrate to the UK and Ireland year on year, the humpbacks that pass through our waters and the corncrakes that take refuge in Rathlin's nettle-filled fields all contribute to something much bigger, providing ecosystem services that are of local and global benefit. This is why taking action to restore nature in the places where we live and work does make a difference. It's why it is more important than ever that we come together to make progress towards tackling a shared global crisis while there is still the chance to do so.

So, how do we get our local representatives and global leaders to act?

We must come together as individuals and communities to remind those in charge of the threats nature faces, and the effect ecosystem collapse has on everything else – including us.

Politicians work in four- to five-year cycles. Most elected officials who hold power don't seem able to look beyond their immediate, elected mandate. For personal gain and short-term political benefits, they so often end up supporting the industries that bring short-term economic benefits – which are often also the industries that destroy our planet, such as intensive farming and big oil.

For far too long, nature and climate issues have felt far

away from individuals in voting chambers – they cannot hear the call of corncrakes and pleas of puffins from their leather benches; they cannot feel the extreme weather affecting the wildlife that calls our islands home; they cannot smell the toxins polluting our farmland; they cannot make the connection between our human world and the wilder, nature-rich one in their legislation, because they've done such a fine job of separating the two.

Our job is to connect these worlds again.

Our job is to make nature every policy maker's key priority, to bring the stories, sounds, smells and senses of the natural world back to people again, to tell them what could happen if we don't.

Our job is to show how we can live differently by changing aspects of our own lives and forging new connections with the wild world outside our familiar one.

Our job is to act radically, to attend rallies, lobby politicians and take part in campaigns like our lives depend on it … because they do.

Our job is to take on the role of warden wherever we are – we are only nature defending itself. Volunteer on a community farm or for a nature conservation charity. Work with wildlife or simply take up some guerrilla gardening.

Our job is to never give up and to never stop speaking on behalf of the species whose calls are becoming quieter.

Every year there are fewer birds singing, insects buzzing and fish in our seas. The UK and Ireland are some of the most nature-depleted countries on the planet, and Northern Ireland, the home of Rathlin and our farm, is twelfth worst in the world for nature loss. According to the State of Nature report (2023), 12 per cent of species in Northern Ireland are at risk of extinction. This doesn't have to be the case. As a society, we have the answers – the technologies and nature-based solutions. We just need the ambition, legislation and will from government, industry, landowners and investors. And most importantly, we need people to be on board – to take nature's side.

I urge you to rewild yourself. Mother Nature is waiting for you to come back. Go out into green and blue spaces – feel the sheer scale of them and the smallness of you and find the space where you slot into it all. This connection to your home will give you a renewed purpose to fight for nature every time the battle feels hard and overwhelming. It will spur you on, and help drive positive change, no matter how big or small it may be.

Thank you for reading this memoir and for letting me take you on a journey in my life that changed it forever.

My name is Ruby Free. I am a campaigner and conservationist, who reads the science, who is angry and scared, but who refuses to give up hope.

Share this book, bend the corners of its pages, pass it along, tell other people the facts.

Species loss is not always noticeable; the nature and climate crisis can be complicated, overwhelming and depressing, but – like seabirds – we have strength in numbers.

My COP28 Speech at the 2023 Global Day of Action rally

I want my grandchildren to see what I see:
abundant forests, wild flowers, a buzzing bumble bee.
I want my grandchildren to see what I see,
breaching whales, diving gannets, biodiversity.
I want my grandchildren to see what I see,
clean beaches, a calm climate, unbleached coral reef.
I want my grandchildren to see what I see,
to watch a nature programme and know it's not history.
I want my grandchildren to see what I see,
Mother Nature in abundance, as it should be.
I want my grandchildren to see what I see,
rich soil, healthy food – just enough for you and me.
I want you, grandchild, to know I honestly truly tried –
I protested and campaigned and changed my way of life.
I want you, grandchild, to know that they lied,
promises from politicians were always pushed aside.
This is the UN Decade of Ecosystem Restoration,
it should be full of action, rewilding, carbon sequestration.
The science warns us, it's bleak, it's stark,
yet the ones who benefit leave us shouting in the dark.
There is hope, there is technology and nature-based solutions,
so stop the ecocide, stop the pollution.
A new era awaits and we have a generational plea,
to make our grandchildren see what we see.

What does Rathlin mean to you?

Liam McFaul, resident and warden
You get maybe a little blasé living in a place like this, and then you go to the mainland and there's something missing, you're searching for something, and you realise, it's what you have at home. It never ceases to inspire you, there's always something new, something different happening – a migrant bird or a golden hare just sitting on the side of the road – and you just think … only on Rathlin.

Honor Poag, visitor
Rathlin to me is a step back in time, to a simpler, more peaceful existence. It is a reminder to care for and protect your environment. It is stunning unspoilt scenery, wild animals and birds, every weather in the space of an hour. It is isolation, quiet, loneliness but the pleasure of a close-knit community where everyone looks out for each other. It is all the sounds of nature that you never hear in a bustling town life. It is my happy place, my home from home, my clear-my-head space.

Eleanor Patience, 'Ballycastle blow-in'
When the opportunity arose to make a daily commute to the island to work for RSPB, I didn't have to think twice. My dad's passing in 2019 had broken me and COVID was the final nail in the coffin. My self-esteem, self-worth and self-confidence were low and Rathlin represented a glimmer of hope that I

was employable. My goodness did the island deliver – from my daily banter with the ferrymen and new colleagues, witnessing dolphins up close for the first time to the privilege of watching the birds feather their nests in preparation for their chicks. The island brought peace and pints and sand between my toes. I'm hugely indebted to this wild place and show my gratitude every day, blowing kisses to the island with the three lighthouses from the Ballycastle shoreline.

Kevin Kirkham MBE, birder, environmentalist and RSPB volunteer

The seabirds cry – the waves they crash – from Bull Point to the Rue – no finer place to feel alive – than on Antrim's Rathlin Isle. A Roonivoolin ramble – high where the ravens nest – over Ushet to Craigmacagan – buteos soaring high – crex crex in the nettles – refreshment at McCuaig's. The West Lighthouse – where the seabirds live – guillemots, razorbills and puffins – onomatopoeic kittiwakes call – albatross-like fulmars cuddle on the grass. Kebble and Kinramer – the golden hare in form – Rathlin is a special place – there's nowhere quite the same – for me it is the shining jewel – in the Causeway's gleaming crown!

Charlie, harbour master

What Rathlin means to me? I love living on Rathlin. It represents a freedom of a way of living. I live on a boat on an island, and that in itself has its challenges. It is beautiful – sometimes the light is so beautiful it takes your breath away. It is not an easy place to live. In the spring and summer, it is ideal. When the sun is shining and the seals are out, you can see the fish swimming around the boat. In the winter, it is tough, weather, light, and isolation. This is part of its attraction for me. You have to be self-sufficient, both yourself and within the community, for repairs, food, and entertainment. It is a creative way to live, and as an artist, this is very appealing.

Kirsty, islander and swimmer
In a simple word Rathlin means home to me. I've travelled the world searching for a place to belong and found that feeling the first day I arrived on Rathlin. The island has a real charm, and when the sun is shining and the birds are singing I wouldn't want to be anywhere else. It's the perfect place to live and raise children away from the commotion of towns and cities. This wild island has certainly taken my heart.

Douglas Cecil, islander
Being born and raised on Rathlin there's no better place to call home. Every day is unique and with it the seasons and weather, the sea dominates all aspects of life and rather than being a barrier it's the gateway to lots of new adventures. From early childhood long summer days spent rock pooling and beach combing, that magic never leaves you even as an adult. There's been plenty of changes over the years but the place that is uniquely Rathlin will always remain the same. From the call of oyster catchers being vocal in their socialising or the dart of the hares through the whins and heather, to seals balancing precariously on the shore at Mill Bay, in the heart of each and every islander – a special place that will always be.

Sorin McMullan, islander
As one of few teens living on Rathlin I can say it can be frustrating balancing a social life on the mainland, but I

wouldn't give it up for the world because it means having the best of both worlds; the fast pace busy life of the mainland and the slower pace peace to come back home to. Living on Rathlin is definitely all about the community, everyone is very close and mostly supportive of one another. However it can be difficult at times, but nowhere is perfect. The community changes as people come and go but they always leave behind their experiences and crazy stories and aren't easily forgotten. I do love our old local traditions of Cèilidh dancing, plays, welcoming new babies to the island with balloons and banners on their first day coming home, New Year's Day swims, card nights and likewise with our newer traditions. I think they are what makes living on Rathlin different from living on any other place in the world.

Rebecca Tanner, Seabird Centre officer in 2022
I had the immense privilege of working at the Seabird Centre for six months in 2022. Rathlin surpasses anywhere else I've lived, not only in its wealth of incredible wildlife but also because of the passionate and knowledgeable people the island attracts, whether residents, volunteers or visitors. My time on Rathlin was marked by experiences which will stay with me for the rest of my life, including hearing Manx shearwaters calling from the north cliffs in the middle of the night, finding Rathlin's first recorded striped hawk moth and spotting my first ever killer whales! But aside from the more unusual wildlife encounters, what really made my time on Rathlin memorable was just turning up to work every day and being able to watch the seabirds for a whole season. Searching for the eggs in spring, seeing the chicks rapidly grow and watching the jumplings take their perilous first leaps off the cliffs. Seabird colonies are an amazing natural

phenomenon and are a huge part of what makes Rathlin so special to me!

Dakota Reid, RSPB volunteer
Rathlin is where I fell in love with conservation work. It was there I realised that spending time outdoors, immersing myself in nature, wouldn't be something I could just do in my free time – this is what I wanted to devote my life to. Rathlin is where I first learnt about moth trapping. It's where I first heard snipe drumming. It's where I met some fantastic friends who were unapologetic nature nerds like me! Not being able to go to Rathlin during Covid was painful, but the return was so, so joyful. Corncrakes, ruby-tailed wasps, puffins, elephant hawk-moths, golden hares and everything in between – it's one of the last oases of wilderness we have in this corner of the world and acts as a reminder of what we can achieve when we take ambitious action for nature. Rathlin for me is a call to action, an inspiration and time spent on the island absolutely rejuvenates my soul.

Wendy Jack, musician and islander
The lyrics from 'Island Bound' reflect the interconnectedness and at times the harsh reality of life on an island. It is a celebration of the hard work and dedication of those who recognise that the health and future of their community is dependent on the health of the environment in which they live and work. It is a profound achievement to create, restore and protect these natural habitats. With vision, hard work and lots of nettles … hope and history has rhymed as the corncrake sounds once again on Rathlin Island.

Island Bound (Abridged version 2023)

Island bound, through the mists of time.
Roots entwined in the hallowed ground.
Island bound, hope and history rhyme.
Dreams combine as the corncrake sounds.
We work together, help one another.
If you fall so do I.
Oh, we live together, making music and laughter.
The future in our eyes.
We are the land … we are the sea.
In health … In liberty.
Hold strong my tether bind us together.
We'll brave this stormy sea.
Worn by the weather, pray it won't sever.
The bond 'tween you and me.
Island bound, through the mists of time.
Roots entwined in the hallowed ground.
Island bound, hope and history rhyme.
Dreams combine as the corncrake sounds.

Carol Cockburn, RSPB volunteer

I have volunteered yearly at Rathlin Seabird Centre since 2021. This is a place I meet so many interesting people, staff and visitors from around the world. I enjoy engaging with them about the wildlife and hearing their stories. The role allows me the best of both worlds in meeting people and walking Rathlin's numerous magical trails which I find therapeutic and restful. If you are interested in nature, history and traditions then Rathlin Island offers it all. One visit is never enough, you will always want to come back – and I do!

Ali McFaul, resident and previous Seabird Centre manager
Thinking of an Island Life

We live on an island – It's like a large boat.
We're moored to the seabed – The sea is our moat.

It's a source of great comfort – fresh food and delight –
but we're never complacent and know of its might.

The island is home to the old and the new –
families who thrive here and singletons too.

Romances have blossomed in adventurous hearts –
and people are welcomed to make their new starts.

The island's a haven for much that is wild –
and a nurturing place for any young child.

So come over the sea and see what you find –
and stay for a while – while you make up your mind.

Mark Latimer, islander

Entering the port of a postcard where life is a caption of tranquillity. Where sounds of mother nature engulf the senses. Hearing the seals, hearing the cows, hearing the birds, sounds that sing songs of peace and harmony. Even sounds of rain, thunder and wind are melodies which inspire. Whatever sound mother nature makes on Rathlin, it's something to perk up your senses and make you smile whilst embracing the heaven that is the island of solitude.

You know (mostly) everyone, and everyone (mostly) knows you, you can be popular or quiet, confident or shy. Discrimination is a word with no meaning or power, Rathlin accepts everyone and won't judge whatever soul you may be.

The sound of a corncrake, whether it be in the middle of the night or during the day, it still lifts up the spirit to hear the sound that penetrates the mind. Hello is a word said throughout the land, the greetings are natural and sincere to all, whether you are a stranger or friend, we welcome you all. A world inside a world where fear doesn't live, happiness rules all and you can be true to yourself. What does Rathlin mean to me? Rathlin is a postcard from heaven where I have the privilege to call home. I can be me without judgement or prejudice.

Jane Wysner, friend and resident
Rathlin Island has been a very special place for me since my family took me there as a young child when I was two. We were the 'rectory' family, so we stayed for a month every summer, for free, while my dad took the services in the little church at the shore. When the old landlord family took the rectory back we stayed in various houses and caravans all over the island, so we got to know it, and the community very well. We watched my dad pump water from the well, and later, when we were older, we were sent to fetch water from an open well. We roamed the fields and coastline absorbing all the nature and wildlife around us. When I got older I brought my own children to Rathlin every summer, to give them the same idyllic childhood I enjoyed there. I can't really describe the importance this place holds for us. It is a magical place that has many interesting physical attributes, but there is also something about the landscape that welcomes you spiritually. It just feels like home.

Tom McDonnell, wildlife photographer and islander (North Coast Nature – Wildlife and Nature Photography)
Rathlin the jewel of the North.
A wonder of Nature.
The sound of the Corncrake.
The song of the cuckoo.
The beauty of the Island.
Nature and Island people together as one.
This is my home, what a place to live.

Daisy May Harris, Seabird Centre officer in 2024
Never did I think I would have the opportunity to go to sleep and wake up to the sound of such coastal biodiversity. It was a young girl's dream that became her very existence.

Rachele Crawford, 2021 Seabird Centre experience manager
Reminiscing about my time as visitor experience manager on the stunning Rathlin Island brings back fond memories. I was captivated by the rugged natural beauty of the island, from the dramatic cliffs and caves to the vibrant seabird colonies that called this place home. Working alongside the dedicated staff and passionate volunteers was a true privilege, as we all came together to share the wonders of Rathlin's rich wildlife and landscapes with visitors from near and far. The camaraderie and teamwork of that close-knit community is something I will always cherish. The breath-taking sunsets, cries of gulls, and fresh, salty air will always hold a special place in my heart.

Acknowledgements

A young, eager woman with a manuscript and a collection of drawings came to Blackstaff Press with a goal; to spread awareness about the plight of UK and Irish species through a unique, joyous experience on a wild little island. This special patch of rock brought me friends – some in human form but mostly ones with wings. It brought me clarity in a time of global confusion. It gave me the closure I needed to pursue the next stages of my life and for that – to Rathlin, I thank you.

Thank you to my Publisher, Blackstaff Press who saw the potential in this book – and me, from day one. It has been an absolute privilege to turn the diary I kept, full of weathered words and sketches into a deliciously detailed memoir and it has been my life's highlight to write about the spectacles of nature that not all of us get the privilege of seeing every day. I'd like to especially thank my Editor, Helen Wright. As a first time Author, your guidance has been sublime. I'll always treasure your knowledge, passion and curiosity for words.

This book took over two and a half years to write, illustrate and edit; it has been a labour of love which wouldn't have been possible without the unwavering support of many friends, family and colleagues. Firstly, I'd like to acknowledge my other half, Craig Holmes. For the endless cups of tea, pep talks and hugs. For taking Isla on walks when I was tied to my desk. For allowing me to share something so personal for us both to the world. I want to thank Craig's kind parents,

Rhonda and Alan who've welcomed me to the farm with such love – I am ever grateful.

Writing this book and the process of publishing it has been one of the hardest, most rewarding things I've ever done. I often debated if it was worth the emotional toll but there were two things that lived rent-free in my mind during the process which kept me going. The first was the nature I adore, species whose voices are quietening. I knew that this story I had within me, and the experiences I held so dear could perhaps inspire some change. From day one, when I began transferring my thoughts and memories to a neat Word document; I pinned a picture of some puffins to the cupboard beside my office desk. These characterful faces I'd had the pleasure of spending much time with, motivated me to carry on. Nature, my muse and forever fascination – thank you.

The second thought that remained ever-present was little Ruby. A girl with an imagination so wide, she could barely sleep at night. I loved reading and writing as a child, but I wasn't always 'good' at it. The joy I felt walking into a book shop to find something new surpassed anything else. And the freeing feeling of getting lost in nature writing brought me such escapism. However, enthusiastically reading lines of a poorly punctuated story aloud in class, children laughed under their breath as my eyes skipped lines on a page. Despite this, by age 9, I created my own series of stapled together adventure books, excited to show people an alternative world full of wildlife. An anxious, undiagnosed neurodiverse girl – if my love for literature and art hadn't been nurtured at home (thank you mum and dad), school would've unfortunately crushed it. Deep down, I knew I could write, so worked at it for years; writing columns, blogs and articles for different organisations, before one day, having the audacity to create this book. Message to younger self: you see the world in a unique way and process information differently, but that doesn't mean you're any less academic, any less deserving of writing that story… So, I thank her for sticking at it, for carrying on.

Thank you to my RSPB NI family for the opportunities they've provided me and their continued support for my work. I'd also like to thank the many other environmental NGOs who've helped me through contributing statistics, research and local insight into this writing process – and thank you to the amazingly passionate and smart individuals within them. Thank you to the wardens, conservation scientists and local communities on Rathlin and across Northern Ireland, fighting for a world with more nature in it. Thank you local campaigners, activists and science communicators for all you do to spread awareness – it takes a team. Thank you to the Friends and Residents of Rathlin who not only welcomed me with open arms in 2021 but contributed so beautifully to this book. I feel honoured to share the personal experiences you hold so dear.

Finally, I am unconditionally indebted to the West Lighthouse, an unassuming white building on a steep patch of cliff, in the middle of the sea. It gave me the opportunity to experience a world I wouldn't have otherwise seen. Thank you to the brave builders who created it over 100 years ago and the lightkeepers and caretakers who've maintained it since, so someone like me could enjoy it.

Beyond Generation Rent

To Kathleen

Beyond Generation Rent

Political Economy, Inequality and the Private Rental Sector

Michael Byrne

polity

Copyright © Michael Byrne 2026

The right of Michael Byrne to be identified as Author of this Work has been asserted in accordance with the UK Copyright, Designs and Patents Act 1988.

First published in 2026 by Polity Press Ltd.

Polity Press Ltd.
65 Bridge Street
Cambridge CB2 1UR, UK

Polity Press Ltd.
111 River Street
Hoboken, NJ 07030, USA

All rights reserved. Except for the quotation of short passages for the purpose of criticism and review, no part of this publication may be reproduced, stored in a retrieval system or transmitted, in any form or by any means, electronic, mechanical, photocopying, recording or otherwise, without the prior permission of the publisher.

ISBN-13: 978-1-5095-6341-8 (hardback)
ISBN-13: 978-1-5095-6342-5 (paperback)

A catalogue record for this book is available from the British Library.

Library of Congress Control Number: 2025938357

Typeset in 10.5 on 13pt Swift
by Fakenham Prepress Solutions, Fakenham, Norfolk NR21 8NL
Printed and bound in Great Britain by CPI Group (UK) Ltd, Croydon

The publisher has used its best endeavours to ensure that the URLs for external websites referred to in this book are correct and active at the time of going to press. However, the publisher has no responsibility for the websites and can make no guarantee that a site will remain live or that the content is or will remain appropriate.

Every effort has been made to trace all copyright holders, but if any have been overlooked the publisher will be pleased to include any necessary credits in any subsequent reprint or edition.

For further information on Polity, visit our website:
politybooks.com

Contents

Acknowledgements	vi
1 Introduction: The Political Economy of the Private Rental Sector	1
2 Conceptualizing Home Rent: The Micro-Political Economy of the Private Rental Sector	16
3 The Revival of the Private Rental Sector: Neoliberalism, Financialization 1.0 and the Concentration of Property Ownership	39
4 The Financialization of the Private Rental Sector: The Rise of Institutional Landlords	65
5 Beyond Generation Rent? Housing Inequality and the Private Rental Sector	90
6 Making a Home in Someone Else's Asset: Insecurity and Power in Lightly Regulated Rental Markets	109
7 The Politics of the Private Rental Sector: Tenants as Agents	129
Conclusion: Transforming the Social Relations of Home	155
References	167
Index	184

Acknowledgements

Many of the ideas in this book began during my involvement with the now defunct Dublin Tenants' Association, so I would like to thank all those who were involved in that project, especially Patrick Bresnihan, Sive Bresnihan, Norma Jean Kenny, Aengus Hennessy, Fionn Toland and Fergal Scully. I would also like to thank my colleagues Michelle Norris and Bryan Fanning for their support and mentorship throughout the writing of this book. Most importantly, I wouldn't have been able to write this book without the constant support of my wife. Thank you, Mags!

This book is dedicated to our daughter Kathleen. Thanks for interrupting me so many times in my 'workshop' to play with me.

1

Introduction
The Political Economy of the Private Rental Sector

In 2015 I started regularly to meet private tenants in Dublin, Ireland, as an activist with the Dublin Tenants' Association. The issues coming up for tenants will come as no surprise: high rents, eviction and poor-quality housing. But there were some surprises too. The tenants were more diverse than one might expect. It wasn't just more vulnerable cohorts who were experiencing these issues, but people from many different backgrounds. From high-skilled migrant workers to lone parents from Dublin's disadvantaged communities, private rented housing seemed to be generating problems across the board. But what surprised me most was the sense of unfairness that characterized tenants' experience of private rented housing. This was especially true of long-term renters, those in their forties, for example, but was shared by nearly all the tenants I met in those years. Tenants were angry and frustrated. They felt powerless. They felt that legislation didn't protect them, that landlords could act with impunity, and that the whole private rented sector was rotten. As a seasoned political activist, I wasn't all that used to finding such a receptive audience for my politics. But in the case of private rented housing, tenants needed no convincing that the housing system systematically favoured property owners, and that any progress for tenants would require radical change.

In those years, the fact that Ireland was experiencing a dramatic decline in homeownership was only starting to be understood. The debate around 'generation rent' hadn't really taken hold. My experience when talking to Dublin's private renters signalled a broader societal shift that was taking place, not just in Dublin but across many advanced economies. Across most countries, the private rental sector (PRS) had been in continual decline throughout the twentieth century, becoming a marginal tenure. Through a combination of state intervention to support the expansion of social housing and homeownership and economic growth, fewer and fewer households depended on a private landlord to meet their housing needs. The issues that had been at the heart of housing politics in the late nineteenth and early twentieth centuries: rack renting, evictions, chronic overcrowding and deplorable conditions, seemed to fade from view as the PRS dwindled. Tenant unions, and their favoured tactics, especially the mass rent strike, gradually became a matter for the history books. As far as housing politics went, the issues that mattered to homeowners and, to a lesser extent, social housing residents, dominated.

In different ways, and with great variation across jurisdictions, the homeownership and social housing tenures offered households the opportunity for residential stability, to put down roots, and to have a stake in their communities and cities (Forrest & Hirayama, 2015). We shouldn't romanticize the achievements of the housing systems of the mid and late twentieth century; they were beset by all manner of exclusions and inequalities (Florida & Feldman, 1988). They did, however, present an attractive alternative to the chronic insecurity and 'landlord tyranny' that many households had previously endured as private tenants. In particular, what Arundel and Ronald (2021) describe as the 'promise of homeownership' offered a compelling vision of mass property ownership, security and the ability to acquire wealth for a wide swathe of the population. This vision, and indeed ideology (Ronald, 2008), proved immensely compelling across many countries. But the 'promise of homeownership', from today's perspective, seems like a broken one. The ability to buy a house is increasingly limited, as house prices soar beyond the reach of many, and investors outcompete would-be first-time buyers. Low-income and precarious households, especially younger households, migrants, lone parents and other more economically vulnerable cohorts, have found themselves locked out of housing markets. The residualization of social housing, typical of the neoliberal turn in housing policy virtually everywhere, has reduced access to this tenure, leaving many with nowhere to turn except the PRS.

Growing demand for PRS housing has in turn created an investment opportunity, leading to a vicious cycle and a housing market that is increasingly perceived as favouring investors over residents (Rolnik, 2019). Although it is not often stated explicitly, at the heart of this process is a *concentration of ownership of residential property*, as a growing proportion of the housing stock is held in fewer hands. In other words, the generation rent phenomenon I first encountered while meeting tenants a decade ago is a symptom of a transformation in property relations within the housing system. This takes the form both of households of multi-property ownership and the mass ownership of rental properties by behemoth financial institutions.

Little wonder, then, that the tenant activism that was beginning in Dublin in those years was also emerging in many countries facing similar issues. In Spain, for example, a new generation of tenant unions was established, starting in 2017, and has grown to be a significant political force. The UK saw a similar development, first with the establishment of Living Rent in Scotland, and later with the emergence of the London Renters' Union and other similar organizations across England. Meanwhile, in societies that had long had much bigger PRSs, especially Germany, the plight of renters also came to the fore, culminating in the now famous Berlin campaign for a referendum on the expropriation of PRS housing owned by financial institutions (discussed in more detail in Chapter 7). This *re-politicization* of the PRS is driven by the poor housing outcomes often associated with the sector, like evictions and rapid rent inflation. But, more than this, it is driven by that sense of frustration I first encountered with the Dublin Tenants' Association: there is something about living in someone else's 'asset' that just *feels* wrong to many people. As well as undermining the ability of tenants to create a real, long-term and stable home, being a tenant can generate a feeling of disempowerment (Madden & Marcuse, 2016).

A lot has been said about these issues over recent years, within the media, politics and academia. Media framings have, perhaps unsurprisingly, been unhelpful. At least in English-speaking countries, there has been a focus on millennials, especially middle-class couples, who have been unable to access homeownership and get 'on the property ladder'. Indeed, it can sometimes feel as though the media diagnosis of the problem is that middle-income millennials have not been able to perpetuate the same way of relating to housing that has gotten us into this mess in the first place. Meanwhile, to the extent that the politics of so-called 'generation rent' is addressed, it is in the form of a simplistic 'boomers vs millennials' narrative. On the one hand, the self-appointed boomer spokespeople blame younger cohorts for their feckless spending on smashed avocado and flat whites; on the other, the righteous defenders of younger generations argue that 'boomers' had everything, especially housing, handed to them on a plate. Both narratives are as false as they are unhelpful.

Thankfully, there has also been a resurgence of academic research that elucidates the nature, drivers and implications of the resurgence of the PRS. Although it is difficult to draw a hard and fast distinction, we can broadly think of this literature in terms of two fields of research: the political economy of housing and housing studies. Beginning with the latter, a huge amount of research has been published over the last decade or so shedding insights into the contemporary PRS. This research includes work focusing on the factors which have caused the decline of homeownership, and to a lesser extent social housing, and the related growth in the proportion of households living in the PRS (Arundel & Doling, 2017; Arundel & Ronald, 2021; Bone, 2014; Byrne, 2020a; Forrest & Hirayama, 2015; Hochstenbach & Ronald, 2020; Kemp, 2015, 2023; McKee, 2012; Ronald & Kadi, 2017). Another strand of research focuses on the subjective experiences of PRS households (Byrne, 2020b; Byrne & Sassi, 2023; Desmond, 2016; Hoolachan et al., 2017; Hulse et al., 2019; McKee et al., 2019; Soaita, 2021; Soaita & McKee, 2020), the nature of 'home' in the PRS (Bate, 2018, 2021; Easthope, 2014; Soaita & McKee, 2019), the social relationships between landlord and tenant, and the related power relationship between them (Byrne & McArdle, 2022; Chisholm et al., 2020; Desmond, 2016; Lister, 2004, 2005; McArdle & Byrne, 2022; McKee & Harris, 2025). A particular concern within this research has been the issue of housing insecurity and its significance for tenants' experiences within the PRS (Desmond, 2016), as well as its impacts on a host of outcomes from mental health to education (Acharya et al., 2022; Bone, 2014; Morris et al., 2017; Power, 2017; Soederberg, 2018). While housing insecurity and eviction is one set of housing outcomes related to inequality, the housing studies literature has also addressed the much wider set of interrelationships between the resurgence of the PRS and inequality (Arundel, 2017; Christophers, 2017; Forrest & Hirayama, 2015, 2018; Hochstenbach, 2022) and 'precarity' (Arundel & Lennartz, 2020; Bone, 2014; Listerborn, 2023; Waldron, 2023, 2024a).

Throughout the following chapters I will delve into the insights from this body of research in much greater detail. Here, I will simply touch upon the strengths and limitations of this work, in order to situate and clarify the contribution of this book and its political economy approach to the PRS. The literature described above has helped immensely to understand the nature of the contemporary PRS, especially with regard to the subjective experiences, social relations and forms of 'home making' and 'home unmaking' that characterize the tenure, particularly in the more lightly regulated PRS markets that dominate in the literature. This has drawn attention to the importance of the experiential dimensions of housing in understanding housing outcomes, such as the ways in which the subjective experience of insecurity and relationships with landlords can exercise a significant impact on the ability to create a 'home', and related outcomes such as mental ill health. Highlighting this subjective, experiential and relational dimension of 'assembling home' (Soaita & McKee, 2019) in the PRS, as well as the all-pervasive question of housing insecurity (Desmond, 2016), is crucial to understanding what, in my view, are some of the most significant issues raised by the resurgence of PRS housing. The housing studies literature has also connected the resurgence of the PRS to wider socio-economic forces. In relation to understanding the drivers of the growth of the PRS, scholars have examined how policy and market forces, such as the processes of neoliberalization and financialization discussed in Chapter 3, have interacted over recent decades to reorganize housing systems, undermine and residualize social housing, and limit access to homeownership, all while generating demand for private rental housing (Kemp, 2015; Ronald & Kadi, 2017). These perspectives help us to see how the everyday experiences of tenants are linked to macro-level socio-economic processes, often, as in the case of financial markets, operating at a transnational scale (Byrne, 2020a). Similarly, in understanding the forms of social stratification and inequality associated with the resurgence of the PRS, research has drawn out the ways in which issues like wealth inequality (Arundel, 2017), income inequality (Bartels & Schröder, 2020) and welfare (McKee, 2012; Ronald & Kadi, 2017) are all tied up with the shifting tenure patterns of contemporary housing systems. In this regard, the housing studies literature touches on questions of political economy, in that it helps us to understand the significance of the interaction between political (i.e. policy) and market forces in explaining and analysing the contemporary PRS.

However, there are some limitations to this literature which it is worth drawing out. First of all, while the housing studies literature certainly engages with political economy, in relation to some important issues this engagement is more implicit than explicit. For example, crucial issues like housing insecurity are often treated as primarily arising from the weak security of tenure that policy and legislation afford tenants in many jurisdictions, rather than foregrounding the economic relationship at the heart of the PRS, which is in turn a property relationship. This is linked to a more fundamental issue, which is the absence of a direct conceptualization of the nature of rent itself, i.e. of the type of economic phenomenon at stake in the renting out of residential property. For example, the nature of the production and appropriation of value that occurs in the context of private rental housing has not been fully addressed. This limits, as I will argue in Chapter 2, a fuller understanding of many of the most important issues addressed in the housing studies literature, including the experience of home, the nature of insecurity and the power relationship between landlord and tenant. A final, and related, limitation of the existing body of housing studies research on the PRS is the absence of a comprehensive framework for the analysis of the PRS that draws together the different aspects and dimensions that have emerged in the recent literature, as well as the older literature.

Some of these shortcomings can be addressed via the literature from the political economy of housing, some of which focuses directly on the PRS (Beswick et al., 2016; Byrne, 2019, 2020c; Christophers, 2022b, 2022a; Fields, 2018; Gray, 2018b; Harvey, 1974; Wijburg et al., 2018), but much of which relates to housing more generally (Aalbers, 2016; Aalbers & Christophers, 2014; Aalbers & Ward, 2016; Adkins et al., 2021; Ball, 1983; Christophers, 2010, 2020; Fuller, 2019; Haila, 2015; Harvey, 2018; Jacobs et al., 2024; Moreno Zacarés, 2024; Schwartz & Seabrooke, 2008; Smith et al., 2022). Much of the contemporary literature here emerged in response to the Global Financial Crisis, exploring the relationship between housing, especially mortgage debt, the dynamics of global capitalism, and the international financial system. In this regard, the political economy literature, perhaps unsurprisingly, pays a lot of attention to the role of housing within the wider dynamics of capitalist accumulation (Aalbers, 2016; Aalbers & Christophers, 2014; Beswick et al., 2016; Gray, 2018b; Moreno Zacarés, 2024). The ownership of capital, investment dynamics and the systemic importance of profits (or rents) derived from housing and from related financial products (from simple mortgages to the most byzantine derivative products) all feature prominently. In terms of explaining housing system change, and especially the decline of homeownership and resurgence of the PRS, this literature has been particularly useful in revealing the crucial role of both neoliberalism and financialization (Aalbers, 2016; Byrne, 2020a; Christophers, 2022b; Fields, 2018). Some political economists within the field of political science have also highlighted the relationship between housing and the financial system and the world of electoral politics and government policy (Fuller, 2019; Schwartz & Seabrooke, 2008).

The approach developed in this book draws on many insights from the political economy literature. Most obviously, it makes explicit, and central to analysis, that in large part housing in the context of capitalist economics is both a commodity and an asset, and is created and allocated via economic relationships which generate value, be it of the 'use' or 'exchange' variety. Thus, the ability of housing to generate value, and for that value to be captured by owners and investors, is made central to analysis. This also foregrounds the fact that property relationships, the private ownership of residential housing and land, are a fundamental feature of contemporary housing systems. Of particular interest in this regard is the importance that the concept of 'rent' plays (Aalbers & Christophers, 2014; Harvey, 1974, 2018; Moreno Zacarés, 2024; Ryan-Collins et al., 2017). Here I do not mean rent in the everyday sense, i.e. the money a tenant pays to a landlord each week or month. Instead, rent within the political economy literature generally refers to the specific form of value that is generated by housing and land as inherently scarce, supply-constrained and immobile resources, and the consequent monopolistic features that characterize property markets (Harvey, 2018). The concept of 'rent', discussed in detail in Chapter 2, is important because it conceptualizes the specific economic form that housing takes, and clarifies what distinguishes housing from other commodities. By conceptualizing rent in this fashion the role of housing markets within wider processes of political economy, such as capital accumulation, can be more clearly addressed.

Focusing on housing as a direct economic phenomenon and relationship can also help to approach issues such as inequality and injustice within housing systems in a particularly valuable way. A straightforward, yet important, example is the simple observation that housing markets privilege exchange value over use value. As indicated in Madden and Marcuse's (2016) *In Defence of Housing*, such a perspective draws out the ways in which many of the problems and contradictions within housing systems are not simply examples of policy failure, but are characteristics of reducing housing to a commodity, i.e. a feature rather than a bug of capitalist housing markets (see also Rosenthal & Vilchis, 2024). More generally, this type of approach has two features which are fundamental to the analysis of housing.

First, understanding issues such as poor housing outcomes or inequality within housing systems is usually strengthened via detailed analysis and critique of related economic processes. For example, during the Global Financial Crisis (GFC) mass mortgage arrears and the consequent foreclosure crisis were directly linked to the economic form taken by specific products, e.g. subprime mortgages, mortgage backed securities and credit default swaps (García-Lamarca & Kaika, 2016; Immergluck & Law, 2014).

Second, and here the Marxian influence within the political economy of housing is particularly evident, the material contradictions that arise within housing systems, and their attendant property relations, can, and often do, give rise to conflicts and antagonism, which can politicize housing and generate social change. Political economists have thus been concerned with understanding resistance, contestation and the politics of housing (Byrne, 2019, 2020c; Fields, 2017a, 2017b; Guzmán & Ill-Raga, 2022; García-Lamarca & Kaika, 2016; Gil & Palomera, 2024; Gray, 2018b; Madden & Marcuse, 2016).

The above literature refers to the political economy of housing in general, rather than the PRS in particular. Indeed, it is somewhat surprising that political economy literature looking specifically at rental housing is rather under-developed. The exception here is in the area of financialization where there has been a wave of scholarship examining the rise of 'financial landlords' in the wake of the Global Financial Crisis, their role in housing issues like evictions, rent increases and gentrification, and tenant-led activism in opposition to them (see Chapter 4 for a full discussion). This literature shares many of the strengths of the wider political economy scholarship discussed above, but it focuses only on one sub-sector within the wider PRS.

Like the housing studies research discussed earlier, the political economy literature is not without its limitations. Perhaps most surprisingly and, as discussed in detail in Chapter 2, a political economic critique of rent in the context of rented housing has never been explicitly developed. While the broader theoretical work on housing and land rent is hugely instructive when it comes to rental housing, the process of creating a home in someone else's 'asset', and being required to pay each month for continued access to that home, requires an explicit theory. Without a critical theorization of rent in this context, analysis of the PRS will be hamstrung; a point I return to below. Furthermore, the political economy literature tends to be most interested either in how housing and the built environment play a role in the wider dynamics of capitalist accumulation (Harvey, 2018; Moreno Zacarés, 2024), or in how different forms of investment interact with and shape the provision of housing. They are typically less concerned with experiences of residents, particularly the subjective experience of housing and home and the direct, everyday relationships that this involves. Indeed, political economy perspectives can sometimes view the 'consumption' of housing as secondary to their core concerns. For example, Moreno (2024: 20) argues that '[a] major problem has been a recurring overemphasis on housing as an object of consumption and exchange, neglecting other aspects of provision, particularly the housebuilding process'. While there is much to be said for an approach that focuses on the production of the built environment, in the case of the PRS, the subjective experience of home, and especially the tensions around everyday practices of home making, are central to the power relation between landlord and tenant (which must, of course, be a central focus of political economy), and to the consequent forms of conflict and antagonism relevant to the politics of the PRS. Finally, the political economy of housing requires a more fully developed and comprehensive analytical approach for the PRS and one, as indicated above, that is founded on a critique of rent in the context of rental housing and can incorporate the subjective, experiential and relational dimensions of home.

In a sense, then, the aim of this book is to incorporate insights from both the housing studies and political economy literatures to develop a comprehensive approach to the critical analysis of the PRS. To do this, the book sets out a novel five-part framework, developed throughout the subsequent chapters and depicted in Figure 1 below. In Chapter 2 I develop a theory of rent in the context of rental housing, which I describe as 'home rent'. I draw on theories of rent in the context of land and housing, especially the work of Henry George, but also urban sociology, Madden and Marcuse's (2016) concept of 'residential alienation' and critical geographies of home to offer a novel theorization. The concept of 'home rent' emphasizes that home, and its value, is *produced* by the resident, but subject to a property relationship, specifically the landlord's monopoly ownership of the tenant's home. The theory of home rent aims to denaturalize private rental housing, to

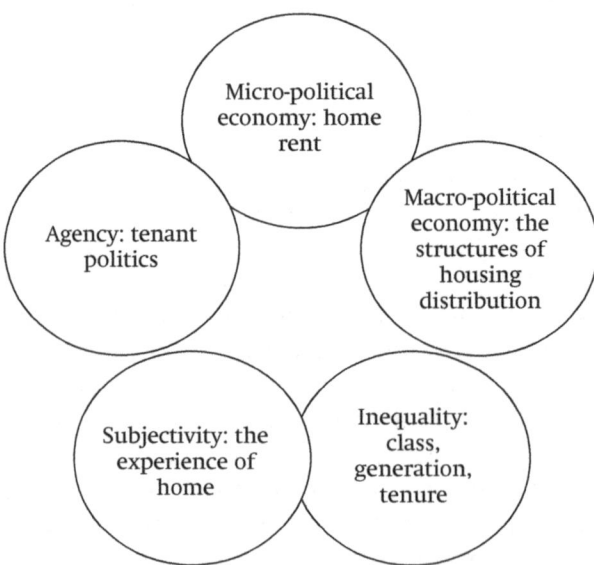

Figure 1 A framework for the political economy of the PRS

recognize the economic activity that is at the heart of this form of housing as a social and property relationship and, consequently, to elucidate the dynamics of power and inequality at stake. A theory of home rent, I will thus argue, offers a fundamental pillar for the analysis of the PRS.

I think of the critique of home rent in terms of what I will call a 'micro-political economy' of the PRS. This refers to that fact that it homes in on the central social relationship between landlord and tenant. However, we also require a 'macro-political economy' of the PRS, which focuses more on the socio-economic structures and processes that shape the nature and extent of the PRS across different times and places. While the micro-political economy of the PRS looks at the dynamic between landlord and tenant, if we are not to naturalize and normalize the PRS, we also need to interrogate the political and economic structures that ensure that some people are landlords, while others are tenants, in the first place. Just as property relationships are at heart of the micro-political economy of the landlord–tenant relationship, a macro-political economy lens reveals how property relations writ large are fundamental to the PRS. This perspective, developed in Chapter 3, recognizes that unequal distribution of residential property, i.e. the concentration of ownership, is a *sine qua non* for the very existence of private rental housing.

A macro-political economy perspective also has *explanatory* power in that it can help us to understand, for example, changes within the housing system. The change of most relevance here is the recent resurgence of the PRS in many countries. Chapter 3 develops this theme by examining the drivers of the resurgence of the PRS, particularly focusing on the interaction of two key political economic processes, financialization and neoliberalization, which together have undermined homeownership, concentrated the ownership of residential property, and thus driven the growth of the PRS. Chapter 3 thus also shows that the growth of the PRS is primarily driven by, and therefore can be explained by, the interaction of political (e.g. housing policy) and economic (e.g. housing and financial markets) forces. Finally, by examining how political and economic forces interact to drive changes in tenure patterns and the distribution of ownership of residential housing, Chapter 3 also attempts to conceptualize the wider political economic significance of the revival of the PRS.

These themes are further developed in Chapter 4, which further explores the financialization of housing to examine the interaction between the financial system and the PRS and the ways in which the latter is imbricated in wider, even global, economic processes that go far beyond the housing system. It thus examines how the political economy of housing in general, and the PRS in particular, calls for an understanding of the role housing plays in contemporary economic developments at different scales. In so doing, we can also examine the crucial role played by the state in enabling and mediating the interaction between PRS housing and larger structures and processes within the economy. The financialization of the PRS, Chapter 4 argues, is also a useful avenue to explore the 'macro-political economy' of the PRS as it is essentially what I call a form of 'super-concentration' of residential property ownership, thus returning to the theme of the distribution of housing.

The remaining chapters focus on the inequalities, power dynamics and politics of the PRS. In Chapter 5 I examine the economics of inequality associated with the revival of the PRS, including wealth and income inequality. The chapter also engages with the question of how to conceptualize the forms of inequality associated with PRS housing. As noted already, media narratives have typically emphasized generational divides, but much of the scholarship has questioned this framing, pointing out the more relevant class-based dynamics (Christophers, 2017; Howard et al., 2024). The relevance of tenure as a useful way of thinking about housing inequalities has also been challenged (Zhang, 2023). In Chapter 5, I engage with these critiques and outline the limits in particular of a simplistic generational perspective, while at the same time arguing that the historical dimension of housing and how it plays out over the *longue durée* mean that social structures associated with class, generation and tenure all interact in shaping the unequal outcomes experienced by tenants.

Chapter 6 delves into qualitative research to discuss tenants' experiences of home and insecurity. It emphasizes the subjective, experiential and relational dimensions of PRS housing, and the importance of grasping these when it comes to 'making home' in someone else's asset (Rosenthal & Vilchis, 2024). Questions like security, privacy and control, all of which are central to the meaning of home, play out within the PRS in ways that are directly related to the economic form of PRS housing. The chapter argues that the experiences of tenants within the home should be just as much a concern for political economy as the experiences of employees in the workplace. Similarly, the everyday social relationships between landlords and tenants are just as pertinent as those between employers and employees. These experiences and social relations, moreover, are in turn related to the frustrations of tenants, and hence the conflicts and antagonisms that arise in the PRS, and which can, in some contexts, become politicized.

In Chapter 7 this theme of politicization is taken up and explored, both by examining some historical examples of tenant activism, such as early twentieth-century rent strikes, but also through some case studies of contemporary tenant movements. A new generation of tenant organizations, especially tenant unions, has sprung up over the last decade in response to the resurgence of the PRS and the issues of affordability, insecurity and poor-quality housing, which are familiar across many jurisdictions. These new organizations aim to politicize the inequality between landlords and tenants, to create a collective tenant identity and to empower tenants through collective organization and action. In this regard, they seek to use political action to overcome the power asymmetry characteristic of PRS housing and the landlord–tenant relationship. This chapter thus introduces the question of tenants' agency, pointing out that tenants, and indeed residents in general, are not merely passengers within the housing system, but instead have the capacity to undertake collective action to transform the housing system itself. Indeed, this sort of collective action played an important role in the development of twentieth-century housing systems.

Taken together, this five-point framework, encompassing micro- and macro-political economy, inequality, subjectivity and agency, offers a comprehensive and novel perspective for the critical analysis of the PRS and its relationship to inequality, politics and change. There are, however, many things that this book does not attempt to do. It is not a housing policy book and there are many important areas of PRS policy that are not dealt with in detail, for example rent subsidies. More generally, the book does not try to cover all of the issues of relevance to the PRS, or even to the political economy of the PRS, as such a task would go beyond the scope of any one book. Instead, the book is organized around the task of developing an approach or framework for the political economy of the PRS, and each of the topics addressed in their respective chapters is selected because they allow the exploration of one or more dimensions of that approach. All of these issues are themselves important aspects of the contemporary PRS, from insecurity to financialization, and I hope that each chapter will contribute to the debate on each of these aspects. But the ultimate aim of the book is that, taken together, they can form the basis of a framework for analysing, explaining and critiquing what happens in the PRS.

It is also important to be clear that the focus of the book is the so-called 'advanced economies' of the Global North, especially Western Europe and the English-speaking world. Of course, the majority of private renters globally live outside these geographies, and therefore the book does not claim to offer a global perspective. The approach developed consequently reflects the focus of the book on advanced economies and the geographic limits of my own expertise. Moreover, while I try to avoid an Anglo-centric perspective as much as possible, I have not been entirely successful. The only international languages I speak are English and Spanish and, therefore, my knowledge of the PRS internationally is limited to the literature available in these languages. For many topics, this has not proved a hindrance, as we have ample scholarship in English covering many jurisdictions on issues such as financialization and affordability. There is, however, a notable dearth of literature in English (or Spanish for that matter) on the subjective experience of home in the PRS and the landlord–tenant relationship beyond the English-speaking world. This topic, in contrast, has been covered quite extensively in the English-speaking countries. My understanding of 'home', and related issues like insecurity, is thus limited to a considerable degree to countries which have lightly regulated rental sectors (i.e. weak security of tenure) and which are 'homeownership societies', where rental is to one degree or another stigmatized or considered marginal within the housing system. This has important implications for the perspective developed in the book, as it emphasizes the power imbalance between landlord and tenant and the implications this has in terms of tenants' experience of home and the politics of the PRS.

The approach I develop, as the reader may already have noticed, is also not primarily focused on what has traditionally been a main area of concern for political economy scholars – the role of housing in the wider dynamics of capital accumulation. I do not investigate, nor do I make any claims about, the structural role that housing plays in overall economic growth. This is not because this issue is unimportant. Rather, it is because my interest is much more in tenants and their experiences. In this sense I am interested in political economy to the extent that it helps us understand housing, not the other way around. My focus is on understanding what is happening in housing systems, why the PRS is growing, the unequal outcomes that arise from this, the experience of home in the PRS, and the forms of collective and political action in which tenants engage. This aspect of the book is related to what we might call the book's normative orientation. On the one hand, like any work of critical social science it aims to make visible and legible forms of inequality and power that operate in taken-for-granted aspects of the social world. Moreover, the approach developed here is informed by a Buddhist ethics that emphasizes the alleviation of unnecessary suffering; in this case the completely unnecessary suffering of private tenants, who are all too often prevented, by both market and policy forces, from having a home where they feel safe and secure. It is also underpinned by a commitment to equality (Baker et al., 2016) and the task of developing housing systems that mitigate, rather than accentuate, socio-economic inequality.

2

Conceptualizing Home Rent
The Micro-Political Economy of the Private Rental Sector

Any attempt to analyse the political economy of the PRS must begin with a critique of rent, as rent is the central economic phenomenon at the heart of this form of housing. Far from a simple monthly payment, rent expresses a complex economic and social relationship. Despite its importance, there has never been a sustained and comprehensive critical account of what I call 'home rent', rent paid in the context of residential housing, and such an account is therefore a glaring absence in our understanding of the PRS, the experiences of tenants and the politics of landlord–tenant relations.

There are a number of reasons why a critical account of rent is a vital starting point for our analysis. Some of these can be elucidated with reference to the limitations of the prevailing ways of understanding the nature of the relationship between landlord and tenant. In political and public debate there are two principal ways this relationship is conceptualized. The first is *policy-centric* and frames the relationship between landlord and tenant in *contractual* terms. Both actors are understood as parties to a contract which sets out their rights and obligations, such as tenancy and rent terms, the tenant's responsibility vis-à-vis the upkeep of the dwelling, the conditions under which a landlord may enter the dwelling, etc. These rights and obligations, whether set out in legislation or in lease agreements, or both, are seen as determining the nature of private rental housing (Byrne & McArdle, 2020). The second prevailing understanding of the landlord–tenant relation is a *market-centric* one, associated with an economistic understanding of rental housing. From this perspective, renting is understood as a voluntary market transaction between rational, self-interested individual actors. The market context within which they engage in this transaction is simply the sum of other similarly motivated individuals engaging in similar transactions, and the balance of supply and demand that arises from this sum of interactions.

These approaches normalize and naturalize the payment of rent and its associated socio-economic relationships. Both approaches cause the power dynamic between landlord and tenant to disappear from view. Moreover, they ignore the fact that the existence of landlords and tenants in the first place requires an unequal distribution of the ownership of residential property. In so doing, they depoliticize rental housing. In order to remedy this limitation, we require a conceptual approach, which captures how the payment of rent arises and is sustained; in other words, the social, political and economic institutions and relationships that underpin the payment of rent, and the landlord–tenant relationship more generally. While common sense, and both the policy-centric and market-centric approaches, suggest that rental housing is natural, normal and inevitable, a systematic and, most importantly, *critical* account must begin by showing that this is not the case, by emphasizing how private rental housing emerges from and is reproduced by certain economic and social relationships and institutions (Rosenthal & Vilchis, 2024). A political economy perspective has particular potential here because, as I hope to demonstrate throughout this chapter, it highlights the key issue of property relationships and power relationships, both of which, as we will see, are crucial to understanding the PRS.

This chapter attempts to home in on the core economic and social relationship that constitutes private rental housing, framing that relationship as one of property and power. In doing so, the chapter focuses on what we might call the 'micro-political economy of the PRS', i.e. on the immediate social relationship between landlord and tenant. This is a crucial facet, as argued, without which a critical account of the PRS is not possible, and therefore a vital starting point. But there are of course also wider social, political and economic structures that shape the PRS, such as the interaction of financial and housing markets, or the legal and policy structures of the housing system in a given country. These aspects, which we might call the 'macro-political economy of the PRS', are addressed in Chapter 3.

The critical political economy of 'home rent' developed in this chapter draws on a variety of disciplines and approaches, in particular classical and feminist political economy, urban sociology, critical geography and housing studies. Indeed, it is perhaps because drawing together these disparate disciplines is necessary for the development of a critical account of rent that such an account has remained absent for so long. The chapter begins with a discussion of the political economy of land rent, explaining why the concept of rent is useful in teasing out the relationship between the economic form of rent and the forms of antagonism, and hence politics, with which it is associated. It then moves on to a discussion of 'place' and 'monopoly rent', primarily via the work of David Harvey (2018) and Logan and Molotch (2007), building on the conceptualization of land rent but adding to it by emphasizing the socially constructed nature of urban places. The next section introduces home as a specific form of place to understand the value of 'home' from the perspective of the resident, and from here introduce the concept of 'home rent'. The remainder of the chapter clarifies what is at stake with this concept and the contribution it makes to the political economy of the PRS.

Rent and the politics of land

The analysis of rent[1] as a specific economic form goes back to the very beginnings of political economy (Aalbers & Ward, 2016). The early political economists (such as Marx and Ricardo) theorized land rent in the context of predominantly agricultural economies. The most lasting contribution of this early scholarship has been an attention to the unique features of land, i.e. the components which distinguish it from other commodities (Harvey, 2018). First of all, and as argued by Polanyi (2001), land is not a true commodity as it is not produced for consumption. At its root, land is merely a parcel of the surface of the earth (Aalbers & Ward, 2016). Moreover,

---

1 For the remainder of this chapter, I use the term 'rent' to refer to the general economic form of rent, understood as income arising from the monopoly ownership of a scarce resource which others wish to access (Christophers, 2023). I use the term 'land rent' to refer to this phenomenon in the specific case of land, 'monopoly rent' to refer to it in the case of urban places, and 'home rent' to refer to it in the context of private rental housing.

and relatedly, land is scarce in the sense that the supply of land is more or less fixed, and more land cannot be produced (with some exceptions such as reclaiming land). Scarcity is also a result of the fact that land is immobile; an increase of the supply of land in one place cannot meet an increase in demand in another (Ryan-Collins et al., 2017). There are two consequences of this. First, that land markets are inherently local. Second, local land markets are relatively insulated from competition, when compared with other commodities. As Logan and Molotch (2007: 32) put it, '[u]nlike widgets or Ford Pintos, more of the same product [land] cannot be added as market demand grows'. Moreover, each local context and each specific parcel of land tends to have relatively unique characteristics which cannot be reproduced (Logan & Molotch, 2007; Smith, 1979). In this respect, land is very much a non-homogeneous commodity, and different parcels of land are typically not substitutable for one another.

As a consequence of the above features, land markets are highly local and subject to inelasticity of supply (Le Grand & Robinson, 2016). This insulates land from market competition, as it is normally understood. It is for this reason that landowners were understood, by the early political economists, as *monopolists* who enjoy exclusive ownership of a unique, non-reproducible commodity or asset which is relatively free from market competition (George, 1879). As Ryan-Collins et al. (2017: 38) note, landowners are in a unique position economically because 'they possess a good that is not subject to the normal laws of market competition' and as such 'benefit from additional unearned income … '.

The term 'unearned income' draws our attention to another key feature of land highlighted by early political economists. Unlike most commodities, there is no productive process as such associated with land and no form of direct investment is necessary in order for land to have some value (Andreucci et al., 2017). Of course, investment may occur which enhances the value of land, for example in terms of agriculture by improving its fertility (Harvey, 2018). But land can have value without any investment of this sort. In this regard, the key dynamics of capitalist accumulation, i.e. the generation of profit through the production of commodities, does not apply in the case of land. Income derived in the form of land rent has thus been interpreted as 'unearned' in that it arises not from any investment or labour, but solely from ownership of a scarce asset (Andreucci et al., 2017). Or as Henry George (1879: 208) put it, in a somewhat more polemical tone, as a landowner '[y]ou may sit down and smoke your pipe...and without doing a stroke of work, without adding one iota to the wealth of the community, in ten years you will be rich'. It is from this aspect of landownership that the term *rentier* gets its notoriety, the idea being that both producers and consumers are forced to pay rent (or to purchase) land from landowners for the simple reason that all human activity must take place *somewhere* (Harvey, 2018).

If the value of land is not generated though the process of active investment and production, what is it generated by? And what accounts for the fact that the value of land varies? In the case of agricultural land, issues like fertility and the types of crops that can be grown will be paramount. But for non-agricultural land, value will be determined by the resources and amenities to which a given parcel of land gives access. These *locational* features may include natural amenities, such as climate, views, access to the sea, or they may include socially produced features such as transport networks (to which I return below). Logan and Molotch (2007: 23) point out that '[t]he owner of a particular parcel controls all access to it and its given set of spatial relations'. For these features to have a value, of course, there must be demand for them, and therefore the level of demand also plays an important role (Ball, 2024). The key point, however, is the value of land, and hence the ability to obtain rent, is locationally determined.

This brief summary of land rent helps us to identify two of the key features of rent as an economic concept, which will be central to our analysis of the PRS below. First, the ability to obtain rent is derived from ownership of a resource or asset and the ability to charge for access to that asset. It thus arises purely, and in a rather brutal form one might say, from *property relations* (George, 1879). Second, the value of land is largely determined *locationally*, i.e. land reflects the value of the place-based features to which it gives access.

We will return to some of these issues presently, but first it is important to highlight some of the political implications of this way of thinking about rent, and the forms of political antagonism and conflict often associated with land rent. This was articulated most clearly and most forcefully by Henry George. Most classical economists, from Ricardo to Mill, bemoaned the ability of landowners to extract an 'unearned increment'. But it was Henry George who raised this idea into a revolutionary cause, and spent much of his life as an activist, journalist and intellectual fighting for it. George was enormously well known in the late nineteenth century. It is often said that his magnum opus, *Progress and Poverty*, was the most widely read book of his time, other than the Holy Bible. An estimated 200,000 people attended his funeral in New York in 1897; to this day the second largest funeral in the history of the US.

What is most interesting about George is his focus on the unjust nature of landownership. He pointed out that private ownership of land arose for the most part via dispossession, as in the famous example of the enclosure of common land in early modern Britain. In addition to this 'original sin', because it is unproductive, land ownership was essentially a form of predation which had no grounding in social justice, and therefore no ethical underpinning. Non-landowners, therefore, were not only right to resent landlords, but should indeed agitate against them. For George and his followers, wealth accumulation through land ownership is a form of appropriation, barely distinguishable from theft, and the relationship between landowners and tenants is one of inequality, power and exploitation (George, 1879; Obeng-Odoom, 2022).

These ideas were highly influential in the late nineteenth and early twentieth centuries, not least in my own home of Ireland. The years following the publication of George's *Progress and Poverty* happened to coincide with the Irish 'Land War' between 1879 and 1882, in which George himself was no mere bystander. His ideas were readily taken up by some of those involved in the Land War, most importantly the revolutionary leader Michael Davitt. Upon his release from prison Davitt travelled to New York in 1880. There, he met George, read *Progress and Poverty* and became an enthusiastic Georgist (McBride, 2006). George also wrote a pamphlet on the Irish Land War in 1881, *The Irish Land Question: What it Involves and How Alone it Can Be Settled*. George (1881: 10) argued that at stake in the 'Irish land question' was a universal issue of political economy:

> What is involved in this Irish Land Question is not a mere local matter between Irish landlords and Irish tenants, but the great social problem of modern civilization. What is arraigned in the arraignment of the claims of Irish landlords is nothing less than the wide-spread institution of private property in land.

George also travelled to Ireland in 1881. He gave talks as well as writing about the Land War for the US newspaper *The Irish World*. During this tour he was arrested while travelling from Dublin to the West of Ireland. In a letter, he noted that: 'The charge against me was being a stranger and a dangerous character who had conspired with certain other persons to prevent the payment of rent' (George quoted in De Mille, 1944: 270).

The point of this brief detour is to highlight the fact that the early political economy critique of land rent was not just analytical, but also political. The critique of land rent not only describes its economic form, but also draws out the material antagonism that arises from this economic form because it is founded on appropriation and unequal property relations. It thus *denaturalizes* land rent by demonstrating that far from an inevitable or natural feature of society, it is a function of a specific constellation of property relations. It also *politicizes* land rent by demonstrating that these property relations have no basis in social justice and, most importantly from the point of view of this chapter, that it is in the material interest of tenants to contest them. George's critique of land rent is also political in the sense that his ideas could be put to work in the political conflicts around landownership and agrarian rent of that time.

From land rent to monopoly rent

The discussion of land helps to clarify why 'rent' is a useful economic concept both analytically and politically. But a discussion of land can only take us so far, as housing and home are different in significant respects from land. While land is simply a natural feature of the earth, housing and home must be constructed and they are typically part of *places* which are also constructed, both physically and socially. The concept of rent, as we shall see, is still of great value here, but it needs to be supplemented with a theorization of place. For this, I turn to David Harvey's (2018) concept of 'monopoly rent' and John Logan and Harvey Molotch's (2008) theorization of 'place as a commodity'. Logan and Molotch's key contribution is to emphasize that places (their focus is on urban places) are more than pieces of land. Most importantly, the *value* of place is not derived solely from the land upon which it sits. To understand the value of place, they argue, we must take a more sociological perspective that illuminates how place is socially constructed through social institutions, relationships, practices and meaning-making. '[T]the attributes of place', they argue, 'are achieved through social action, rather than through qualities inherent in a piece of land, and places are defined through social relations … ' (Logan & Molotch, 2007). This social construction of place, as we will see below, is central to how we should understand rent, the politics of cities and rental housing itself.

In delving into this theoretical insight, it is important to emphasize that in contrast to common sense assumptions, it is not simply that social action occurs *in* places, but that social action constitutes place as such. This takes place via countless forms of social action, but Logan and Molotch identify some of the most decisive. First, public investment in the form of transport networks, infrastructure (e.g. energy, telecommunications and water), and public services such as schools and hospitals. These forms of investment enhance the value of a place by enhancing the quality of life accessible in that place (through schools, parks or healthcare, for example) and by connecting a place (via transport) to other places which contain valuable resources (e.g. employment). Public investment is typically paramount in making places liveable, but we should not neglect the ways in which private investment adds to the value of place. Providing services and making goods available (i.e. the retail aspect of a place) adds enormous value from the point of view of residents, as does the creation of employment. Moreover, some of the services and infrastructure discussed under public investment above can also be provided privately, such as private transport services, private hospitals, private schools and private provision of water.

A less obvious form of 'investment' takes the form of everyday, largely informal social practices and relationships that shape any given place (Madden, 2024). This may include: making a place interesting through cultural practices or the organization of social activities; making a place safe (through a culture of neighbours looking out for each other, for example); making a place friendly (by being good neighbours); and so on. This may also include more economically tangible social capital, in the form of relationships with neighbours that may generate employment opportunities. This aspect of the social construction of place is of course extremely complex and the list of forms of social action that play a role here is potentially endless. There are, however, a number of points here worth emphasizing. First, these actions often occur informally and in a bottom-up fashion; the value they generate emerges from everyday practices. If actions are performed regularly over time, they become embedded in and characteristic of place, such that the value thus generated can be recognized by residents and potential residents. Second, many of these practices involve 'meaning making', i.e. they operate through the investment of places with symbolic meaning. The value of a given place, then, is characterized by a significant *subjective dimension*, something we will return to below in our discussion of home. Finally, it is important to note that social construction of place can occur, and typically does occur, over long time horizons, for example in the case of historical city centre areas or cultural districts. In other words, the subjective experience of a place, the forms of investment that make it valuable, the cultural and social practices embedded within it, etc. can all develop over decades or even centuries (Harvey, 2012). Place can thus crystallize a complex array of 'use values' which residents derive from it, and it is this which, according to Logan and Molotch (2007), is at the heart of the complex emotional, psychological and affective attachments that people often have to place. These 'place attachments' (Blunt & Dowling, 2006) are typically very different from the casual and instrumental relationship consumers typically have with standard commodities.

From our point of view, the most important point is that part of the value that is created via the forms of social action and public/private investment discussed already is absorbed, so to speak, by the place in which it is located. In this sense, and this is a theme we will return to below, places have the ability to *capture value* associated with forms of investment or activity which occur in them, near them or in places that are easily accessible from them (Madden, 2024).

Although thus far I have emphasized that place should be conceptualized as distinct from land in that the value of place is largely socially produced, there are important similarities in the nature of place and land as idiosyncratic commodities. Most importantly, each place is *unique*. The uniqueness of place arises from its socially constructed nature. The complex, historical and bottom-up nature of the social actions that constitute a place and its value cannot be reproduced (Harvey, 2012). Of course, some places are more unique than others. But at a conceptual level no two places can be said to be identical. This is, of course, very different from the vast majority of commodities in a capitalist society, which are characterized by a high degree of homogeneity and interchangeability. This uniqueness means that places can be seen as even more 'supply constrained' than land. For example, if we take a unique city-centre neighbourhood with strong cultural associations, vibrant social relationships and a legacy of decades or even centuries of investment in infrastructure (sewers, etc.), it is not possible for a competitor to produce a rival place with the same characteristics. Thus, while land is merely locationally unique, places are unique locationally *and* in a much wider sense.

Returning to the concept of rent, the uniqueness of place grants it its own monopolistic quality, because real-estate owners can capture the value generated by place by charging rent for access to that place, or indeed via selling properties that allow access to that place (Haila, 1990, 2015). This form of rent, which Harvey (2018) calls 'monopoly rent', is a specific form of rent in which monopolistic real-estate ownership makes possible *the appropriation of the socially produced value of place*. The crucial insight here relates to this capacity of 'place', and ownership of place in particular, to make possible the appropriation of value, which is produced collectively. Artist-led gentrification is a classic example. In such instances, artists move into low-rent neighbourhoods making them more attractive and lively (at least to some), generating a form of place-based value. This, in turn, manifests in higher rents and house prices, which wind up excluding the very people who generated the value in the first place. In cases such as this, rents and property prices 'capture' the socially produced value which triggered the process of gentrification. We see here once more how, just as in the case of land, rent arises from the unique and monopolistic nature of place mediated by property relationships, specifically real-estate ownership.

These insights into the nature of monopoly rent will prove central to how we understand rent in the context of the PRS. Before turning to that, however, it is worth addressing how the nature of monopoly rent is associated with particular forms of urban conflict and politics. Just as George allows us to see how the nature of land rent is related to the politics of land and the conflict between landowners and tenants, so too Logan and Moloch shed light on the politics of place. They argue that the politics of place emerges from two features of the latter's distinctive nature. First, and most obviously, residents in a particular place have access to and consume the same, or similar services, such as local transport. This gives them a certain shared interest, in line with Castells' (1983) politics of collective consumption. For our purposes, however, the second feature is much more significant. Both real-estate owners and residents are in a position to impact the value of place through collective action. The 'value' at stake here can take two different forms. For the real-estate owner, value can be enhanced via increases in land and property prices (or rents). For the resident, value is more typically enhanced by increasing the 'use values' they derive from the place they live. In relation to owners, Logan and Molotch (2007:32) talk about 'place entrepreneurs' as a 'dynamic social force':

> [P]lace entrepreneurs attempt, through collective action and often in alliance with other business people, to create conditions that will intensify future land use in an area. There is an unrelenting search, even in already successful places, for more and more.

Place entrepreneurs form 'growth coalitions' that seek to enhance real-estate values by shaping urban development, such as infrastructure and public investment, a phenomenon captured very effectively in the Italian neo-realist classic *Hands over the City*. The fact that the value of place is derived from social and public investment gives owners both opportunity and motive to shape local politics.

From the perspective of residents, the use value they derive from place (including access to housing, access to other services like health and education, cultural and recreational amenities, retail services, etc.), which is also shaped by various forms of public and social investment, can also be protected and enhanced through local politics. This leads to various local organizations, such as resident associations, which seek to shape urban development. Of course, residents' interest, if they are owner occupiers, can also coincide with those of 'place entrepreneurs', as they may seek to protect house prices through, for example, lobbying against social housing.

There are three conceptual insights that flow from this. First, the fact that the value of place is determined through a wide variety of social and political investment and activity means that the politics of place takes the form of competing attempts to shape place and the value, both of the exchange and use variety, that can be derived from it. Second, 'consumers of place', such as residents, are not passive. They have *agency* which allows them to act to shape the place they live in. Third, the nature of any given place is in part determined by these political actions and conflicts. The politics of place is, thus, part of its constitution:

> The reality of place is constructed through political action, with the term *political* encompassing both individual and collective efforts, through both informal associations and institutions of government and the economy. (Logan & Molotch, 2007: 48)

Home as place: Ontological security and social reproduction

We can now move on to consider what kind of place 'home' is. This question, I will argue, is central to understanding the form of rent associated with residential housing. However, and remarkably, the concept of 'home' has been largely divorced from the political economy of housing (Saunders & Williams, 1988). This is due to a focus within economics and political economy on housing in its narrowest sense, i.e. on access to dwellings, and what people have to pay for it, be it for sale or for rent. Moreover, housing is often conceptualized in a manner which emphasizes its land-based or land-related features, such as supply inelasticity, monopoly, immobility, etc. These features are important in theorizing home and rent, but leave much out. The peculiar place we call home requires a clearer conceptualization which captures its complexity and the unique form of value it has for the resident, a value which, as argued below, is principally found in the subjective experience of home and practices of social reproduction. This, in turn, can serve as a foundation for a comprehensive theory of home rent. Given its neglect within the political economy of housing, to approach this task, we must turn to literatures from housing studies and critical geography, as well as feminist political economy.

To conceptualize home, we must recognize that it is a centrepiece, nexus and anchor of a set of experiences and resources which are central to human needs and wellbeing (Saunders & Williams, 1988). It is a *place* (or 'locale', to use the terminology employed by Giddens, see Saunders & Williams, 1988) which is entirely unique in terms of the extent to which it is linked to human wellbeing, its centrality to the most intimate and important social relationships, and the intensity of attachment it is characterized by. It is no exaggeration to say that it is at the heart of selfhood. It is vital that we take seriously just how significant and distinctive a 'commodity' home is. Two concepts, ontological security and social reproduction, help to clarify the forms of value that are characteristic of home. To begin with the former, ontological security captures the centrality of home to wellbeing at its deepest level. Central to ontological security is a sense or subjective experience of the reliability of things and place over time (Byrne, 2020b; Easthope, 2004). As Walshaw (quoted in Easthope, 2004: 581) argues, ontological security refers to the ways in which '... home [is a] safe haven in which individuals can be themselves and from which they can derive an enhanced sense of emotional security'. Ontological security is thus derived from home when it endures and is stable and predictable over time (Hiscock et al., 2001).

There are a few further aspects of ontological security that are important to note as they are of direct relevance (as developed below) to the theorization of home rent. First, ontological security is linked to the '*interrelationships* between the physical dimensions of housing (such as safety and security) and the psycho-social dimensions of home such as privacy, emotional security and identity' (Hulse & Milligan, 2014: 638, my emphasis). In other words, it emerges from both everyday practices of meaning-making and the objective physical structure of the dwelling. As Saunders & Williams (1988: 83) argue, '[t]he home ... is not reducible either to the social unit of the household or to the physical unit of the house, for it is the active and reproduced fusion of the two'.

Second, and relatedly, ontological security does not emerge spontaneously from the mere fact of inhabiting a dwelling, but rather through practices of 'place making', or rather 'home making', through which we construct a sense of ownership, autonomy, stability, privacy and safety. This occurs through the changes we make to a dwelling in terms of design and furnishing, through the organizations of our belongings within the dwelling in a particular fashion and through the routines of work, recreation, socialization and rest we establish within a dwelling (Byrne & Sassi, 2023). The resident is thus active in an ongoing way in producing home. This means that 'home making' is a kind of labour which produces the 'homely qualities' that transform a house into a home (van Lanen, 2017), and therefore generate the value of home for the resident.

Moving on to social reproduction, the second key concept through which we can grasp the value of home, feminist political economists have long argued that the home is not just a place of consumption, or a private space divorced from economic activity, but is itself a workplace (Dyck et al., 2005; Power, 2019). A huge proportion of social reproduction or care work takes place within the home, and the home itself is typically designed, organized and structured around activities associated with social reproduction (Madden, 2025). Care work is typically embedded in the physical structure of the home and in the particular organization of belongings, for example the layout of the kitchen, how food and other necessities are stored, and routines around children's bedtime (Saunders & Williams, 1988). Care work is thus intertwined with a particular dwelling (Blunt & Dowling, 2006). Indeed, Power and Mee (2020: 484) have conceptualized home as an 'infrastructure of care' and as a 'hub of care practices and relations'.

It is worth noting that social reproduction and care work are typically highly routinized forms of social activity. Everyday forms of care work, such as cooking dinner or putting children to bed, take place through routines to make them more manageable. The home, when it is a place that is stable over time, provides a place in which these routines can be established and maintained. Social reproduction is also embedded in many of home's place-based features, i.e. it relies on localized networks of social relations, transport and services (Mee, 2009). Relationships with neighbours, access to certain shops, schools and healthcare facilities are all central to practices and routines associated with care. As Logan and Molotch (2007: 104) note, a 'place of residence is the potential source of an informal network of people who provide life sustaining products and services'. As in the case of ontological security, everyday activities of social reproduction produce value for the resident which is *invested in the home*, as the home becomes a place that sustains and supports care work, a process which arises directly from the activity, and labour, of the resident.

In emphasizing ontological security and social reproduction as important resources, which residents will typically strive to derive from their experience of home, there is a risk of falling into an overly romanticized perspective. Feminist scholars have for many decades demonstrated that the ideas of home as a 'safe haven' separated from the public sphere have a gendered and ideological dimension. Most notably, they make invisible forms of gendered and intra-household conflict, inequality and violence, which are all too prevalent (Blunt & Dowling, 2006; Ecker, 2016). Gender-based or domestic violence is the most obvious example but home has a wider 'dark side' (Gurney, 2023). Indeed, it is precisely because home offers privacy and security that it is so often the location where abuse is perpetrated. In terms of privacy, for example, the way the home allows us to escape constraining social forces like judgement and shame may enable practices that are not socially accepted or are illegal, including forms of violence and abuse. Thus, one person's privacy may be another's abuse. It is therefore important to recognize that while residents typically strive to 'make home' in order to meet their needs in terms of ontological security and social reproduction, they often do so in the context of forms of inequality and oppression, such as those relating to gender (Madigan & Munro, 1991; Watson, 2023).

Summing up our conceptualization of home, we can say that *the value of home consists in the fact that it is a particular kind of place, which acts as a resource, bundling together a physical dwelling with the intangible value residents create in their efforts to create ontological security and to take care of themselves and others, and it gives stability to this resource over time.*

It is worth emphasizing that home thus differs markedly from land, as well as from the urban places theorized by Logan and Molotch. Home has a specific value for residents that is not just locational, but subjective, experiential and embedded in the relationship between the individual or household and the dwelling (Saunders & Williams, 1988). This form of value, because it is subjective, experiential and relational, is specific to a given resident and their home. In other words, the value of home a resident creates for themselves, via the everyday practices and investment of meaning described above, cannot be transferred to another. Some improvements that a resident may make, via decoration for example, can remain after they have departed. But, typically, when an eviction occurs, the value of the home, the routines of care work and the sense of security and stability the resident has generated through their activities, or labour, are essentially destroyed. This is another sense in which social reproduction theory is relevant to understanding the value of home. The form of value invested in home is a 'use value', but one that is primarily useful, or valuable to, the resident. It is not 'alienable' and therefore does not take the form of exchange value in that it will not, generally, manifest in the price of a house or the rent another household would be willing to pay for access to the same dwelling. This is obviously because for another household the house would have none of the subjective sense of safety and stability, or the established practices of care work, or indeed the place-based social relationship, which have been created by the existing residents.

Home and rent

Clarifying the value of home allows us to conceptualize 'home rent', i.e. the specific economic form that rent takes in the context of residential property. As emphasized, the value of home for the resident is complex and multifaceted, incorporating both the unique locational values of a house (and the land it sits on), as well as the unique value associated with ontological security and social reproduction. The key point is that for an existing tenant, rent is the sum of money that must be paid for continued access to their home. From the tenant's perspective, the value of home is in large part created independently of any action of the landlord, in that it is derived from the qualities of a given place, which are essentially socially determined, and from the home, which is created and sustained by the tenant. Of course, the landlords do create a portion of the value through the services they provide, i.e. property management and maintenance, and through the initial investment, for example in furnishing a dwelling. However, the home has additional value for the tenant, which is essentially independent of any investment or action on the part of the landlord.

Madden and Marcuse (2016) capture this dimension with their concept of 'residential alienation', which they describe as the 'painful, at times traumatic, experience of a divergence between home and housing'. If home is the result of processes and practices of home making which are an 'extension and expression of our capacity to create' (Madden & Marcuse, 2016: 59), residential alienation arises when one's home is controlled by someone else, i.e. a landlord. In this context, the resident's 'housing is the instrument of someone else's profit, and this confirms their lack of social power' (Marcuse & Madden, 2016: 59). This dimension relates to a simple, but ultimately the most consequential, fact of PRS housing: *tenants must make their home in someone else's asset*. The tenant must pay rent each month for continued access to the home they have themselves created, a home which becomes invested with value and meaning, not simply as a result of the physical structure of the dwelling, but as a result of the sense of security and stability created *by the tenant*, the routines and practices of care work established *by the tenant*, and the place-based relationships with neighbours and service providers (from local doctors to schools) *created by the tenant*. The landlord acts, then, as a kind of gatekeeper to all of these resources and forms of value, embedded in home and place, obtaining rent via the ability to charge for access to them.

We can also consider home rent as monopolistic in nature. As housing sits on land, it shares many of the monopolistic features of land, namely that any given house is in a relatively unique, and fixed, location which cannot be reproduced. But homes are of course unique in another sense. As the cliché puts it, there's no place like home. Once a tenant has occupied a dwelling for a period of time, it becomes a unique place due the establishment of daily routines and its investment with meaning, an obvious example being the family memories created there. Thus, home is unique and it is in no way possible for the tenant to substitute it with an identical commodity. As anyone who has conducted qualitative research with tenants in the PRS can attest, for long-term tenants, being evicted can be an extremely emotional form of loss (Desmond, 2016; Fullilove, 2001; McArdle & Byrne, 2022). Of course, tenants may be more or less invested in a given home and therefore a specific home may have more or less value for them (obviously the value of any class of commodity or resource will vary). However, the conceptual point remains that homes, even more so than locations, are unique and not substitutable.

From this perspective home rent is at its core a social relationship. But this social relationship does not, of course, arise out of thin air. It is based, first and foremost, on a relationship of property and of power. The ability to capture rent is entirely predicated on the private ownership of residential property, and the ability, on the basis of that ownership, to control access to, and exercise power over, that property. The political economic foundation of home rent is property relations, i.e. it arises purely from the legal recognition of the landlord's right to charge for access to a dwelling and to terminate access to a dwelling, usually under certain conditions. I will return below to the issue of legislation and tenancy rights, but for now note that, while in different jurisdictions the landlord's right to raise rent or the circumstances under which the landlord can evict are curtailed to different extents, every PRS in the world involves the landlord having a right to obtain rent and the right to terminate a tenancy where rent is not paid, for this feature is inherent to the PRS and a pre-condition of its existence. In this sense, the property rights that underpin home rent also represent a power relation between landlord and tenant. This power relation, which again can be more or less asymmetrical, is ultimately the basis upon which rent can be appropriated by the landlord. Because of the ability to exercise power over a fundamental resource for the tenant, the landlord–tenant relation can be considered as having a dimension of coercion, or what Ahmari (2023) calls 'private tyranny', i.e. a form of coercion, which arises within market relationships.

The property and power relations associated with home rent are also established and sustained by a wider institutional context, consisting primarily in the domains of policy, markets and culture. Legislation establishes tenancy terms and the rights and obligations of landlords and tenants; for example, the nature of rent setting or the grounds for tenancy termination or eviction. It should be noted that this legislative context may constrain what landlords can do, for example via rent regulation but, just as importantly, if not more importantly, this legislation empowers landlords to set rent and to evict tenants due to the non-payment of rent. In other words, the power of landlords arises in part from housing policy. The market context, in particular the balance between supply and demand, also plays a decisive role in determining the degree of asymmetry in the relationship between landlord and tenant. A tight rental market reduces the alternative housing options available to a tenant, making them more dependent on their current landlord (Soaita, 2024).

Finally, the cultural context also plays an important role. In homeownership societies, where renting has traditionally been seen as a transient tenure, landlords may feel empowered by the lack of social recognition of the tenant's home. This cultural dimension is important, as in many contexts compliance with legislation in the PRS is poor and, therefore, the behaviour of landlords (and indeed tenants) is more constrained by cultural norms about what is and is not legitimate behaviour (Rex & Moore, 1967). We will return to some of these issues in later chapters. For now, I want simply to emphasize that the power relationship between landlord and tenant is neither natural, inevitable, nor spontaneous, but in fact is constructed, shaped and maintained by a series of social, legal, cultural and economic institutions.

Conclusion

Before clarifying the contribution of the approach developed throughout this chapter, it is important to add a number of caveats with regard to the conceptualization of home advanced here. First, the conceptualization of home advanced should not be taken to imply that housing is always stable, safe and predictable. For many, insecurity is a characteristic feature of housing, as I discuss in detail in Chapter 6. Moreover, forms of violence and abuse can and do occur in the home, often shaped by gender-based and intergenerational forms of inequality and power. My intention, rather, is to argue that the act of home making strives to create stability and predictability over time in order to generate the subjective state of ontological security, and the wellbeing derived from it, and to support and sustain social reproduction. But this cannot always be achieved, especially in the face of social, economic or political conditions, which undermine those 'homely qualities'. Even individuals and households who are deprived of a secure home will typically attempt to engage in home-making practices to the extent that they can (Lenhard et al., 2022; van

Lanen, 2022). Second, and as noted already in the previous chapter, some countries have strong tenant protections, which constrain landlords and empower the home-making practices of tenants. These include rent regulation and protections against eviction. Moreover, in some contexts the dynamics of supply and demand favour tenants, which can also mitigate the power asymmetry between landlord and tenant. Thus, the landlord–tenant relation is heavily mediated by political and policy factors, and this can have a decisive impact on the relationship between landlords and tenants and the latter's experience of home. Nevertheless, even in those countries that have the strongest tenancy protections, landlords can evict due to non-payment of rent, and thus, ultimately, home rent is linked to this property-based form of power. Moreover, these jurisdictions also typically allow tenancy terminations under certain conditions, such as sale of property, albeit subject to some constraints. They also often have black markets within the rental sector, for example via informal subletting. Finally, the countries with the strongest tenant protections, such as Sweden, Germany and the Netherlands, also have the largest tenant unions, indicating that even in these contexts tenants recognize the power imbalance between landlords and tenants.

In concluding this chapter, and having set out a critical account of home rent, it will be useful to clarify the utility of this approach, i.e. what it does and does not allow us to understand. The approach developed here is first and foremost an account of what home rent is; in other words, what kind of economic phenomenon is at stake here, what is the nature of home rent as a form of commodity. This account of home rent is focused on how the value of home is generated and what this means for understanding how it operates as an economic phenomenon. This approach is an important corrective to the more commonplace approaches, which assume that rent is simply a given feature of housing systems and need not be subject to any scrutiny. Relatedly, the theory of rent advanced here emphasizes the nature of the *economic relationship* at the heart of home rent, i.e. the relationship, between landlord and tenant. Thus, it emphasizes that home rent is a social relationship, and specifies what the nature of this relationship is and how the nature of this relationship relates to the type of value at stake.

The first contribution of the approach here is, therefore, to shed light on the nature of home rent as an economic phenomenon, form of value and social relationship. As indicated at the outset, however, the approach is also critical in orientation. It does not treat the payment of rent for access to residential property as normal or natural but, instead, uncovers the dimension of power that is central to the appropriation of home rent. It therefore centres the inherent power asymmetry that is at the heart of the landlord–tenant relationship. This power asymmetry is in turn founded on the unequal distribution of residential property between landlord and tenant but also within the wider system. Finally, I place 'home' at the centre of the analysis, rather than simply housing. In so doing, I capture the idiosyncratic nature of home as a commodity, both the unique nature of any given home, and the ways in which home arises out of the activities of the resident.

The theory of home rent adds to our ability to analyse and understand the PRS in two principal ways. First, by emphasizing that home rent is predicated on socially constructed value, my approach puts tenants' *experience* of home and home making in private rental housing at the heart of its analysis. This facet, discussed in further detail in Chapter 6, is important because, as amply demonstrated in the literature, much of the experience of private renting is characterized by the struggle to create a home within a house that is someone else's asset and property, and within an asymmetric social relationship (Madden & Marcuse, 2016). Second, and relatedly, the theory of home rent allows us to properly understand the forms of antagonism and conflict at stake in the PRS, because these arise from the experience of 'making home' within the unequal property and power relations of the PRS. Although the literature on the politics of housing is typically concerned with homeownership and how, for example, it affects electoral outcomes or policy preferences (Dancyger & Wiedeman, 2024), political conflict has, in fact, been central to the PRS in many jurisdictions for much of the last century. Indeed, some of the most radical developments in housing systems, including the imposition of rent controls and the expansion of social housing in many European countries in the early twentieth century, arose in direct response to tenant activism (Gray, 2018a). The issue of tenant politics is the subject of Chapter 7 but, for now, let us simply highlight that without understanding the ways in which the power relationship between landlord and tenant are inherent in the material nature of private rental housing as a form of commodity, the politics of rent strikes and tenant agitation cannot be fully understood.

It is also useful to clarify here what my theorization of home rent does not do, and indeed does not aim to do. First of all, it does not attempt to explain the level of rent prices or increases/decreases in average rents. Many of the older theories of land rent attempt to do precisely this. Other political economy approaches have argued that rents in the PRS do not correspond to the balance of supply and demand in the neoclassical sense and that market competition does not operate in any straightforward sense (Gilderbloom & Appelbaum, 1987). My emphasis on the monopolistic nature of home, housing, place and land supports this line of thinking. However, it is also the case, as Ball has argued, that in many senses supply and demand are the chief drivers of rents:

> The relative attractiveness of scarce land and willingness to pay for those attributes in the face of competition from many others after the same thing determines rents in aggregate and across space ... In modern urban terms, as cities grow through, say, population increases, rising living standards or greater business demands, rents in them grow too. Conversely, if those factors decline, rents go down as well. (Ball, 2023: 6)

I thus remain agnostic about the factors that shape prices, which appear to relate both to the unique features of land and housing but also to the dynamics of supply and demand.

Second of all, the approach outlined here does not attempt to determine the role of home rent within the overall economic system, or indeed its relationship to other forms of capitalist accumulation. I emphasize this because this has been a chief concern of much political economy writing on contemporary housing in general and the PRS in particular. Much of this literature is in one way or another based on David Harvey's (Harvey, 2018) emphasis on investment in the built environment, and associated rents (in the broad sense), and the role they play in capitalist development and, especially, averting crises of overaccumulation. Moreno Zacarés (2024), for example, in a recent intervention, argues that residential accumulation is a distinct logic from capitalist accumulation, with which it is, moreover, in tension. This trend is particularly evident in the literature on the financialization of the PRS in the 2010s and 2020s, which, it is argued, plays a role in the overall dynamics of financialized capitalism (discussed further in Chapter 4). While there is much merit in such approaches, and the role of housing investment in the overall dynamics of capitalist accumulation is an important issue for investigation, my theory does not aim to capture this relationship and remains essentially agnostic about what role the PRS plays, at any given time, in macro-economic dynamics. My focus is instead on the nature of the economic relationship at the heart of the PRS, the experiences of tenants, and the politics of the landlord–tenant relationship.

This, however, is not to say that more structural or macro-economic issues are not relevant to the political economy of the PRS. On the contrary, it is to this level of analysis we must look in order to explain and analyse why some people are tenants and some are landlords in the first place. Political and economic structures and processes interact to shape the housing system and, most importantly, the ownership of residential property, and therefore constitute the taken-for-granted categories of landlord and tenant. The next chapter addresses this 'macro-political economy of the PRS', as well as examining the recent resurgence of private rental housing across many countries of the Global North.

3

The Revival of the Private Rental Sector

Neoliberalism, Financialization 1.0 and the Concentration of Property Ownership

How can we explain the revival of private renting in the twentieth century, and how should we interpret its significance for the politics and economics of housing? This chapter addresses these two questions. The resurgence of renting must be understood as a consequence of two major processes of political economy: neoliberalization and financialization. The impact of these two processes is characterized by a remarkable contradiction. Financialization, at least in the sense of expansion of mortgage credit, and neoliberalism are ostensibly both forces that should promote the extension of homeownership. From Thatcher's 'property-owning democracy' to Bush's 'ownership society', neoliberals have always presented marketization and the promotion of homeownership as two sides of the same coin (Béland, 2007). Margaret Thatcher's 'right to buy' programme, the poster child for neoliberal housing policy, was very much in this vein. The expansion of mortgage markets from the 1970s and 1980s, similarly, promised to remedy the scarcity of credit which had for many decades made homeownership inaccessible for much of the population. These two processes, it should be noted, are also intimately related: the marketization of housing is a fundamental prerequisite for the expansion of mortgages. The contradiction resides in the fact that the long-term consequence of the neoliberalization and financialization of housing regimes has been to systematically undermine homeownership and drive, instead, a renaissance of the PRS. Instead of a secure home and a 'stake in society' for the masses, they have delivered affordability crises, volatility in housing supply and prices, and the return of many housing issues once thought to have been relegated to the early twentieth century: evictions, insecurity, rack renting and abusive landlord practices (Gray, 2018a). This change has created a sense of generational unfairness (Ansell, 2023) and put housing back at the centre of politics in many countries (Fuller, 2019).

This chapter will attempt to show that a political economy approach has two valuable contributions in terms of how we make sense of these developments. First, I argue that a political economy approach allows us to *explain* the revival of the PRS by examining the intersection of political and economic forces, specifically the intersection of housing policy, financial and monetary policy, housing markets and financial markets. In so doing, I highlight the *explanatory potential* of a political economy approach as a framework for analysing how, and why, housing system change occurs in general, and specifically with regard to the PRS. I develop this argument via an exploration of three case studies: Ireland, the UK and Spain. Although I make reference to other examples, examining the decline of homeownership in these three classic 'homeowner societies' sheds light on many of the fundamental processes that are more widely evident in advanced economies.

Second, I argue that a political economy perspective can allow us to *analyse* the significance of the revival of private renting, and the related decline in homeownership, because it highlights how housing is embedded in wider political economy structures and processes. Here, I delve into the literature on the 'homeownership society' to examine the ways in which mass homeownership was much more than a housing phenomenon; it also underpinned many important political and economic phenomena in post-war advanced economies. Against this background, I discuss the ways in which the revival of renting suggests new ways in which housing impacts politics and economics.

Both of these contributions, however, draw our attention to a third important insight, and one which goes to the heart of the political economy of the PRS. In both explaining the growth of the PRS and in interpreting its significance, a central question is the distribution of the housing stock, i.e. the extent of concentration of ownership of residential property. This 'distributional question', although seldom acknowledged, is core to the political economy of the PRS, as the size and nature of the sector is obviously intertwined with the extent to which the ownership of housing is concentrated or, in contrast, is more equally distributed. This in turn highlights the fact that for the PRS to exist in the first part the taken-for-granted categories of 'landlord' and 'tenant' need to be produced, i.e. political and economic structures must interact to generate a housing system in which some individuals, households and firms own multiple properties, while others own none. We are reminded here of Marx's famous quote from the Grundrisse:

> In the shallowest conception, distribution appears as the distribution of products, and hence as further removed from and quasi-independent of production. But before distribution can be the distribution of products, it is: (1) the distribution of the instruments of production, and (2) ... the distribution of the members of the society among the different kinds of production. [...] To examine production while disregarding this internal distribution within it is obviously an empty abstraction ...

In a similar fashion, a 'shallow' (but all too prevalent) account of the PRS tends to assume the existence of landlords and tenants, as if this was a phenomenon requiring no scrutiny, and focuses solely on the conditions according to which the latter are able to access the housing owned by the former. But, of course, before a tenant can rent a home from a landlord, there must be a set of social structures that distribute residential property between households. Understanding these social structures, and the extent to which they lead to a concentration of ownership of housing is thus a crucial task for the political economy of the PRS, if it is to provide a perspective that is both comprehensive and critical.

In sum, while Chapter 2 theorized what I call the 'micro-political economy of the PRS', via a focus on the direct political and economic relationship between landlord and tenant, this chapter 'zooms out' to provide a 'macro-political economy of the PRS' via a focus on the political economic structures that shape the distribution of residential property, paying special attention to the interaction of housing policy and housing and financial markets.

Before delving into this discussion, a brief note on concepts. Financialization has been defined as 'the increasing dominance of financial actors, markets, practices, measurements and narratives, at various scales, resulting in a structural transformation of economies, firms (including financial institutions), states and households' (Aalbers, quoted in Wijburg & Aalbers, 2017:2). While I return to this in Chapter 4, for the present chapter, the focus will be solely on the importance of mortgage credit, which, as we shall see below, played a decisive role in the 1990s and 2000s. Neoliberalism is a much-contested concept, but at its core is a focus on the role of the state in expanding marketization and commodification (Madden & Marcuse, 2016).

The revival of the private rental sector in the advanced economies

Although the resurgence of the PRS and concerns around declining access to homeownership and social housing are widespread, there are significant differences between different jurisdictions. In Britain, Denmark, Ireland, Spain, the US, Australia and New Zealand the growth in the share of private renting households has been substantial (Kemp, 2023). But, beyond this, many countries have experienced a modest, but important, increase in private renting. Martin et al. (2018) found that seven of the ten advanced economies they looked at had experienced a revival of the PRS. Another recent study found that 23 of the 28 European countries examined had witnessed a decline in homeownership and an increase in private renting between 2005/7 and 2016/18 (Hick et al., 2022, see Figure 2 below). The authors, focusing on the under-sixties, find that virtually every country has experienced a decline in homeownership over the period. France, Malta, Czechia, Slovakia and Poland are the only countries to experience an increase in homeownership, and in most cases this is modest. What is also notable is that there does not appear to be any discernible pattern in terms of the level of decline experienced by different types of housing regime.

As Hick et al. (2022) show, Anglophone countries show a consistent pattern, with both the UK and Ireland seeing a decline in the proportion of homeowning households of more than ten per cent (the sharpest declines except for Slovenia). Southern European countries show a clear tendency of decline, but only Spain has witnessed a very significant fall (almost 10%), with Italy and Portugal falling by around five per cent. The Social Democratic countries exhibit a similar tendency, with one country, Denmark,

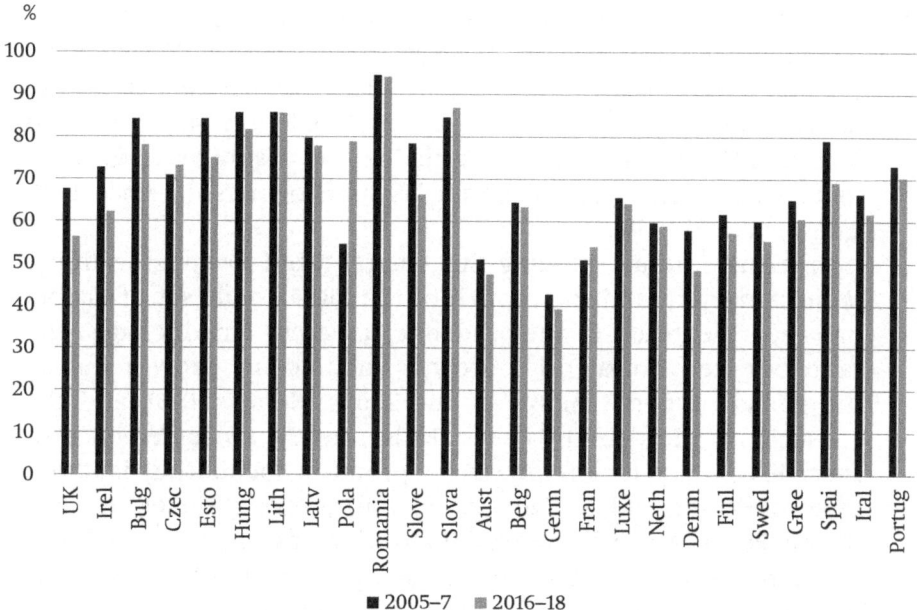

Figure 2 Change in homeownership rates, under-60s
Source: Based on Figure 6 in Hick et al., 2022: 16. I am grateful to the authors for providing this data.

showing a large decline of almost ten per cent, while in both Sweden and Finland homeownership fell by less than five per cent. The continental countries were most stable, with Germany and Austria showing declines of a little less than five per cent, Belgium very stable, and France showing a modest increase. In Central and Eastern Europe, Slovenia witnessed a significant decline of homeownership (more than 10%), as did Estonia (close to 10%), but Poland saw a dramatic increase, and the other countries remained relatively stable, although almost all countries experienced a decline, and for Hungary and Bulgaria this was close to five per cent. Stepping back, the most notable pattern, from both the European data and more broadly, appears to be that countries such as Denmark, Ireland, UK, Spain and the US all experienced significant declines in homeownership, and these are all countries that experienced financialized housing market bubbles in the early and mid 2000s.

Beyond Europe, Australia, like most Anglophone countries, has experienced a significant increase in private renting, with around a quarter of households renting privately today, an increase from approximately one fifth at the start of this century (Hulse, 2023: 19). Similarly in the US, in the year 2000, 67.5 per cent of households were homeowners. This had declined to 63.4 per cent of households by 2016, and has remained at a similar level since then. This is the lowest level for US homeownership since 1965 (Schwartz, 2023: 54).

In terms of understanding the general trend towards declining homeownership and the growth of the PRS, a wide set of factors has been identified in the academic research. This includes: the ratio between house price growth and income growth during the global property bubble of the 2000s (Kemp, 2015), and the consequent affordability issues (Hulse, 2023); the emergence of specialized 'buy to let' mortgage products facilitating landlord investment (Ronald & Kadi, 2017); the deregulation of the PRS (Kemp, 2015); the decline or residualization of social housing (Bone, 2014); restricted access to mortgage credit since 2008 (Forrest & Hirayama, 2015); declining employment, income and job security (Arundel & Doling, 2017); and changes in processes of urbanization, such as migration (Dewilde, 2017), the locational and job density patterns associated with a service sector economy and the preferences of contemporary urban dwellers (Hulse & Yates, 2017). Other demographic factors should also be noted, including the tendency for people to have children later, spend more time in education, and live in one-person households. Housing systems are, as can be seen, thus shaped by an incredibly wide array of phenomena, and it is very hard to develop a framework that would capture their full complexity. Nevertheless, in the next section, I focus on the examples of the UK, Ireland and Spain to give a sense of how these various processes play out and interact to undermine homeownership and grow the PRS. In so doing, I pay particular attention to how policy and market forces interact to shape housing system change via the concepts of neoliberalization and financialization.

Neoliberalism, financialization and the revival of the PRS in the 1990s and early 2000s

Ireland, as a country that historically enjoyed extremely high rates of homeownership, is a good place to start. The Irish state has, since its foundation, provided a myriad of interventions to ensure high homeownership rates. These included direct public construction for the homeownership sector, grants and tax relief, and the public provision of mortgage finance, especially to low- and middle-income households (Norris, 2016). State intervention drove homeownership rates in Ireland to among the highest in Western Europe; eighty per cent of Irish households were homeowners by 1991. In the late 1980s, however, most of the direct homeownership supports were withdrawn or substantially reduced during a series of austerity budgets (Norris, 2016). Public provision of mortgage credit, public construction of dwellings, and grants and tax reliefs all became very minor parts of Irish housing policy (Norris, 2016). Funding for social housing was reformed and reduced and the sector became highly residualized (Byrne & Norris, 2021). This also had an impact on homeownership as, since the Housing Act of 1966, social housing tenants enjoyed a 'right to buy', which became one of the main avenues for working-class households to become homeowners. By the 1990s, two thirds of the social housing dwellings built in the twentieth century had been transferred to homeownership in this way (McCabe, 2011). As is common in many jurisdictions, the rolling back of social housing provision occurred in tandem, with the rolling out of rent subsidies for private tenants (Byrne & Norris, 2021).

The period thus saw the neoliberalization of the Irish housing system. As Norris has argued, the post-war Irish housing regime can be characterized as one of 'socialized homeownership', given the extensive nature of homeownership supports. From the 1980s, however, the state 'walked away', leaving housing to market forces to a much greater extent (Ó Broin, 2019). Consequently, the provision of private mortgages by banks became central to accessing housing for most households (McCabe, 2011). In this sense, the processes of neoliberalization and financialization of housing went hand in hand. This proved decisive, as the rapid expansion of the mortgage market ensured that households could continue to access homeownership, despite the rolling back of state supports.

Between the late 1990s and 2008, as the country's 'Celtic Tiger' economy took off, Ireland experienced a remarkable property bubble. Nominal house prices increased by approximately 300 per cent between 1996 and 2006 (Kitchin et al., 2012), with lending for the construction and real-estate sector growing from €5.5 billion in 1999 to €96.2 billion in 2007, and outstanding residential loans rising by 281 per cent between 2000 and 2006 (Ó Riain, 2013). This expansion of credit for real estate and housing was made possible by Ireland's integration into the Economic and Monetary Union following the Maastricht Treaty. This gave Irish banks access to the European interbank lending market, through which to fund the expansion of credit (Kelly, 2014), and led to lower interest rates, the elimination of exchange rate risk and the deregulation of capital markets (Norris & Byrne, 2015). This also developed on the back of the more global liberalization of the financial system in the 1980s and 1990s, leading to much greater flows of capital across borders. Irish banks' net borrowing from abroad went from ten per cent to sixty per cent of GDP between 1999 and 2007 (Ó Riain, 2013). This 'wall of credit' entered the Irish market in the form of development finance, investment loans and mortgage credit. The latter was made possible via higher 'loan to value' (LTV) and 'loan to income' (LTI) ratios as well as longer maturities, giving more households access to larger mortgages (Downey, 2014).

The early period of financialization in Ireland thus made possible the continuation of the already extremely high rate of homeownership, in spite of the withdrawal of the key government supports for the sector, as argued by Norris (2016). As such, homeownership remained relatively stable from its high of eighty per cent in 1991 to 2001. The intertwined processes of financialization and neoliberalism appeared to have delivered on the 'homeownership promise' (Arundel & Ronald, 2021), albeit at the cost of exceptionally high levels of personal debt and an unsustainable financial system.

As the property bubble reached its peak, however, this gave way to a resurgence in the rental sector. Irish census data shows that homeownership declined between 2002 and 2006 – from 79.7 to 77.2 per cent of households. However, the most intense period of change came between 2006 and 2011, when the private rented sector expanded from 11.2 to 18.6 per cent of households and owner occupation fell from 77.2 to 70.8 per cent (Byrne, 2020a). There are a number of factors that explain the resurgence of the PRS in the midst of what appeared at the time to be a homeownership and house price bubble. Housing affordability was one such factor. Although LTV, LTI and mortgage maturities grew throughout the period (Downey, 2014), the gulf that opened up between wage growth and house price growth inevitably led to affordability issues. Ireland did not experience a subprime sector on the scale of the US. As such, a growing proportion of low-income households were left behind and became a new generation of renters. Due to the retrenchment of social housing, this option was no longer available to those who could not purchase on the market. The 'Celtic Tiger' years also saw a rise in the proportion of younger households and significant net immigration, which further fuelled demand for private renting.

The house price bubble made investment by small-scale landlords attractive, and the introduction of buy-to-let mortgages in the early 2000s was central to a new wave of landlord investment, which sustained the property bubble. The proportion of outstanding mortgages held by homeowners fell by 6.7 per cent between 2004 and 2006, while the proportion held by landlords expanded by 6.3 per cent (Norris & Coates, 2014). What this suggests is that buy-to-let investors piled in at the peak of the housing bubble, thus creating an 'affordability gap' by driving house prices beyond the reach of first-time-buyers. The buy-to-let boom was also facilitated by the neoliberalization of PRS policy in the 1980s. In the mid 1980s, rent controls were discontinued, following a court case which found them unconstitutional. Tellingly, they were not replaced by any alternative form of rent regulation, leading to the full marketization of rent setting. In terms of tenancy legislation, the Irish rental sector was virtually unregulated until the introduction of the Residential Tenancies Act in 2004. Even after that date, it remained one of the most poorly regulated rental sectors in Western Europe, until a new phase of reform began in 2016. A poorly regulated sector, with little or no limit on rental growth, made for an attractive investment. Thus, while the intersecting processes of neoliberalization and financialization initially appeared to herald an era of the market-led expansion of homeownership, it quickly reversed into the first significant decline of the sector in modern times.

As in Ireland, the recent resurgence of the UK's PRS stands in contrast to past experience. For much of the latter half of the twentieth century the sector experienced what appeared to be an irreversible decline (Kemp, 2015). Post-war UK housing policy revolutionized housing through investment in social housing as well as supports for homeownership (Ball, 1983). The 1980s was, however, a period of radical change. Unlike in Ireland, where governments rarely embraced neoliberal ideology explicitly (Byrne & Norris, 2021), the neoliberalization of the UK's housing regime was a more or less explicit policy goal, most famously in the example of Thatcher's 'right to buy' policy, which privatized and residualized the social housing sector (Kuenzel & Bjornbak, 2008). Again, echoing Ireland, the PRS underwent radical deregulation in the form of the 1988 Housing Act, which deregulated rent setting and weakened security of tenure by introducing shorter tenancy periods. In consequence, the PRS began to grow, if modestly, in the 1990s, from around nine to ten per cent of households (Kemp, 2015).

A more significant period of PRS growth began as a result of the property bubble in the 2000s. Between 1997 and 2007 UK house prices were among the fastest growing in Europe, with average nominal house prices tripling between 1998 and 2007 (Kuenzel & Bjornbak, 2008). The house price bubble was also strongly correlated with credit growth and financial deregulation (Whitehead & Williams, 2011), as house price increases were primarily driven by increases in mortgage credit, itself a result of significant credit market liberalization from the 1980s (Kuenzel & Bjornbak, 2008), as well as higher LTVs and longer maturities. Securitization also played a role in increasing mortgage credit (Hay, 2009). As in Ireland, despite easier access to mortgage credit, the pace of house price increases inevitably led to an 'affordability gap', which pushed homeownership beyond the reach of many. Real income per household increased by 27 per cent between 1995 and 2007, while real house prices increased by 168 per cent (Kuenzel & Bjornbak, 2008). As Kemp argues, 'as house prices surged out of the reach of many prospective first-time buyers, they had little choice but to rent from a private landlord' (Kemp, 2015: 611). The reduced role of social housing also played a role in terms of the higher proportion of low-income families renting privately (Kemp, 2015). In a pattern very similar to our discussion of the Irish case, '[a]n apex for Britain as a "nation of homeowners" was ostensibly reached around the mid 2000s' (Ronald & Kadi, 2017: 789). Homeownership peaked in 2003 at 70.9 per cent of households, and had declined to less than 67 per cent by 2010 (Whitehead & Williams, 2011).

In the mid 1990s, specialized buy-to-let mortgages were introduced by credit providers, and these became enormously popular.

In 2000, just one per cent of all mortgages were buy-to-let. By the mid 2000s that figure stood at ten per cent and in 2007 it was twelve per cent (Kemp, 2015). The value of outstanding buy-to-let mortgages increased from £12.2 bn to £43.8 bn between 2002 and 2007 (Ronald & Kadi, 2017). As in Ireland, then, a two-pronged dynamic led to the shift from homeownership expansion to decline, arriving around the peak years of the property bubble. One the one hand, house price increases brought about an affordability gap that halted the expansion of owner occupancy. On the other hand, rapid capital gains, a deregulated rental sector, and the availability of specific financial products for landlords (buy-to-let mortgages), meant that the demand for rental property created by the 'affordability gap' could be met by a remarkable increase in small-scale landlord investment (a similar dynamic has been identified in Australia, see Adkins et al., 2021). In both instances, the foundations of these developments were laid by the intertwined processes of neoliberalization and financialization in the 1980s and 1990s.

The interaction between neoliberalization and financializaton played out in a similar fashion in the Spanish context, although with some notable differences. As in the UK and Ireland, the Spanish rental sector had been in steady decline since the end of the Second World War. During the 1960s and 1970s, state supports to the development sector reduced house prices significantly below the market rate. In addition, while private banks were largely absent from providing mortgages to the working class until the late 1980s, the Spanish Mortgage Bank, a public institution, fulfilled this role (Palomera, 2014). From the 1990s, Spain moved away from the direct financing of construction and mortgage loans and towards tax incentive- and tax relief-based measures to stimulate supply and demand. By 2003, it was estimated that tax deductions for property purchase and other supports represented between twenty and fifty per cent of the total price paid by households (Palomera, 2014).

The rental sector's decline was also driven by legislation dating back to the Urban Rent Law of 1946, which introduced rent freezes and indefinite leases, both of which undermined investment in the sector (Pareja-Eastaway & Sánchez-Martínez, 2011). Although both of these measures were rescinded in the Boyer Decree of 1985, subsequent legislation introduced in 1994 established five-year standard tenancy periods and pegged rent increases to the Consumer Price Index (Pareja-Eastaway & Sánchez-Martínez, 2011). Thus, throughout the bubble period of the 2000s, Spain's rental sector was comparatively more strongly regulated than in either the UK or Ireland. This regulatory environment made investment, particularly by non-professional, small-scale landlords, less attractive. Taken together, these factors led to an exceptionally high homeownership rate of 80.7 per cent in 2001. However, unlike in both Ireland and the UK, homeownership expanded throughout the property boom of the following six years.

Despite the fact that Spanish house prices tripled between 1995 and 2007 (Norris & Byrne, 2015), by 2007, on the eve of the financial crisis, 87 per cent of Spanish households were homeowners and just 7.6 per cent lived in the PRS (Palomera, 2014). The expansion of homeownership throughout the bubble was certainly not due to the affordability of housing; in 1997 the average house price was equivalent to four times the average annual gross salary, by 2007 it was nine times (Pareja-Eastaway & Sámchez-Martínez, 2017). Unsurprisingly, credit availability was the key factor, made possible by Eurozone integration, access to inter-bank lending and the low interest rate environment. Outstanding residential loans expanded by 204 per cent between 2000 and 2006 (Norris & Byrne, 2015).

The most significant difference in the Spanish case, when compared with the Irish and UK examples, is that credit standards deteriorated to an even greater extent, ensuring an ever-growing number of households could access homeownership. In 2009, Spain was the country with the highest ratio of long-term household mortgage debt to disposable income in the world (García-Lamarca & Kaika, 2016). There are two interrelated factors which help us to understand the specificity of the Spanish experience. First of all, in Spain there was a much clearer trend towards the 'subprimization' of the housing market (García-Lamarca & Kaika, 2016). Unlike the US, this did not primarily take the form of a dedicated subprime industry, but rather the catastrophic decline in credit standards among mainstream mortgage providers (López & Rodríguez, 2011). The average maturity of mortgages increased from ten to 28 years between 1990 and 2007 and the number of mortgages issued to lower-income groups rose, as did mortgages issued to non-European migrants, particularly in the last phase of the boom (López & Rodríguez, 2011). During the most intense phase of the property bubble, between 2003 and 2007, approximately one million migrants from the Global South were granted mortgages (Palomera, 2014). Research has documented the common use of 120 per cent mortgages, as well as spurious 'cross-guaranteeing' of mortgages (Palomera, 2014). Second of all, securitization played a much more significant role in funding the Spanish mortgage bubble. Spanish securitizations account for almost half of all securitizations in the euro area (López & Rodríguez, 2011). They expanded by 65 per cent per annum during the 2000s and Spain's share of Eurozone mortgage covered bonds more than doubled between 2003 and 2007 (European Central Bank, 2009).

As in the case of the UK and Ireland, an affordability gap was created as income-to-house-price ratios increased throughout the early 2000s. Unlike those countries, however, Spain covered over this affordability gap via an even more intense expansion of mortgage debt.

After the crash: The Global Financial Crisis and the private rental sector

In the case of all three countries, the expansion of credit was entirely unsustainable and led to a series of post-crisis dynamics, which further undermined homeownership. As we shall see, however, these played out in different ways in different national contexts. Most notably, as the Spanish homeownership bubble had been more prolonged, when the crash came it led to particularly high levels of mortgage arrears and foreclosures, much like the US case. Nevertheless, in all three cases the ultimate consequence of the neoliberalization and financialization of housing was the decline of homeownership and the revival of the PRS.

Once house prices declined precipitously as part of the Global Financial Crisis (GFC), many Spanish homeowners found themselves in negative equity and mortgage arrears, leading to a massive wave of foreclosures and evictions due to mortgage arrears. During the peak of the crisis, between 2008 and 2014, almost 600,000 foreclosures occurred (García-Lamarca, 2022). Indeed, the mortgage evictions crisis was one of the defining political issues in Spanish society throughout the 2010s (Colau & Alemany, 2014), and pushed many former homeowners into the rental sector (Pareja-Eastaway &

Sánchez-Martínez, 2011). This strongly resembles the US experience, aptly captured by Schwartz (2023: 54):

> The mortgage crisis that began in 2007 marked an inflexion point in the growth and composition of rental housing. Prior to the crisis, home ownership rates were rising across nearly all population groups and rental housing was increasingly dominated by people with low incomes. Of course, the surge in risky, subprime lending proved to be unsustainable, and the resulting collapse sparked a global financial crisis and economic recession that caused millions of homeowners to lose their homes to foreclosure. From 2007 through 2012, more than eight million homeowners lost their homes to foreclosure, short sales and other related means ... Most households who lost their homes to foreclosure and related causes became renters ... Foreclosures probably accounted for an increase of several million renters since the onset of the crisis.

Changes at the level of housing policy have also played an important role in the post-crisis growth of the Spanish PRS. The rental sector was significantly deregulated through the Urban Letting Act of 2013 (Observatori DESC, 2013), reducing the tenancy period from five to three years and deregulating rents. The period after which a tenant can be evicted due to rent arrears was also substantially shortened (Byrne, 2020a). Although in more recent years there has been re-regulation of the PRS (discussed in Chapter 7), the post-crisis reforms afforded few protections to tenants and were much more favourable for investors. Combined with this, the growth in rent, particularly since 2014, and the decline in property values led to high investor yields (Pareja-Eastaway & Sánchez-Martínez, 2011). The combined effect of these factors has seen a remarkable and unprecedented turnaround in the tenure balance within Spain. The number of rented dwellings has increased by 51.1 per cent in a decade (2001–2011) (Pareja-Eastaway & Sánchez-Martínez, 2017). In terms of tenure change, the proportion of households in homeownership fell from 87 per cent in 2008 to 77.8 per cent by 2016, and those renting privately rose from 7.6 to 13.8 per cent over the same period. Thus, the prolonged homeownership bubble in the Spanish case ultimately led to the same outcome as in Ireland and the UK, albeit delayed until after the crash of 2008.

In Ireland by 2012 average house prices had declined by 41.6 per cent and house building had declined by 90.9 per cent (Norris

& Byrne, 2015). The banking sector entered into a period of freefall, as deposits contracted, mortgage lending collapsed, bad loans increased, and banks faced rising difficulties in borrowing on international markets. The years since the crisis have solidified the set of dynamics that undermined homeownership and fuelled further growth in the rental sector in Ireland. As noted, the proportion of households renting privately grew from 8.1 per cent in 1991 to 18.8 per cent in 2016. Mortgage arrears and repossession were an important part of Ireland's post-crisis housing system. The credit bonanza, combined with the rise of unemployment, declining incomes and an average decline in house prices of almost fifty per cent, inevitably led to very high levels of mortgages arrears: fourteen per cent of homeowner mortgages were in arrears in March 2015. The outworking of this has been very different in Ireland when compared with Spain; however, a mixture of government intervention and forbearance on the part of banks (the majority of which were state owned during the crisis period) resulted in much lower levels of repossession (Waldron & Redmond, 2014).

Another factor contributing to lower rates of homeownership was that the issuing of mortgage credit plummeted during the crisis years, making it extremely difficult for first-time buyers. Difficulties in accessing mortgages were further compounded by new macro-prudential lending rules introduced in 2015, which capped LTV and LTI ratios. Indeed, stricter mortgage underwriting standards in most countries meant many households who would previously have been able to access mortgage credit were unable to do so (for example, for the US case see Schwartz, 2023; for the Netherlands see Haffner, 2023). Housing policy in the post-crisis years, as in the case of Spain, also played a role in the growth of the PRS. For example, funding for social housing was cut by 88 per cent between 2008 and 2014, ensuring that the PRS became the only option for those who could no longer access homeownership. This was accompanied by an increased emphasis on rent subsidies, which represent an enormous demand subsidy for the PRS and thus a key part in the growth of tenancies and in investment in the sector. Legislation was also introduced (in 2013) to establish Real-Estate Investment Trusts in Ireland, and governments at the time courted institutional investors in an explicit attempt to draw on international capital to expand the PRS (Hearne, 2020).

Meanwhile, rapidly rising rent levels led to attractive yields in the PRS (Byrne, 2020a).

The UK was also beset by crisis across its property and banking sectors, with house prices falling by as much as thirty per cent from peak to trough (Ball, 2011). Mortgage advances also fell, as the banking sector moved to restrict new lending by strengthening credit standards (Whitehead & Williams, 2011). As Ronald and Kadi (2017: 6) note, 'the GFC appears a catalyst rather than a break on investment in private rental property'. This has been reflected in the continuing growth of the absolute number and proportion of households renting and the continued decline of homeownership. The 'apex' of homeownership in the UK was around 2003, at which point 71 per cent of households were homeowners. By 2014, this had declined to just 63 per cent. Meanwhile, by 2017 eighteen per cent of households were renting privately (Ronald & Kadi, 2017).

For landlord investors, the UK's post-crisis context has been favourable. On the investment side, rent increases and reduced house prices led to growing yields on rental property in the wake of the financial crisis. Buy-to-let mortgages have financed a large proportion of investment since the crash. As Ronald and Kadi (2017: 795) note, '[t]he 2008 GFC was an important moment in the advance of small-time landlords, and, although many overleveraged landlords had their properties repossessed during the initial downturn, as house prices continued to fall, more BTL buyers stepped in to acquire property cheaply'. The continuing decline of social housing has also played a role as, like in Ireland, the growing role of subsidized private rental accommodation. This has been a long-term process, with rent allowances doubling in cash terms between 2003 and 2013 and forty per cent of state housing benefit payments now going to the PRS (Ronald & Kadi, 2017), thus fuelling demand.

As seen throughout the discussion, in all three cases the revival of the PRS and the decline of homeownership has been the long-term consequence of the intersecting processes of neoliberalization and financialization. While this process began during the peak of the property boom in the UK and Ireland, as a lack of affordability locked out potential first-time buyers and BTL landlords piled into the sector, in the Spanish case (and indeed in the US) the more dramatic deterioration of credit standards allowed homeownership to expand right up until the crash. Housing policy and market forces thus take different forms, and interact in different ways, in different national contexts. Nevertheless, the medium to long-term impacts have been the same. The processes of neoliberalization and financialization have thus been at the centre of housing system change over the last number of decades.

This chapter focuses on Ireland, the UK and Spain. A brief discussion of some other cases can highlight some of the ways the processes identified are also occurring in other types of housing systems, as well as some important differences. Howard et al.'s (2024) comparative analysis of Australia and the Netherlands is instructive here. Australia's housing system is in many respects similar to that of Ireland and the UK, although with a much more residualized social housing sector, but it was largely unimpacted by the global property crash. The Netherlands' housing system is, in contrast, characterized by comparatively low rates of homeownership, and a large social rental sector. Nevertheless, as argued by Howard et al. (2024), there are some interesting common tendencies across the two countries. Both have experienced a revival of their PRS in recent decades, although in Australia this trend has been well established since the beginning of this millennium (Hulse, 2023), whereas in the Dutch case it is mainly associated with the post-GFC era, and especially post-2015. However, in the Dutch case, homeownership rates have not declined to the same extent as they have in Australia (and indeed the three case studies discussed in this chapter), but homeownership has become more 'polarized along the lines of class and age' (Howard et al., 2024: 8). This draws our attention to the fact that some of what we are seeing in terms of housing system change is not just about changing proportions of homeowners, but also about the composition of the different tenures, and related questions of inequality (discussed further in Chapter 5). Importantly, in the Dutch case the retrenchment of the social rental sector has played an important role in the revival of the PRS:

> Over the 2009–2001 period (but especially post-2015), private rental growth among young adults in the Netherlands was, to a greater extent, driven by decreasing access to social-rent than by homeownership opportunities being closed off. As private renting among young adults in the Netherlands increased from 19 per cent in 2009 to 36 per cent in 2021, homeownership rates declined by 6 percentage points

and social-rental rates declined by 9 percentage points. (Howard et al., 2024: 11)

This analysis provides an important corrective in terms of recognizing the role of social housing retrenchment and residualization in understanding the way housing systems are changing and the 'generation rent' phenomenon. It also shows how different processes and patterns of tenure change are driving the revival of the PRS across divergent housing systems, as the residualization of the social rental sector has played a very limited role in the Australian context. Nevertheless, Howard et al. (2024: 8) also highlight some key common tendencies across the two cases:

> In both countries, young adults now face highly financialized, owner-occupied housing markets underpinned by high levels of mortgage debt, commodified private rental sectors, and residualizing social-rental housing sectors fraught with barriers to access (albeit to substantially different disagrees).

This also readily applies to Spain, the UK and Ireland, as we have already seen. Thus, while the specific dynamics vary across different jurisdictions, there are some points of remarkable similarity.

Attention to how policy (housing policy as well as financial/monetary policy) and markets (both housing and financial markets) interact thus has valuable *explanatory power* when it comes to understanding how housing systems change, as well as the relative importance of private rental housing within the housing system. Indeed, we can see throughout the above analysis that housing policy interacts with and shapes the housing market and tenure patterns in many ways, from homeowner supports to rent subsidies to social housing retrenchment. Financial and monetary policy also play a crucial role, for example through mortgage credit standards. Questions of scale are also relevant here. In the case of financialization, regional and international financial dynamics were fundamental in the boom/bust era (and continue to be today). The way these super-national processes play out varies in different jurisdictions, but with Spain and Ireland especially effected by European Economic and Monetary Union, while the UK's integration into global flows of finance took place via somewhat different avenues. Moreover, global financial dynamics interact with local housing markets in different ways, and, more importantly, are mediated by national housing policies, such as the extent of PRS regulation (Aalbers, 2017). Neoliberalism, too, can be seen as a global process, both in the sense that neoliberal ideas and policies have been adopted across much of the world, but also in that international (e.g. EU) and global (e.g. the IMF and World Bank) institutions have been instrumental in this. But neoliberalism also plays out in different ways in different national contexts, shaped by national politics and path-dependent features of housing systems, for example (Brenner et al., 2010).

A political economy approach has powerful explanatory potential in large part because its analytical focus is precisely at the point of interaction, at different scales and in different local contexts, of these political and economic processes.

Beyond the 'homeownership society': The political economic implications of the revival of the PRS

As we have seen, a political economy approach emphasizes the embeddedness of housing within wider political and economic structures and processes. In so doing, it can not only help to *explain* why housing systems are changing, but also allow us to *analyse* the meaning and significance of this change. The relationship between housing and wider questions of political economy is amply demonstrated in the vast literature on homeownership. Examining this literature helps us understand the broad impact of housing systems, but also allows us to examine what is at stake in the transition away from the homeownership norm, and to begin to consider what we might be *transitioning to*.

The opening up of the 'promise of homeownership' (Arundel & Ronald, 2021) to a relatively wide swathe of the population in advanced economies developed in the decades following the Second World War. Here it is important to recall that before this development, and the parallel mass construction of social housing which occurred in many countries, the PRS was the overwhelmingly dominant form of housing for poor, working-class and middle-class households (Harloe, 2021). There are many factors which explain the remarkable expansion of homeownership during the period, summarized by Kemp as follows:

> The growth in the owner-occupied market in the advanced economies during the second-half of the twentieth century was the result of a highly favourable and unpreceded conjunction of trends. These included: full male employment for the three decades up to the early 1970s; the more recent rise in female participation in the labour market, which increased the income of two adult households; improvements in social safety nets after 1945 …; rising real incomes …; increasing availability of mortgages for home purchase… and, not least, the promotion of home ownership by governments in the advanced economies. (Kemp, 2023: 2)

Mass homeownership was thus supported by a wide variety of historically specific factors, leading to what Forrest & Hirayama (2018) describe as an idiosyncratic 'generation own', i.e. a historical anomaly in the context of capitalist society characterized by a huge increase in upward 'residential mobility'.

This, in turn, shaped a variety of political and economic processes which characterized the post-war era. In economic terms, homeownership shaped consumer trends in everything from domestic appliances to interior design products, thus fuelling demand for mass consumption (Ronald, 2008). Homeownership also played a huge role in the dynamics of wealth distribution, with widespread homeownership supporting much greater wealth equality (Arundel & Ronald, 2021). Mass ownership of a valuable asset also fuelled economic demand and consumption, as asset ownership allows equity release and other forms of borrowing, which has been described as a form of 'asset-price Keynesianism' (Brenner, 2006). Mass homeownership is also of course closely tied with the growth and maturation of mortgage markets, and the banks and other financial institutions that provide them (Forrest & Hirayama, 2018). Finally, mass homeownership shaped the development of welfare systems, as demonstrated in the literature on 'asset-based welfare' (Watson, 2009).

Housing systems and tenure patterns also play a role at a more cultural level. Mass homeownership gave rise to all manner of cultural phenomena, from gardening to dinner parties. In particular, homeownership in many advanced economies has become a quintessential expression of middle-class identity (Ronald, 2008), and to some extent adulthood itself. Housing is perhaps the ultimate 'status' or 'positional' good (Foye et al., 2018), allowing individuals and households to express their social status as well as their identities through the choices they make about the neighbourhood they live in, the type of dwelling they purchase, and how they design and decorate. The way homeownership supports the expression of middle-class identity is of course intimately connected with the mass consumerism noted already, e.g. with home interior and related products (Florida & Feldman, 1988).

The relationship between homeownership and middle-class identities is linked to the question of class (Saunders, 2021). In many societies around the world, and especially in quintessential homeowner societies, which include the English-speaking and Mediterranean countries, homeownership played a crucial role in the political integration of the working class into post-war capitalism. Access to homeownership was often viewed by post-war politicians as an antidote to 'Bolshevism' and workers' militancy (Forrest & Hirayama, 2015; Harvey, 1976). Ownership of property would give workers a stake in the system, it was thought, and support their identification with post-war capitalism (Edel, 1982; Saunders, 2021). In some post-colonial countries, mass homeownership became associated with the politics of post-independence land redistribution, as struggles around land ownership often played a central role in anti-colonial independence movements (for the case of Ireland, see Norris, 2016; for a somewhat different case, that of Singapore, see Lee, 2000). Post-war social housing played a similar role, although it is more typically framed as a mechanism to protect the working class from capitalism rather than to incorporate them into it (Forrest & Hirayama, 2015). Nevertheless, the provision of secure and reasonably affordable social or public housing provided a degree of stability and certainty which mitigated class conflict in many post-war societies, especially in Europe (Arundel & Ronald, 2021). While the focus of this section is homeownership, it should also be noted that 'first-generation' rent controls (which often took the form of rent freezes) and other forms of rent regulation also formed part of the politics of post-war housing (Harloe, 2021).

The post-war housing system can thus be seen as part of a wider political settlement or 'class compromise', and as part of the development of welfare states and labour market regulation (Ruggie, 1982). What Arundel and Ronald (2021: 1112) call the 'promise of homeownership' was characterized by 'a model of owner-occupation based on three key tenets: being *widespread* in access, being *equalizing* in wealth distribution, and providing household economic *security* over the life course'. It should not be forgotten, of course, that the promise of homeownership only extended so far and came with its own exclusionary dynamics, for example of class and race. As Florida & Feldman (1988: 199), writing in the US context, note:

> Housing opportunities were similarly segmented. Relatively high-quality suburban housing accrued to working-class groupings that were privy to the accord. Both the growing ranks of managerial personnel – the 'new middle class' – and unionized, primary sector workers used the housing finance system to purchase suburban homes. Owner occupancy conferred additional economic advantages to such groups in the form of tax breaks, retained equity and appreciation of home values, further exacerbating inter- and intraclass differences. Groups that were peripheral to the accord were limited to much lower quality, multifamily, inner-city rental or public housing. Residential patterns both reflected underlying class fragmentation and added a spatial dimension to it.

Nonetheless, post-war 'Fordist housing systems' did extend homeownership, and social housing in some cases, to a much greater proportion of the middle and working class (Florida & Feldman, 1988).

Fordist political economy, at a general level, was a reflection of the heightened levels of class conflict and working-class political organization in the early and mid twentieth century, the extension of franchise to the working class in the period proceeding this, and conflict between the West and the Soviet Union (Aglietta, 2000; Florida & Feldman, 1988). The post-war housing system, and especially mass homeownership, thus played a decisive political role in terms of the political integration of the working class (Florida & Feldman, 1988), the dominant forms of political ideology (Ronald, 2008) and voting behaviour (Saunders, 2021).

This overview of the centrality of housing and homeownership to the development of post-war capitalism in the advanced economies shows how a political economy approach, which emphasizes housing's embeddedness within wider political and economic structures and processes, is vital if we are to grasp the full impact and significance of housing systems. This also implies, of course, that it can help us understand the significance of housing system *change*. In other words, when we speak of declining homeownership, we are speaking of much more than the question of access to and experience of housing. This begs the question, to which I now turn, of how we might understand the political economic significance of the revival of the PRS. To begin this discussion, I draw on the work of Forrest and Hirayama (2015; 2018) and Ronald and Kadi (2017) and their respective, and largely compatible, concepts of the 'late-homeownership' and 'post-homeownership' societies.

Forrest and Hirayama (2015) argue that the post-war expansion of homeownership represented an 'inclusive political agenda' (Forrest & Hirayama, 2015: 238). Current housing dynamics, in contrast, represent a shift to an 'unapologetically exclusive' political agenda (Forrest & Hirayama, 2015: 238; see also Adkins et al., 2021) under which the 'relentless logic of commodification has served to undermine a key element of the social cement of contemporary capitalist societies: homeownership' (Forrest & Hirayama, 2015: 234). Rather than the state-supported widening of access to homeownership, neoliberalism results in a *concentration of property ownership* among wealthier households, who draw on existing housing wealth and can acquire additional properties as landlords (Forrest & Hirayama, 2015; see also Ronald & Kadi, 2017). Thus, '[t]he homeownership systems which have emerged from the [global financial] crisis are ones which favour the financially privileged ...' (Forrest and Hirayama, 2015: 237). Echoing this point, Arundel and Ronald's (2021: 1120) analysis of the US, UK and Australia suggests that contemporary housing market dynamics 'enhance inequality and insecurity'. The fundamentally more unequal nature of contemporary housing systems has implications for social and political cohesion:

> In the post-war Keynesian era ... homeownership systems were set within frameworks of public policies aimed at enhancing social cohesion and protecting people's lives in volatile, uncertain and precarious economic conditions. In this sense, problems in homeownership sectors not only impact on housing situations but are a potential catalyst for undermining wider, established social contracts. (Forrest and Hirayama, 2015: 241)

In their later work, Forrest and Hirayama (2018) argue that the transition to what they call the 'late homeownership society' is bound up with new forms of social stratification in which the ownership of residential property plays a decisive role. With access to homeownership increasingly limited, family deployment of existing real-estate assets, accumulated during the post-war era, takes on a new significance in terms of shaping social stratification and inequality. This takes place primarily via inter-generational transfers (the so called 'bank of mom and pop') and inheritance. What is occurring, thus, is not simply a return to the pre-war era of mass private renting. Rather, '[e]merging late homeownership societies, while associated with a revival of private landlordism, are built upon the housing wealth accumulated in the exceptional era, and the consequent structure of housing and social inequalities is very different' (Forrest and Hirayama, 2018: 268). They argue that this stratification can be conceived according to three categories. First, 'real-estate accumulators', who 'maintain or further accumulate valuable multiple property assets over generations'. Second, 'housing wealth dissipaters', who 'experience a decline of residential property assets accumulated during the post-war, high-growth period'. Third, and finally, 'perpetual renting families', who were never able to access homeownership in the first place and are 'increasingly excluded from the social mainstream' (Forrest and Hirayama, 2018: 268).

Ronald and Kadi (2017), focusing on the British case, develop a parallel concept of the 'post-homeowner society'. While Forrest and Hirayama emphasize a shift from 'generation own' to 'generation rent' in terms of stratification, Ronald and Kadi (2017) note that the emergence of 'generation rent' is linked to a new 'generation landlord'. The dynamics of the post-homeownership society are characterized by the exclusion from property ownership for some, and the concentration of property ownership (including multiple-property ownership) for others:

> Generation rent is only one side of housing and welfare restructuring in the British post-homeownership context. It has been mirrored by the emergence of a new class of non-professional landlords largely made up of an earlier generation of homebuyers who have become focused on the acquisition of extra property to let. (Ronald and Kadi, 2017: 790)

One of the most salient impacts of the transition to the post-homeownership society relates to what is known as 'asset-based welfare', i.e. the tendency in recent decades for households, typically supported and encouraged by policy, to treat housing assets as a hedge against future risk, especially in relation to loss of income in later life. Declining access to homeownership thus has profound impacts on the relationship between housing and welfare (Ronald and Kadi, 2017). But the implications of the post-homeownership society go beyond this. Given homeownership's role in notions such as that of the property-owning democracy, the rise in renting may have implications for political legitimacy. Furthermore, it may lead to the erosion of 'middle-class adult identities and economic autonomy' and thus 'represents a challenge to the reproduction of middle-class status' (Ronald and Kadi, 2017: 799).

The concepts of both the 'late homeownership' and 'post-homeownership society' draw our attention to the fact that a necessary corollary of the revival of the PRS is the growing *concentration of ownership of residential property*. It follows that the sector is intimately bound up with property relations between socio-economic groups, and hence with inequality. Such a perspective must be a central component of any political economy of the PRS because it, crucially, allows us to denaturalize the categories of landlord and tenant. Rather than assuming these as givens with any housing system, as if they simply appeared out of thin air, we must see landlords and tenants as two mutually constitutive forms of relationship with the ownership of residential property. These relationships, in turn, are produced and reproduced by the political economic structures that shape the ownership of residential property, and draw the line between the 'housing haves' and the 'housing have nots'. The political economy of the PRS cannot be simply an examination of the issues that affect tenants, such as rents and supply, but must also address the political and economic structures, chiefly relating to housing policy and housing/financial markets, that generate the distribution of residential property and therefore explain why some households are landlords and some are not, i.e. how the stratification of housing access and ownership is produced. From this point of view, one of the chief impacts of the transition towards the 'post-homeownership society' is that housing systems are increasingly driving social and economic polarization, exclusion, alienation and political conflict, the diametric opposite of the role they played during most of the latter half of the twentieth century. The 'macro-political economy of the PRS' is therefore a crucial counterpart to the 'micro-political economy' approach developed in the previous chapter, with its emphasis on the direct social relationship between landlord and tenant associated.

Chapter 5 delves further into the multiple forms of inequality that are associated with private rental housing, and the challenge of conceptualizing the relationship between class-based and tenure-based forms of inequality, but also the generational inequalities that have been subject to much public discussion and debate. Before getting to this, however, the next chapter examines another, and perhaps even more contentious, way in which political economic forces are re-shaping the distribution of residential property, this time in the form of the 'super-concentration' of ownership associated with the rise of 'financial landlords'.

4
The Financialization of the Private Rental Sector
The Rise of Institutional Landlords

Throughout this book we have seen that there are a wide variety of issues and challenges faced by private tenants in the contemporary context. But, amongst these issues, perhaps none has garnered as much public interest, and been subject to as much debate, as the rise of what have been called 'financial landlords'. The arrival of Wall Street firms as major landlords in countries across the world, a development as sudden as it was unexpected, has polarized opinion, challenged policy-makers, and, at least to some degree, changed the nature of PRS housing.

Some of these issues have already been raised in Chapter 3, where we saw how the expansion, and later contraction, of mortgage markets led ultimately to the erosion of homeownership and the consequent resurgence of private renting. In more recent years, however, the PRS has been more directly subjected to processes of housing financialization as a host of financial institutions rapidly expanded into rental housing (Gabor & Kohl, 2022). Globally, investment in what is known as the 'multifamily' sector, primarily rental apartments owned by institutional landlords, has grown from €30 billion in 2010 to over €200 billion as of 2020 (Guironnet et al., 2024). Against this backdrop, we have also seen a flurry of political and policy activity. Activist campaigns against, for example, Blackstone, one of the world's most well-known private equity firms, have taken place in numerous countries, including Spain and the US. Meanwhile, policy-makers in countries like Ireland and Denmark have introduced specific new legislation to limit the reach of this new breed of landlord.

Examining the financialization of the PRS is a complex task. Despite a recent spate of literature, a full understanding of the extent of its influence on housing systems has yet to emerge. Moreover, the concept of housing financialization itself, as well as its application to the PRS, requires further development (Christophers, 2015). Nevertheless, it is an important topic in two senses. First of all, it represents an important recent development in many housing systems and, as a consequence, in the lives of tenants. Second, examining the interaction between the financial system and PRS housing allows us to explore the ways in which the latter is imbricated in wider, even global, economic processes that go far beyond the housing system. It thus allows us to explore how the political economy of housing in general, and the PRS in particular, calls for an understanding of the role housing plays in contemporary economic developments at different scales. In so doing, we can also examine the crucial role played by the state in enabling and mediating the interaction between PRS housing and larger structures and processes within the economy (Gabor & Kohl, 2022).

In this sense, examining the financialization of housing allows us to draw out some of the ways in which private rental housing interacts with wider political economic dynamics. Housing is at the same time a driver of financialization and a 'victim' of financialization, in the sense that financialization has a series of negative impacts on housing systems (Gabor & Kohl, 2022). Indeed, for many scholars and activists today, financialization is the principal force challenging the right to housing (Hearne, 2020; Rolnik, 2019), undermining affordability (Lima, 2020), leading to gentrification (Aalbers, 2019; August & Walks, 2018; Raymond et al., 2021), causing homelessness (Lima et al., 2023), and generally intensifying the commodification of housing (Aalbers, 2016; Beswick et al., 2016; Gil García & Martínez López, 2023; Hearne, 2020; Janoschka et al., 2020). However, it has also been argued that the financialization of housing literature in general (Christophers, 2015), and PRS financialization in particular (Nethercote, 2020), is hampered by a lack of conceptual clarity with regard to the key processes involved. As finance (specifically credit) has played a role in housing throughout much of modern history, it is not always entirely clear what makes the present era of financialization qualitatively different. Moreover, although it is argued that 'financial landlords' intensify the commodification of housing and treat housing as an asset, thus potentially undermining tenants' experiences of housing, it is not always clear why, and in what way, they differ from conventional landlords. This chapter cannot definitively resolve these challenges, but it will draw on the existing research from a variety of countries in an attempt to clarify the nature and form of PRS financialization, the different firms and business models involved, and the impact it can be said to have on tenants.

Introducing the financialization of housing

The story of the rise and rise of financial landlords is part of a wider story of housing financialization, which, according to some commentators, is one of the most significant political economic phenomena of recent decades (Aalbers, 2017). The relationship between housing and the financial system, and with it the role of housing within the wider economy, has undergone profound transformation over recent decades. Housing assets, and income streams arising from housing, are now routinely traded on global financial markets, the volume of mortgage debt in relation to GDP has grown across the advanced economies, and developments in housing markets have become key economic bellwethers.

The experience of the global property boom and bust which occurred during the 2000s had a profound impact in terms of scholarly examination of the relationship between housing and the wider economic system. The fact that subprime mortgages played the role of catalyst in the greatest global capitalist crisis since the Wall St Crash could hardly fail to draw attention to the interrelationship between housing and the global capitalist system. Against this background, a variety of themes have emerged which have helped researchers understand how the everyday housing experiences of residents are plugged into global flows of capital. Over the course of the 1990s and 2000s, for example, mortgage debt came to play a role in underpinning consumer demand via the wealth effects of housing equity and how it facilitates growing household indebtedness (Brenner, 2006; Crouch, 2009). Relatedly, growing mortgage debt and house price growth acted as the driving motors of economic development in a number of jurisdictions, in particular the UK (Crouch, 2009), the US (Brenner, 2006), Ireland (Ó Riain, 2013) and Spain (López & Rodríguez, 2011; Norris & Byrne, 2015).

One important strand within this literature has identified the central role of housing, and real estate in general, in the expansion of the global financial system over the last number of decades. Aalbers (2017), the leading scholar within the field of housing financialization, draws on the work of Jordà et al. (2016) in noting that over the period between 1870 and 2010, private debt-to-GDP ratios remained relatively stable at around fifty to sixty per cent of GDP. However, from the 1980s this began to increase, reaching 118 per cent by 2010. Most notably, Aalbers (2017: 5) argues, this rapid increase in private debt was largely predicated on the growth of mortgage debt, as between 1914 and 2010 mortgage debt grew from twenty to 64 per cent of GDP. Housing has thus played a role in the general process of financialization, understood as: 'the increasing dominance of financial actors, markets, practices, measurements, and narratives at various scales, resulting in a structural transformation of economies, firms (including financial institutions), states, and households' (Aalbers, 2016). Financialization has occurred through, and impacted on, housing and urban development in myriad ways, including mortgage securitization (Gotham, 2009), the privatization of public housing (Fields & Uffer, 2016), new forms of entrepreneurial urbanism (Byrne, 2016c) and the general expansion of the volume of capital flowing into housing investment (Aalbers, 2016).

It is important to note that housing and finance have a long relationship. However, throughout much of the early and mid twentieth century, and focusing on the advanced economies, housing finance was typically offered through relatively conservative channels separated from the mainstream banks, such as building societies and savings and loan institutions (and their equivalents) (Aalbers, 2017). The state also often played a very significant role in relation to mortgage provision, from offering subsidized loans to shaping the allocation of credit (Norris, 2016). As Aalbers (2017: 3) argues, '[h]ousing finance was sheltered from the volatility of financial markets, and even if it was not provided by the state (which was the default option in most countries), it was still heavily regulated and, indeed, controlled by state institutions'. Aalbers' analysis also dovetails with the analysis developed in Chapter 3 of this book, in that he argues that the financialization of housing can be situated within the wider context of the transition from 'Fordist' to 'Post-Fordist' housing regimes, i.e. the shift from a regime in which 'housing, including housing finance, was considered too important to be regulated and controlled solely by markets' to the neoliberalized and financialized regimes we see today (Aalbers, 2017: 3).

In theorizing the financialization of housing, a dominant perspective is that associated with David Harvey's (2018) Marxist view of the relationship between urban development and capitalist political economy (Aalbers, 2017; Beswick et al., 2016; Fields, 2018; Gil García & Martínez López, 2023). While Harvey's work is complex and theoretically dense, and therefor hard to summarize in the limited space available here, the key point with regard to housing financialization is that urban development plays a role as an outlet for capitalist investment, particularly in the context of crises of overaccumulation. Where there is a lack of available opportunities for profitable investment, urban development serves to create and to store 'surplus value'. Finance capital is crucial to this process because it is essentially by lending money with interest for urban development that capital flows into the built environment. While this perspective strikes me as overly deterministic and reliant on Marxist theory (as opposed to empirical analysis), it can serve as a useful way to situate the financialization of the PRS which has occurred since the Global Financial Crisis (GFC) of 2008. As the next section will show, the entry of financial institutions into the rental housing sector emerged against the backdrop of falling house prices, distressed mortgage debt and the decline of homeownership. This confluence of factors created the opportunity for a new form of housing financialization, while the collapse of mortgage markets generated the need for financial institutions to find new outlets for investment (Beswick et al., 2016). Thus, the financialization of the PRS can be understood as emerging out of a period of capitalist crisis, and specifically how that crisis interacted with dislocation within housing systems and property markets.

The financialization of the private rental sector

Turning specifically to the financialization of the PRS, the literature covers a variety of types of landlord, with various terms such as financial landlord, build to rent, transnational landlord and 'global corporate landlord' all being used. It also covers a variety of business models and forms of investment. This can result in a degree of confusion with regard to what is meant by institutional landlords, but also what exactly is at stake in processes described as forms of PRS financialization. For the purposes of this chapter, I differentiate between five of the most salient phenomena described in the research. First, the growth of institutional investors in the US 'single-family rental' (SFR) market. While the US has a long tradition of institutional landlords, their rapid penetration of the SFR market, characterized by detached, semi-detached or terraced housing (i.e. not apartments), and typically in a suburban setting, represents one of the most novel, and indeed controversial, developments in the US PRS over the last decade or two. Second, the acquisition of existing apartment units by a new generation of institutional landlords in the wake of the GFC, including private equity firms, hedge funds, pension/insurance funds and real-estate investment trusts (REITs). This section focuses especially on the acquisition of distressed property assets in Spain and Ireland. Third, the mass acquisition of former public or social rental housing, focusing on Germany and Sweden. Fourth, a particular investment strategy or business model occurring across numerous jurisdictions, focused on acquisition of second hand, often older, stock and its renovation. This phenomenon overlaps with the acquisition of former public or social rental housing. The final form of PRS financialization examined, in contrast with those described thus far, relates to new apartment/multi-family supply in the form of 'build to rent', which has grown in the US, UK, Ireland and other countries.

Before delving into our exploration of these diverse forms of financialization, a note of caution is warranted. Institutional landlords have attracted a lot of attention both within academic research and political debate, but they still represent a minor part of the PRS. Even in a country like Sweden, which has a very large share of institutional landlords by comparison with the European average, the share of apartments owned by institutional landlords is fourteen per cent. In Germany, the Netherlands and Denmark, which again have unusually high levels of activity in terms of institutional landlords, the share of institutional investors in the PRS is between three and five per cent. In many countries, they account for less than one per cent of PRS units (Holm et al., 2023). Relatedly, it is clear from the existing research that in many countries the recent expansion of the PRS has been driven by small landlords, typically owning less than five units and often just one or two (for France see Guironnet et al., 2024; for the Netherlands see Hochstenbach, 2022; for the UK see Ronald & Kadi, 2017). It is thus important to situate the recent rise of institutional landlords within this context, and to understand that, for the moment at least, their impacts are somewhat limited. Nevertheless, it is also the case that institutional landlords typically invest in specific locations, and often acquire assets at scale in these locations (Gabor & Kohl, 2022; Holm et al., 2023; Pfeiffer et al., 2021; Wijburg et al., 2018). As such, while the proportion of PRS units owned by institutional investors may be reasonably low *at a national level*, this can disguise the role they play within particular locations or market segments. For example, and as we will see below, in Ireland institutional landlords play an outsized role in relation to a specific housing typology (one- and two-bedroom apartments) and in relation to new housing supply in particular locations (neighbourhoods in Dublin with links to high-paid employment) (Daly, 2022; McCarthy, 2024). Similarly, in some US 'Sun Belt' cities, institutional investors play a very important role within the specific single-family dwelling market segment (An, 2023; Pfeiffer et al., 2021). Institutional investors can also play an outsized role in shaping processes of urban development, such as gentrification (Aalbers, 2019; Raymond et al., 2021). Moreover, while the literature drawn on below from the fields of housing studies, political economy and economic geography tend to adopt a critical perspective on institutional landlords, there is also evidence of some benefits of institutional landlords and the related build to rent sector, emphasizing their role in increasing the supply of PRS housing and, potentially at least, generating more professional and tenant-oriented outcomes (e.g. Carvalho et al., 2023).

US single-family rental

The United States has a long history of institutional PRS housing, going back to the post-war era (Nethercote, 2020). However, until recently this was confined to what is known in the US as the 'multifamily' sector of the housing market, referring to apartment complexes primarily in urban areas. Until more recent years, the market segment known as single-family rental (SFR), comprised of suburban detached, semi-detached and terraced housing, was dominated by smaller landlords, which are often referred to as 'mom or pop' landlords. This form of housing was understood as unsuitable for institutional investment because the housing units are dispersed rather than concentrated, and are not easy to either acquire or manage at scale (Fields, 2018). However, in the wake of the GFC institutional players rapidly entered the market, bringing about a significant, and controversial, change in the country's PRS. There are two aspects to the change that took place. Firstly, the proportion of single-family homes which are rental grew from thirteen per cent in 2006 to seventeen per cent in 2015, representing an increase of over 34 per cent in less than a decade (Colburn et al., 2021). Second, the form of landlord that drove this shift was not the traditional 'mom and pop' investor, but a new breed of large firms driven by Wall Street money (Christophers, 2022b).

One of the biggest institutional landlords is Invitation Homes, originally established in 2012 by the private equity firm Blackstone. The firm entered the market rapidly, taking advantage of the crisis conditions that pertained immediately after the GFC. In the latter part of 2012 and in 2013, it acquired 40,000 dwellings, spending $100 million on SFR properties (Christophers, 2022b). Fields and Vergerio (2022) note that today the four biggest SFR operators together control over 200,000 homes. As Colburn et al. (2021) note, '[i]n less than a decade, this industry went from non-existent to a new asset class with over $30 billion of invested capital'. Although the first phase of SFR financialization was concentrated in the immediate post-crisis context (roughly 2008 to 2016), investment has continued in more recent years. Indeed, in 2021 institutional landlords purchased 21 per cent of all single-family homes for sale (Nethercote, 2024). The significance of institutional landlords in the SFR market has been further enhanced by a series of mergers that took place as the industry matured, leading to the dominance of a small number of firms. Invitation Homes and American Homes 4 Rent, the two largest companies, together owned around 130,000 SFR dwellings by the end of 2018 (Colburn et al., 2021).

Despite this rapid growth, it is important to highlight that institutional landlords represent just 1.2 per cent of the national SFR market. It is thus important not to overplay the significance of this phenomenon. Nevertheless, because institutional investment in SFR tends to be highly geographically concentrated, it can have a significant impact on particular markets. For example, institutional ownership of SFR properties in 2015 represented fourteen per cent of stock in Pheonix and twelve per cent in Tampa, two of the biggest markets (Pfeiffer et al., 2021). This geographical concentration relates to the specific business strategy of institutional SFR landlords, as well as the context within which they emerged. This novel form of financialization took shape against the background of the mass foreclosure of owner-occupied housing following the subprime crisis, and the related collapse in house prices and proliferation of distressed property assets. As already noted in Chapter 3, between 2007 and 2012 more than 8 million homeowners experienced foreclosure in the US (A. Schwartz, 2023). Private equity firms, hedge funds and REITs could take advantage of access to cheap Wall Street capital to acquire the resulting distressed properties in bulk (Christophers, 2022b; Fields, 2018). Foreclosures, and consequently institutional investment in SFR, were concentrated above all in the Sun Belt region, where property prices fell the sharpest (Immergluck & Law, 2014), thus explaining the concentration of SFR financialization in these areas.

The state also played an important role in the financialization of the SFR, as argued by Christophers (Christophers, 2022b), due to its failure to support homeowners experiencing foreclosure, while at the same time ensuring conditions, via bank bailouts and loose monetary policy, which supported the expansion of the financial system. The US government also engaged specific policies to stimulate the financialization of SFR housing. In 2012, the Federal Housing Finance Agency initiated a scheme of selling 2,500 foreclosed government-owned SFR properties. Fields (2018: 126) quotes one of the scheme's architects describing it as an attempt 'to gauge investor appetite for a new asset class, that is scattered-site single-family rental housing'.

It is also important to note, as argued by Fields, that the financialization of SFR led to both the transformation of this form of housing into a target for institutional investors, but also to the creation of a new financial asset via the securitization of rental income streams. In 2013, Blackstone's SFR subsidiary Invitation Homes developed the first SFR-backed security, i.e. a financial asset underpinned by the rental income of thousands of SFR homes (Fields, 2018). Over subsequent years this form of securitization has continued to develop.

Wider housing and financial market conditions also played a role here. The tightening of mortgage issuing criteria, as well as the mass foreclosure mentioned already, made access to homeownership more challenging, thus enhancing demand for private rental housing (Christophers, 2022b). This also meant that mortgage markets, which had been one of the most important avenues through which financial actors could invest in residential housing, all but dried up. The development of SFR thus created a new avenue through which capital could flow into residential housing:

> [As] we know from the role of subprime lending in the US housing bubble, finance capital's pursuit of accumulation through real estate also creates cycles of speculation-fuelled crisis. The detrimental effects of such crises on the urban landscape subsequently offer new opportunities for value extraction, propelling the uneven reproduction of urban space. Thus, we can understand the process by which the US mortgage crisis created the conditions for the SFR asset class, with homeowners' dispossession setting in motion a new round of financial accumulation situated in the rental sector. (Fields, 2018: 121)

The impact of this form of financialization is wide and varied. While some commentators have argued that the role of institutional investors in SFR helped to stabilize local housing markets in the wake of the GFC (Colburn et al., 2021), others have pointed towards some of the negative impacts which may be experienced by tenants. Here it is worth noting that SFR financialization has been concentrated in areas which have particularly weak tenant protections (Pfeiffer et al., 2021). Raymond et al. (2018) found that institutional SFR landlords are 68 per cent more likely than small landlords to file eviction notices, even when controlling for property, neighbourhood and tenant characteristics. There is also research indicating that financial investors are more likely to engage in harsh cost-cutting measures to boost profits, which can in turn lead to poor tenant experiences (Colburn et al., 2021). This is also suggested in research which highlights the importance of technology and the automation the property management of SFR, which can lead to a lack of responsiveness on the part of institutional SFR landlords (Fields & Vergerio, 2022). Higher-than-typical rent increases are another impact that undermines tenants' access to and experience of rental housing. Although research is limited, it has been argued that institutional SFR landlords dampen access to homeownership, due to their outsized ability to acquire properties in a way which households cannot compete with (An, 2023). Thus, while institutions own a small minority of single-family dwellings nationally, the pace at which they have grown, their extremely high level of consolidation, and the concentration of their investments in particular local markets, all mean they can have important impacts on the nature of PRS housing. It is also worth noting that the financialization of SFR housing is also emerging in embryonic form in some other jurisdictions, notably in the UK (Nethercote, 2024).

Post-crisis acquisition in Ireland and Spain

The acquisition of foreclosed SFR dwellings in the US is just one form of PRS financialization, and one which relates to that country's particular post-crisis context. For some European countries, notably Ireland and Spain, the process has developed in a very different form. Both these countries, as discussed in Chapter 3, experienced acute property and financial crises in the years after 2008, and both went on to implement enormous bank bailout packages and experience related sovereign debt crises (Beswick et al., 2016; Byrne, 2016b; Janoschka et al., 2020; Norris & Byrne, 2015). As part of the unwinding of their financial real-estate crises, large volumes of distressed financial assets, chiefly loans secured by real estate, were 'deleveraged' (Gabor & Kohl, 2022). This took place through two primary avenues. First, private banks selling distressed property loan portfolios, or in some instances actual real-estate assets which had been repossessed. Second, through 'bad banks', a form of public asset management company established by both the Irish and Spanish governments, respectively, to acquire distressed property-related assets from across their ailing bank systems and to dispose of them in bulk sales (Byrne, 2016b). Spain's bad bank, SAREB, was created in 2012 to take control of debts and repossessed homes from across the Spanish banking sector (García-Lamarca, 2021; Gil García & Martínez López, 2023; Guzmán, 2023; Janoschka et al., 2020). Ireland's NAMA (National Asset Management Agency) played the same role, acquiring €72 billion in real-estate-related debt, equal to a remarkable 47 per cent of Irish GDP at the time (Byrne, 2016a). The main purchasers of distressed real-estate assets in both countries were international, mainly US, private equity firms and hedge funds. In Spain, for example, Blackstone, once again revealing itself as a key player in the global financialization of the PRS,[2] purchased the entire defaulted mortgage portfolio (consisting of 94,000 loans) from the bank Catalunya Caixa (Beswick et al., 2016). Indeed, between 2013 and 2018, Blackstone created 'a real- estate empire in Spain, consisting of more than 120,000 assets that include rental flats, mortgages, offices, hotels, and land ready for real-estate development. Blackstone has now become the most important single actor in the Spanish real-estate market ...' (Janoschka et al., 2020). In the Irish case, NAMA sold ninety per cent of its assets to US financialization institutions such as Blackstone, Cerberus, Oaktree and Loan Star Capital (Byrne, 2016a).

Interestingly, both Ireland and Spain also facilitated the repurposing of distressed residential real-estate assets as PRS housing via legislation providing for the establishment of Real-Estate Investment Trusts, which has played an important role in re-financializing housing via the rental sector in Spain and Ireland (Beswick et al., 2016; Gabor & Kohl, 2022; Waldron, 2018). REITs, which own and manage rental property and issue shares, have played a key role in the expansion of financialized PRS over recent years across many jurisdictions (Fuller, 2021). By issuing shares, REITs allow the incomes arising from spatially fixed residential property to be traded globally in the form of liquid financial assets (Fuller, 2021; Wijburg et al., 2018). Moreover, while REITs typically invest in property on a long-term basis, shares in REITs can be short-term investments.

The form of PRS financialization at play here relates primarily to apartment blocks in urban and suburban areas, and their mass acquisition by financial institutions over a few short years. The Irish example is instructive as there were effectively no institutional landlords prior to the GFC. Today, the country's largest landlord is IRES REIT, which was founded by Canada's largest landlord, CAPREIT, in the early 2010s to acquire assets from NAMA and other deleveraging institutions, and rapidly accumulated a portfolio of several thousand properties. Today, Ireland's institutional PRS is

---

2 At one point, Blackstone was the largest owner of repossessed homes in both the US and Spain (Beswick et al., 2016).

one of the most active markets in Europe. Remarkably, given the absence of institutional landlords prior to 2010, around ten per cent of Ireland's PRS nationally is held by institutional landlords, and twenty per cent in Dublin (Residential Tenancies Board, 2024).

Thus far, there has been little systematic empirical research examining the specific impact that this form of financialization has for tenants. While both countries have suffered significant rent increases in recent years, the specific role of institutional landlords, and how they may be different from other types of landlords, remains unclear. In the Irish context, there has been some research to show that institutional landlords have acquired a significant degree of market power (Daly, 2022) and that they increase monthly rents by 4.1 percentage points more than other landlords following purchase (McCarthy, 2024). Research into Blackstone's operations in Spain identified strategies such as increasing utility costs and strategic running down of tenancies through deliberate non-maintenance (Gil García & Martínez López, 2023; Janoschka et al., 2020). García-Lamarca (2021), also examining Blackstone, but this time in Barcelona, finds that Blackstone-owned rental companies set higher-than-normal rent prices and deliver poor tenancy management service to tenants.

Acquisition of formerly public rental housing

Spain has also experienced another one of the main forms of PRS financialization that has been identified in the literature: the acquisition of public or social rental housing (Janoschka et al., 2020; Yrigoy, 2021). But this form of financialization has been much more important in Germany, where it represents not just the most significant avenue of PRS financialization, but also one of the largest forms of housing redistribution ever undertaken. Unlike most of the cases evident in the international literature on PRS financialization, the German case does not originate in the years following the GFC, but instead in the context of the neoliberalization of German housing in the 1980s, 1990s and 2000s. The German housing system throughout the twentieth century and especially in the post-war era was characterized by a high degree of publicly subsidized housing, often provided by state and local government-owned companies, as well as trade unions and corporations. This form of housing, which was a core part of the country's 'social market' model (Wijburg et al., 2018), enjoyed significant public subsidies on the basis that it was let out at below market rents for thirty years (at least) after construction, under a principle known as 'common public interest' (Fields & Uffer, 2016). Berlin enjoyed particularly high levels of such housing. In the post-war era, in West Berlin over eighty-five per cent of new housing was publicly subsidized, and often provided by state-owned companies. Meanwhile, in East Berlin 'private landownership was almost entirely abandoned and state-led housing associations provided housing, which became part of Berlin's state-owned housing stock following reunification' (Fields & Uffer, 2016). By 1991, Berlin's local authority owned approximately 480,000 rental units through various publicly owned companies, 28 per cent of the city's housing stock (Fields & Uffer, 2016).

This aspect of housing policy was, however, abandoned in the 1980s, and during the 1990s and 2000s much of this housing was effectively privatized (Wijburg et al., 2018). At the national level, 'between 1999 and 2011, around 1.4 million housing units were sold off, almost 3.5 per cent of the entire housing stock in Germany', with the majority of these units sold to institutional investors such as UK- and US-based private equity firms and hedge funds (Wijburg et al., 2018). In the case of Berlin, this was motivated by the city's fiscal difficulties. The city has sold more than 200,000 rental units over recent decades (Fields & Uffer, 2016), representing a significant redistribution of the capital's housing stock. For example, in 2004 the US private equity firm Cerberus purchased one of Berlin's municipal housing companies, which owned 65,000 rental dwellings, thus becoming the 'largest landlord of Berlin overnight' (Wijburg et al., 2018). Vonovia, which is today one of Europe's largest landlords (it owns 400,000 units in Germany in addition to 60,000 units in Sweden and Austria, see Grander and Westerdahl, 2024), arose from the consolidation of two other institutional landlords, one of which acquired more than 60,000 rental properties from German Federal Railways (Wijburg et al., 2018).

Vonovia, as noted, has also become one of the largest landlords in Sweden, in fact becoming the country's largest landlord for a time in 2019 after it acquired two large institutional PRS landlords (Grander & Westerdahl, 2024). In a manner not dissimilar to the German example, post-war Swedish housing featured widespread housing provision via municipal housing companies (MHCs).

However, since the 1990s these have been increasingly subjected to commercial pressures and 'for-profit principles', one consequence of which has been the sale of former municipal housing to private actors, i.e. another example of mass privatization:

> From 2001 to 2018, about 20% (or more than 165,000 units) owned by MHCs in 2001 were sold (see Figure 1) The majority 95,700 of these 165,000 units (58%), were sold to private rental housing companies while 69,000 units (42%) were sold to sitting tenants. (Grander & Westerdahl, 2024: 6)

While much of this housing was originally purchased by local (i.e. Swedish-based) and relatively small companies, these have been increasingly acquired and consolidated by large international 'financial landlords', marking a rapid transformation of the country's PRS (Gustafsson, 2024).

The privatization and financialization of former public or social rental housing in countries like Germany and Sweden has been very controversial. In the German case, there have been widely supported calls for the return of much of this housing to the public sector (see Chapter 7). This is no doubt in part because the process of PRS financialization in these cases centres on former affordable and social housing, and the related role governments played (Wijburg et al., 2018). Beyond this, the actions of the institutional landlords who have acquired such housing have also come in for scrutiny, with research suggesting they implement strategies including failing to maintain properties, increasing rents and displacing tenants (Fields & Uffer, 2016; Grander & Westerdahl, 2024; Gustafsson, 2024; Wijburg et al., 2018). Due to the highly concentrated nature of their property ownership, these strategies can also play a role in processes of gentrification. One strategy that has been pursued in both Germany and Sweden involves upgrading properties to increase rents, often exploiting aspects of rent regulation. This type of investment strategy features more widely in the PRS financialization literature, as discussed in the next section.

Renovations and property upgrading

A fourth form of financialization, renovation-led investment strategies on the part of institutional landlords, has been identified in research on Denmark (Christophers, 2022a), Canada (August & Walks, 2018) and the US (Raymond et al., 2021), as well as featuring in the literature on the financialization of former public rental stock in Germany and Sweden. In the Swedish case, 'renovations of rental apartments have physically and socio-economically transformed the private rental housing stock' (Gustafsson, 2024). Renovations are used to upgrade apartment standards, and thus take advantage of a particular aspect of Swedish rent regulation, which is generally very stringent, that allows for increasing rents following renovations:

> These widely criticized 'luxury renovations', comprise various alterations that increase an apartment's standard of living, ranging from the installation of a dishwasher to a full upgrade. As Sweden's rental legislation allows for rents to be lifted when living standards are improved, these renovations lead to often substantial rent increases, not uncommonly more than 40 per cent. (Gustafsson, 2024)

Although Sweden has a longer tradition of domestic institutional landlords, in more recent years a significant proportion of the rental stock has been acquired by international or internationally backed 'financial landlords', such as Vonovia. Of course, rent increases may represent a desirable investment strategy in general, but Gustafsson (2024) argues that financial landlords are particularly interested in the uplift in capital gains, and therefore share price, that institutional landlords enjoy on the basis of increased rental income revenue. This is an interesting example of a specifically financial logic introduced via these novel investors. Christophers (2022a: 711), discussing the Swedish case but also the wider international picture, also notes that only very large financial institutions with enormous access to capital have the scale and financial power to acquire, *en masse*, residential housing and invest in renovations:

> Not many property investors are of a sufficient scale and financial wherewithal to undertake the combination of acquisition and largescale renovation that Blackstone pursued in say Sweden or the US; perhaps no others, meanwhile, would be big enough, not to mention motivated enough, to undertake the strategy in Sweden, the US, Berlin and Copenhagen more-or-less simultaneously.

Acquisition and renovation strategies are tailored to both national legislative contexts, for example the specific rent regulations in the Danish and Swedish cases, but also to local neighbourhood contexts and the path-dependent legacies of the built environment (Christophers, 2022a).

In the Danish case, Blackstone's investment strategy has come in for particular scrutiny. Between 2017 and 2019, Blackstone acquired approximately 2,300 rental units in Copenhagen, focusing on older housing in need of renovation (Christophers, 2022a). As in Sweden, rents are tightly regulated in Denmark, but undertaking renovations can release properties form the rent-controlled segment, thus enabling significant rent increases. This can also lead to the displacement of tenants (Christophers, 2022a). Blackstone's activity was strongly criticized in Denmark, including by the country's prime minister: 'An American private equity fund is purchasing our houses. Does greed know no boundaries?' (then Prime Minister Mette Frederiksen, quoted in Christophers, 2022a: 709). The Danish government went on to introduce a law, which became known as the 'Blackstone Law', that prohibited rent increases on renovated apartments for a period of five years.

On the other side of the Atlantic, Toronto's rental market experienced renewed activity of institutional investors in the years following the GFC:

> After 2009, a series of new and aggressive financialized landlords entered the Toronto market, rapidly acquiring buildings and applying sophisticated asset management strategies. From 2000 to 2015, approximately 42% of the apartment stock in Toronto traded hands, with ownership increasingly consolidated in the hands of REITs, institutional investors, private equity funds, and (still a few) large family-owned companies. (August & Walks, 2018: 128)

August and Walks' research finds two principal investment strategies associated with this wave of investment:

> In affordable buildings in lower income post-war suburban neighbourhoods, revenues are squeezed from tenants and buildings ... rent increases, increased ancillary costs, and building-wide efficiencies. With a reliable supply of desperate renters, landlords can be sure that tenants will absorb increased costs or be replaced, and can nudge rents higher upon unit turnover. The second general strategy is gentrify-by-upgrading- aggressively 'repositioning' buildings in coveted areas with strong markets, transforming them from affordable into luxury buildings ... In some cases, such a repositioning strategy is coupled

with a short-term investment focus and the flipping of repositioned buildings. (August & Walks, 2018)

The impact of renovation as a strategy of institutional investors is rather well documented, perhaps offering the best evidence of the potential negative impacts of financial landlords. As noted already, renovation can lead to higher rents for tenants, which can in turn trigger displacement (August & Walks, 2018; Gustafsson, 2024). Due to the concentration of ownership of rental stock in certain local markets, this can also be part of neighbourhood-wide gentrification processes (Gustafsson, 2024; Wijburg et al., 2018). Raymond et al. (2021), for example, found that neighbourhoods with investor purchases of multifamily residential properties had 33 per cent higher odds of an eviction spike in the same year, and that such neighborhoods have 166 fewer Black residents and 109 more White residents over a six-year period compared with adjacent neighborhoods without such purchases. Overall, then, it seems clear that '[l]imited access to affordable housing is a primary consequence of renovations' (Gustafsson, 2024).

Build to rent

Thus far all the forms of PRS financialization discussed relate to the acquisition of existing rental housing. The fourth, and final, form which will be addressed relates to the construction of purpose-built private rental housing, known as the build to rent (BTR) sector. This is perhaps the least-well-researched phenomenon related to PRS financialization. The relationship between BTR and the wider process of financialization has been primarily explored in contexts in which BTR represents a new development, notably the UK and Ireland (Nethercote, 2020). In such countries, private housing, including apartment housing, was traditionally constructed for the owner-occupier market. BTR represents a novel departure because it involves large-scale construction of multi-unit, or 'multifamily', properties which are constructed and designed to be owned and managed by a single large landlord. This contrasts with cases like Germany, where the new generation of financial landlords has not been significantly involved in new supply (Holm et al., 2023), and the US, where the BTR sector has been well established for several decades.

As Nethercote (2020) notes, in the case of BTR:

> [I]nstitutional investors retain single-ownership of large apartment complexes as revenue generating assets, with rental income and capital gains providing for potentially thin but secure margins. BTR typically entails large (50–100+ units) purpose-built rental accommodation with advocates emphasizing professionalized onsite management, hotel-style amenity and services, and more convenient and flexible tenancies ...

Proponents of the BTR sector argue that by enabling international capital to invest in new PRS housing, the BTR sector enhances the supply of housing (for the Irish case see Lyons, 2024). Because in many countries there is a widespread view that lack of supply is at the root of their housing crisis, supply-based arguments can be effective. In both Ireland and the UK, government action in the years following the GFC aimed to facilitate the emergence of the BTR sector backed by international capital. The UK government, for example, created a €1.1 billion BTR fund to this end in 2013, and legislated for the establishment of REITs in 2007 (Nethercote, 2020). In Ireland, as already noted in Chapter 3, legislation enabling REITs was passed in 2013, with the specific goal of attracting international capital in the context of the country's crippled banking system and property market (Waldron, 2018). As noted, the financialization of the PRS in the Irish case initially centred on the acquisition of distressed assets. However, as this market dried up, around 2018, institutional landlords moved towards the BTR sector and expanding their portfolios through new construction. This was also facilitated by government action, including specific reforms to planning requirements that allowed BTR operators to construct smaller apartments (Waldron, 2018). The Irish government's *Strategy for the Rental Sector*, published in 2015, highlighted the benefits of institutional provision of rental housing in terms of its capacity for 'tapping new sources of finance from institutional investors such as pension funds and Real-Estate Investment Trusts' as well as 'an enhanced level of professionalism to the estate management of housing projects, seeing residents as long-term "customers" rather than tenants' (see Byrne, 2021).

BTR is typically higher-end rental housing, and often includes 'hotel-like' amenities as well as the 'in-housing' of retail services. BTR developments may include concierges, gyms, meeting rooms and workspaces, laundry facilities and similar. They often brand themselves as providing a more professional and tenant-focused experience, often associated with a certain kind of lifestyle which might appeal to young urban professionals (Byrne, 2021a). Consequently, rents in the segment are typically at the higher end of the market.

The shift towards BTR can been situated as part of a wider shift towards what is sometimes known within the industry as the 'living' sector. In addition to BTR, this term encompasses a variety of forms of accommodation or housing which span the lifecycle of individuals and households, including in particular purpose-built student housing (Hubbard, 2009; Revington & August, 2020), co-living (Casier, 2024; Harris & Nowicki, 2020), and care homes (Horton, 2021). In the UK's care sector, for example, it is now estimated that REITs own and manage up to 30,000 beds (Horton, 2021). Institutional investors can thus be involved in the provision of accommodation from 'cradle-to-grave'. As with BTR, institutional investment in sectors such as co-living and care homes has been associated with a 'hotelization' and standardization of these types of accommodation, in an effort to produce what Horton (2021) conceptualizes as 'liquid homes'.

There appears to be limited research on the impact of the BTR sector on either tenants or cities. It seems likely that, given the focus on high-end rental property, the sector will tend to lead to higher average rents and is unlikely to contribute to the provision of affordable housing. From the tenants' perspective, however, it is reasonable to assume that such landlords can deliver a higher-quality tenancy management service, and can allow relatively high-income tenants to avoid the low level of professionalization and sometimes poor-quality dwellings that typify rental sectors in many jurisdictions. Beyond this, it is clear that further empirical work is needed in understanding specifically how the provision of new housing supply via this avenue impacts on rental markets, affordability, gentrification and housing insecurity.

Analysing the financialization of the private rental sector

The above discussion has shown that there are diverse forms of PRS financialization, with a host of different types of firms and varying business models. These firms operate, moreover, in national and local contexts which vary significantly, as they are determined by distinct housing policy regimes and also the path dependencies created by the form of built environment and the nature of housing markets (Holm et al., 2023). The literature on institutional landlords and PRS financialization is, moreover, reasonably new and neither the concept of PRS financialization nor the specific nature of the actors involved has been fully elaborated. Nevertheless, thanks to the literature that does exist, which covers a wide range of countries and case studies, we can identify a number of core trends associated with PRS financialization, which will help to clarify the nature and significance of this phenomenon. In this section I identify six of the core features of PRS financialization that emerge from the literature.

First, the internationalization of the PRS via direct ownership and management of PRS housing by international firms. This can be contrasted both with the historical norm in rental sectors across the advanced economies, traditionally owned by domestic operators, be they institutional or household. It can also be contrasted with the financialization of homeownership and mortgage markets in the 2000s. That process was in large part driven by international flows of credit and capital, for example in the mortgage-backed securities market, but the provision and ownership of housing itself remained at a more national and local level. In this case, banks or other forms of mortgage providers acted as intermediaries between households and international capital markets. In the case of the PRS, housing is being operated directly by firms which are owned or financially backed by international financial institutions, including REITs, private equity firms, hedge funds, pension funds and insurance firms. This necessarily involves the deep integration of a portion of the PRS into financial markets, and thus a more intense linkage between 'high finance' and 'the everyday life of households' (Gabor & Kohl, 2022; Santos, 2025). PRS housing is now subject to a variety of forces, such as currency exchange values, interest rates set by the Federal Reserve and the ECB, and the performance of competing assets such as government bonds, in a way which has simply not been the case heretofore. We know from the experience of financialization during the 2000s that the integration of housing and financial markets can lead to unappreciated risks to housing systems, financial systems and, most importantly, residents (Gabor & Kohl, 2022; Horton, 2021). Currently, the types of risks that might be generated by PRS financialization appear to be unknown.

Second, the financialization of the PRS is clearly closely related with a wider phenomenon already discussed in Chapter 3: the concentration of ownership of residential property. Given that for financial institutions scale is crucial (Byrne, 2016a; Gil García & Martínez López, 2023), they are associated with a particularly acute form of concentration (Santos, 2025). A notable related issue has been the tendency towards consolidation, via mergers and acquisitions, over recent years, which is evident both in the US SFR market, but also in markets such as Germany and Sweden. As the reader may have noticed already, one company that has featured throughout this chapter is Blackstone, which has at different points held vast amounts of residential property not just within specific countries, but across them (Christophers, 2022a). In terms of the €167 billion of residential assets held by private equity firms in Europe, the top four (Blackstone, Patrizia, Lone Star Fund and Amundi) account for an incredible fifty per cent of assets under management (Gabor & Kohl, 2022). Thus, while there is a general concentration of housing ownership among better-off households, the financialization of the PRS in the form of institutional landlords represents a kind of 'super-concentration'.

Third, the state has played a key role in the financialization of the PRS in many, although not all, jurisdictions (Beswick et al., 2016; Fields, 2018; Gabor & Kohl, 2022; Gil García & Martínez López, 2023; Holm et al., 2023; Waldron, 2018; Wijburg et al., 2018). State intervention has played a role both in terms of transformations to the housing system (as well as planning) and to the financial system. In terms of housing, and as discussed throughout this chapter, there are a number of ways in which governments have reformed or intervened in housing policy and provision to facilitate the financialization of the PRS. Perhaps the clearest is the case of countries like Germany and Sweden, which experienced direct sales of public and social housing to institutional landlords (although this also happened to some degree in other jurisdictions, such as Spain, see Gil García and Martínez López, 2023). In other cases, reforms of tenancy legislation have played a role. Gill and Marinez argue that the deregulation of both security of tenure and rent regulation in Spain, which took place via legislative reform in

2013, was aimed at facilitating financialization. In the case of US SFR, while there was direct state action in the form of bulk sales of SFR dwellings aimed at igniting the industry (Fields, 2018), it is also the case that what the state *didn't do* was equally important, specifically the lack of support for homeowners in distress to avoid the tsunami of foreclosures which made possible the rise of institutional SFR landlords (Christophers, 2022a). State action at the level of the financial system has been at least as important, if not more so. Examples include legislation to establish REITs, which occurred in the UK, Spain and Ireland between 2007 and 2013 (Fuller, 2021; Nethercote, 2020); and the mass sale of distressed property assets by public 'bad banks', especially in Ireland, Spain and the US (Byrne, 2016b). There are thus a variety of ways in which government action can underpin the financialization of the PRS. However, it should also be noted that the process has in some countries occurred without significant state action, with Denmark being a salient example. The state, of course, can also play a role in curbing institutional investment.

Fourth, as Aalbers (2017b) has argued in relation to housing financialization in general, the financialization of the PRS is a variegated process because while the actors themselves are often international, and financial markets are rather well integrated globally, housing itself tends to be heavily shaped by national and local housing policies, as well as local housing market conditions and path dependencies (see also Guironnet et al., 2024; Holm et al., 2023). Thus, while a similar set of factors relating to financial systems are salient across many jurisdictions, such as the low interest rate environment that pertained in the 2000s and 2010s, local factors have led to very different forms of PRS financialization. For example, in Germany and Sweden the business models of financial landlords were shaped by the selling off of public rental stock as well as the comparatively strong tenant protections that pertain in both countries. Recent work by Holm et al. (2023: 162), which compares seven European cities, is worth quoting at length, as it highlights not just the specific trajectories of PRS financialization in different countries, but also the potentially significant impacts this can have:

> Malmö and Berlin thus represent unitary rental systems with a strong history of state intervention into housing. In both cases financial

> investors entered local housing markets through mass-privatisation of state-owned or social housing. Usually, this privatisation was achieved by the sale of entire housing blocks to private investors ... At the same time, 'en bloc' takeovers were only possible exactly because of the very history of unitary rental systems in which the state has traditionally taken a considerable role in the housing provision. Ironically, this more centralized management structure has eased the acquisition of large packages of properties through financial investors when states decided to reduce their role in housing provision and sell their stocks ... As the building blocks sold to financial investors are geographically concentrated in specific neighbourhoods, these transactions have transformed institutional investors into key real-estate actors in these areas and often even in the respective metropolitan housing markets ...

In contrast, in a context like that of the UK, 'financial investors have entered residential real estate by developing alternative asset classes' and 'niche markets':

> These niche markets target the needs of very specific demand groups. Elderly homes, student accommodation, short-term-renting or co-living arrangements are typical examples here ... In these cases, institutional investors have not become dominant real-estate market actors, but they have introduced new products and new logics into existing local housing markets by developing new housing products, promoting new construction and transforming their land use. In most of the cases the focus is on expensive units rather than on providing affordable stock. (Holm et al., 2023: 162)

In other contexts, for example that of the Paris region, national and local conditions, as well as the fact that France was only impacted by the GFC to a limited extent, have resulted in relatively low levels of institutional investment (Guironnet et al., 2024).

Fifth, while the full impact of PRS financialization is not known, and even the full extent of this phenomenon is not fully understood (Holm et al., 2023), the existing literature indicates that the most significant impact of this shift in the PRS will be a reduction of affordable rental housing. However, the way institutional landlords contribute to unaffordability again differs in different contexts. In cases like Germany and Sweden the mechanism is rather straightforward, i.e. via the privatization of formerly affordable public and social housing. In cases like Ireland and the UK, the relevant issue is that institutional landlords are mainly focused on the higher end of the rental market. Meanwhile, in many countries the salient phenomenon centres on renovation, upgrading and gentrification. More generally, there is evidence that institutional landlords set higher rents, likely related to the scale of their investment, its geographical concentration, and hence the market power they may exercise in setting rents (Fields & Vergerio, 2022; McCarthy, 2024).

Sixth, institutional landlords may be associated with further undesirable outcomes from a tenant's perspective, including higher levels of evictions (Gomory, 2022), displacement (Gustafsson, 2024), unresponsiveness and poor maintenance (August & Walks, 2018; Fields & Vergerio, 2022; Janoschka et al., 2020; Wijburg et al., 2018), and gentrification (August & Walks, 2018; Raymond et al., 2021; Wijburg et al., 2018). However, the extent to which these outcomes are characteristics of institutional landlords in general, or to which they are more systematically associated with institutional landlords compared with other types of landlords, is not entirely clear from the existing research.

Seventh, and finally, institutional landlords, again due to their scale, may introduce new dynamics in terms of their interaction with the policy and political process, i.e. lobbying (Card, 2024; Holm et al., 2023). Firms who own very large volumes of rental housing obviously have an extremely strong incentive to attempt to influence the policy process, and of course are likely to have substantial financial firepower with which to pursue those aims. This is not the only way that PRS financialization might shape politics, however. There is increasing evidence that the financialization of the PRS is producing a backlash, with tenants organizing and campaigning and policy-makers sometimes directly targeting institutional investors. This will be discussed in greater detail in Chapter 7.

5
Beyond Generation Rent?
Housing Inequality and the Private Rental Sector

The critical analysis of housing inequality is almost as old as the discipline of political economy itself. Perhaps its earliest expression is Engels' (1993) famous work *The Condition of the Working Class in England*, in which he highlighted not just the poor housing conditions experienced by Manchester's workers, but also the inequality in access to and experience of housing they experienced when compared with the 'bourgeoisie'. Engels thus noted that inequality in the relations of production, and consequently the labour market, generate structural inequalities in terms of housing. The commodification of housing plays a central role in this process, moreover, because it means that housing is allocated according to price, and therefore unequally distributed according to incomes. This inequality has a spatial dimension in that housing is segregated according to class in different neighbourhoods. Thus, the parallel commodification of both labour and housing associated with the development of industrial, capitalist cities inevitably generates housing market segregation of city-dwellers along class lines in different neighbourhoods. Du Bois, another early analyst of housing and urban inequality, further nuanced the analysis of housing inequality by pointing out how structural racial inequality interacts with class and the nature of housing markets to produce racialized modes of housing inequality (see Loughran, 2015).

Today, housing inequality is back on the top of the agenda, both in terms of housing research and in wider public debate. 'The housing question', as Rosenthal and Vilchis (2024: 12) argue, 'is, at its centre, a question about inequality: who gets to be housed and who doesn't, who profits from housing and who falls into poverty paying or failing to pay the rent'. Some of the reasons for this were already touched upon in Chapters 2 and 3. In Chapter 2, I argued that rent is inherently founded on an unequal property and social relationship, in which the extraction of value and power play key roles. Inequality, from this point of view, is part of the DNA of the PRS. Moreover, in Chapter 3 we have seen the restructuring of housing systems, and specifically the revival of the PRS, is associated above all with the concentration of property ownership, which in turn is linked to new forms of social stratification. In other words, housing inequality appears to be integral to the ways our housing systems are changing.

Housing inequality, however, has also risen to the top of the agenda because its negative effects on residents are increasingly evident, as is the growing evidence that it disproportionately affects private renters. Housing unaffordability, and especially high rents, have been subject to political discussion across many of the advanced economies, and a number of countries have recently introduced, or strengthened, rent regulation (Marsh et al., 2023). Homelessness is also of growing concern, as are issues like poor-quality accommodation and overcrowding. Many of these issues became particularly salient in the context of the Covid-19 pandemic – as people had to spend much more time at home, the highly unequal experiences of home that characterize contemporary housing systems were thrown into sharp relief (Brown et al., 2020; Byrne, 2020b, 2021b). It is not surprising, indeed, that a new generation of tenants' movements has sprung up across a whole host of countries (see Chapter 7 for further discussion). There is now a wide body of research shedding light on the relationship between the various facets of housing inequality and the experiences of private renters. One of the main aims of this chapter is to summarize this research and distill some of its most valuable insights both in terms of the extent, but also the nature, of housing inequality today as it relates to the PRS.

In popular framings, especially in the English-speaking world, housing inequality has been interpreted as a generational clash, as evident in the widely used term 'generation rent' (Howard, 2025; McKee, 2019). This simplistic 'boomers versus millennials' analysis, however, obscures more than it reveals, as will be discussed below. But within academic debates there has also been considerable confusion about how best to conceptualize contemporary housing inequality and how it relates to the PRS. In particular, academics have critiqued both generationally based analyses of housing inequality, as well as those based on tenure (e.g. based on the dichotomy of owning versus renting). The complexity of the relationship between tenure, generation and indeed class is made all the more confusing by the fact that we currently lack a clear framework for defining and analysing housing inequality (James et al., 2024).

This chapter focuses on the economic dimension of housing inequality, which encompasses affordability (the unequal distribution of housing costs as a proportion of income), wealth inequality (the unequal distribution of residential property assets) and income inequality (the unequal distribution of income arising specifically from the payment of rent). Each of these forms of inequality relate to housing as a good (i.e. something which is produced and exchanged for money) and an asset (a store of value over time). Housing inequality, of course, goes beyond this, and the crucial issue of housing insecurity in the context of the PRS is addressed in detail in the next chapter. The second part of the chapter engages with contemporary academic debates on how best to conceptualize the relationship between housing inequality and the PRS, particularly in relation to debates around the validity of tenure and generation as analytic lenses. It argues for a conceptual framework that centres on class, generation and tenure, and how they relate to each other.

Housing inequality: Affordability, wealth and income

Housing affordability and inequality

The relationship between economic inequality and the PRS is typically addressed through the lens of affordability. There is ample evidence that housing affordability, in terms of both house prices and rents, has been declining in recent decades. A recent OECD study, for example, claims that as house prices have risen in the past two decades in most OECD countries, households are spending an increasing share of income on housing costs (OECD, 2021, see also Housing Europe, 2022). While some studies suggest caution in relation to a generalized decline of housing affordability

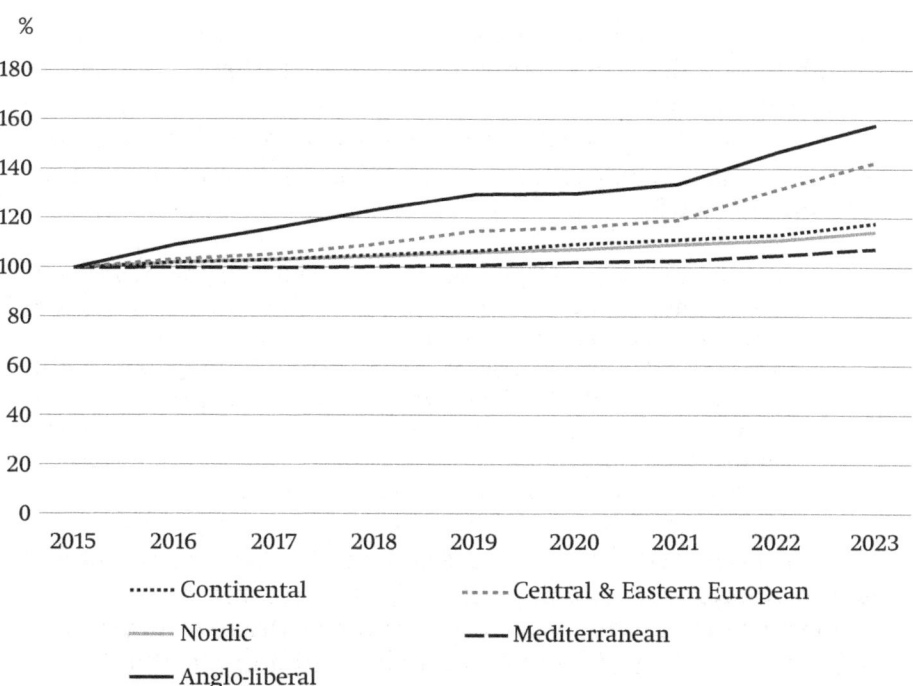

Figure 3 Average rent increases, selected EU countries
Source: Eurostat, 2023. Selected countries have been loosely grouped together based on housing regime: Nordic (Denmark, Sweden, Norway), Continental (France, Germany, Austria, Netherlands, Belgium), Mediterranean (Spain, Greece, Italy, Portugal), Central and Eastern European (Poland, Slovenia, Slovakia, Estonia, Lavia, Lithuania, Czech Republic, Hungary, Bulgaria, Romania), Anglo-liberal (United States and United Kingdom)

(Hick et al., 2024), what is clear is that there are some cohorts who are certainly experiencing declining housing affordability, and private renters are chief among them.

Affordability is, of course, an important concern because it tells us a lot about how easy, or difficult, it is for households to access housing. Where housing is unaffordable, it means that it is likely that many households will (a) be unable to access quality housing in a location they chose, and (b) struggle to afford non-housing expenses (Meen & Whitehead, 2020). Thus, 'deteriorating housing affordability might lead to the intensification of poverty' (Hick et al., 2024: 2), as high housing costs are linked to being unable to purchase food, healthcare, transport, childcare, etc, and also increase the likelihood of homelessness (Dong, 2018), with low-income households especially vulnerable (Pittini, 2012). There is also a well-developed literature demonstrating that unaffordable housing negatively impacts the health and mental health of families and children (Arundel et al., 2022; Gold, 2020; Sandel et al., 2018).

One of the most detailed recent studies on rents and affordability in the European context (Hick et al., 2024: 13) shows a number of concerning trends:

> [M]arket renters have higher odds of cost overburden than mortgaged homeowners in most countries ... [In] most countries, the relative position of market renters has deteriorated between the years 2010 and 2018. The relative risk of market renters being poor is much greater than mortgaged homeowners ... and there is some increase over time ... Thus, in terms of *both* housing cost burdens and low incomes, the position of market renters has deteriorated relative to mortgaged homeowners over the period 2010–2018.

This is supported by other research in the European context, which shows that housing unaffordability is far more severe among private renters than other tenures (Disch & Slaymaker, 2023; Inchauste et al., 2018), and that 'declining affordability of housing for private renters is a common trend across Western European countries' (Dewilde, 2018). Indeed, in two thirds of countries, housing affordability for low-income private renters declined between 1995 and 2013. The countries in which the proportion of low-income urban renters with housing affordability problems grew the most between those years include Ireland (40% to almost 70%), Spain (just under 45% to almost 80%), Portugal (just over 20% to 75%) and Belgium (30% to just over 60%) (Dewilde, 2018, see Figure 2).

Evidence from the US also clearly shows the declining affordability experienced by low-income renters. Average rents are 21 per cent higher in 2022 than they were in 2001 (inflation adjusted), while the income of renters has risen by just two per cent over the same period (JCHS, 2024). Moreover:

> At last measure in 2022, a record-high 22.4 million renter households spent more than 30 percent of their income on rent and utilities ... Among cost-burdened households, 12.1 million had housing costs that consumed more than half of their income, an all-time high for severe burdens ... As a result, the share of cost-burdened renters rose to 50 percent ... (JCHS, 2024: 2)

Schwartz (2023: 49) notes that '[n]early half of all renters in 2021 were cost burdened, spending more than thirty per cent of their income on rent and related expenses', and that '… median renter incomes increased by only 0.5% from 2000 to 2018, while the median gross rent … increased by seventeen per cent [adjusting for inflation]' (Schwartz, 2023: 50).

Housing affordability is thus unequally experienced by renters (as opposed to owners), with research showing how this leads to poorer outcomes in terms of health, wellbeing and housing quality issues (Baker et al., 2014; Bentley et al., 2011; Kearns et al., 2000; Mason et al., 2013). Low-income tenants are particularly vulnerable to rental unaffordability, but other forms of socio-economic inequality, such as race, ethnicity and migration status, also feature prominently. In the US, '[r]ent burdens fall heaviest on Black and Latino renters, whose share of rent-burdened households is more than 10 percentage points higher than that of White renters' (DeLuca & Rosen, 2022: 346). Migrants and ethnic minorities are also more likely to be in rental housing, and suffer from issues of affordability and access (Lukes et al., 2019). Research from Australia (Andersen et al., 2018) and New Zealand (Goodyear, 2017) shows that the same applies to indigenous peoples.

The issue of affordability is linked to inequality between renters and owners, as the former are increasingly experiencing a disproportionate housing cost burden. Returning to the themes of the last chapter, we might say that, if the growth of the PRS is associated with declining access to both social housing and homeownership, the economic penalty associated with that exclusion is growing (Christophers, 2020). Housing and housing costs are unequally distributed, in other words, according to income and tenure, and the PRS plays an especially important role in understanding how this plays out.

Wealth inequality and the PRS

The relationship between the PRS and economic inequality goes significantly beyond questions of affordability. As has been extensively documented, the distribution of residential property is a key determinant of wealth inequality (Arundel, 2017; Kholodilin & Kohl, 2023). Due to the increasing importance of real-estate assets, and the sustained growth in prices in most advanced nations, housing wealth has become increasingly central to wealth distribution:

> A distinguishing feature of the past half century is the unprecedented accumulation of wealth into residential property through a long wave of house price appreciation that was, for the first time, global in reach and synchronized across jurisdictions. This turned residential property into the largest capital asset in the investable economy, exceeding the combined value of equities, commercial property, agricultural land, forestry and all gold ever mined. Housing accounts for the majority of the twenty-first century rise in total private wealth, for the lion's share of total return on aggregate wealth. (Smith et al., 2022)

Housing wealth not only plays a key role in wealth inequality, it has also come to play a much more central role in the wider political economy, as 'residential real estate is one of the defining characteristics of the current regime of accumulation' (Smith et al., 2022). This is evident both in the proportion of wealth held in the form of real-estate assets, but also in the role of residential real-estate assets in fuelling demand and economic growth (Brenner, 2006). Indeed, it has been argued that in many countries houses 'often appreciate by far more in a given year than it is possible for middle-class wage earners to save from wages' (Adkins et al., 2021), or, more provocatively, 'homes earn more than jobs' (Ryan-Collins & Murray, 2023).

In general, societies with lower levels of homeownership have higher levels of wealth inequality, as for many households housing will be the only form of wealth they hold. Renters typically have very low levels of wealth (Kholodilin & Kohl, 2023). Over recent decades, unsurprisingly, the decline of homeownership combined with the growth of house prices has played a decisive role in increasing wealth inequality (Arundel, 2017). Smith et al.'s (2022) analysis of Australia, the US and the UK between 2011 and 2017 shows that as homeownership has declined in these typical 'homeowner societies', wealth inequality has increased overall, and housing wealth has become more concentrated among wealthier and older households. Today, the top ten per cent of wealthy households in the UK, Australia and the US hold 39, 45 and 53 per cent, respectively, of all housing equity. Similarly in Ireland, another former homeowner society with declining rates of homeownership, the wealth divide between homeowners and renters is stark: 97 per cent of Ireland's wealth is held by homeowners, while just three per cent is held by renters (despite the fact that renters account for 30% of households). The average net wealth of Irish homeowners in 2020 was €303,900, while for renters the figure was just €5,300 (Moreno, 2024). Housing wealth inequality is not just about homeowners and renters, however. Adkins et al.'s (2021) research, moreover, highlights the concentration of housing wealth specifically among landlords, who are both disproportionately high-income households, as well as owners of multiple residential properties (on the Dutch case, see Hochstenbach, 2022).

What should not be forgotten in all this is that housing wealth is linked to a much wider variety of forms of inequality. As Zavisca and Gerber (2016: 350) note:

> Like other sources of wealth, housing could influence life chances by facilitating consumption smoothing, longer time horizons, and intergenerational transfers. Homeownership could amplify the stratifying effects of wealth that have been identified in domains such as education, living standards, and happiness.

The absence of wealth exposes households to greater future risk and uncertainty, and undermines their ability to respond to adverse events. The wealthlessness of renters can thus be seen as a form of precarity.

Income inequality and the PRS

While we have seen that greater levels of wealth inequality arise from the decline of homeownership, income inequality is another, albeit much less studied, aspect of the relationship between the PRS and economic inequality. Focusing on income inequality homes in more specifically on the immediate economic relationship between landlords and tenants, rather than focusing on inequality between homeowners and non-owners, as the housing wealth inequality literature typically does. One way to open up this interesting, and novel, research area is by examining the impact of rent controls on economic inequality, as Kholodilin and Kohl (2023) do in a fascinating cross-national study. Their research finds that the introduction of 'hard' (or first-generation) rent controls around the early and mid twentieth century is associated with declines in economic inequality. In other words, lower rents mean lower economic inequality.

Kholodilin and Kohl (2023) identify three ways in which rent levels effect economic inequality. First, rent levels are linked to house prices, and thus wealth inequality. This is especially the case when considering landlords, who typically own their own primary residence, as well as additional rental properties. Second, as landlords are mainly high-income households (Hochstenbach, 2022; Ziegelmeyer, 2015), the receipt of rental income adds to income inequality. Third, because housing costs are typically the largest aspect of household expenditure, and because poorer households tend to spend a greater proportion of their income on housing (known as 'Schwaube's law'), the payment of rent significantly reduces the non-housing disposable income of tenants. Thus, rent controls limit 'the incomes and real-estate values of generally richer landlords while simultaneously increasing the disposable incomes of generally poorer tenant households' (Kholodilin & Kohl, 2023: 170). Kholodilin and Kohl (2023) find that rent controls are negatively associated with rent increases, rental expenditure, income inequality and wealth inequality, and that this 'is mainly because landlords are rich and tenants are poor, such that rent control acts as a channel for redistribution'.

If lowering rents can redistribute rental income away from (richer) landlord households and towards (poorer) tenant households, it follows that rent increases over recent years have redistributed income from tenants to landlords, and thus increased income inequality. Although this is not a widely studied area, recent evidence suggests that this is indeed the case. For example, in the US, where economic inequality is higher than at any point in the past four decades, research has shown that income inequality is linked to rental unaffordability. Looking at the period 2008–12, an increase in the Gini coefficient of 0.1 in a given area was associated with 4.4 percentage points more severely rent-burdened low-income households (Dong, 2018). Similarly, looking at 28 European countries, Dewilde and Lancee (2013) show that an increase in income inequality increases the likelihood of affordability issues for low-income renters.

Recent research from Germany is instructive, as Germany has a particularly low homeownership rate (44%, see Kindermann & Kohls, 2018) and therefore a large rental sector. In terms of income inequality, Dustmann et al. (2022) show that increases in housing costs have amplified income inequality over time. The share of income spent on housing has increased for lower-income households, while for high-income groups it has declined, such that 'the increase in ... inequality in net household incomes between 1993 and 2013 is almost three times as large once housing expenditure is considered' (Dustmann et al., 2022: 1733). Thus, 'the increase in real housing expenditure vastly amplified the effect of the loss in real income' (Dustmann et al., 2022: 1710). Moreover, Bartels and Schröder (2020) show that the proportion of households who are landlords increased significantly between 2002 and 2017 across Germany (to about 10% of households),[3] and the average annual income from rental property also grew over this period, to €12,000 for landlords in cities with more than 100,000 population. Therefore 'the importance of rental income for explaining overall income inequality increased' (Bartels & Schröder, 2020: 14). This pattern also applies to Europe more widely (Kholodilin & Kohl, 2023). While €12,000 may not seem like a large amount of money, consider that the *average* household disposable income in Germany was €25,000 in 2017 (Bartels & Schröder, 2020). Moreover, households who are landlords are more likely to have high incomes outside of their rental income, and thus we can see that the rental income they receive can make a significant contribution to income inequality, and this contribution increased between 2002 and 2017.

More recent research further highlights the property-based inequality between landlords and tenants in the German context. Kadekle (2023) finds that landlords are much more likely to be older (over 50), but more importantly also have higher incomes and more wealth. The household income of landlords, including rental income, is almost twice that of tenants in Germany, and 1.6 times the overall population average, and 'the odds of landlords being in the wealthiest decile are over 100 times higher than tenants' odds' (Kadelke, 2023: 70). Moreover, on average, tenant

---

[3] Contrary to the sometimes common view of the German PRS being dominated by institutional landlords, small-scale landlords control two-thirds of its rental housing stock (Kadelke, 2023).

households in Germany transfer more than a quarter of their income to their landlords (Kadelke, 2023). Thus, Kadelke (2023: 70) argues that 'landlords and tenants sit on opposite sides to each other ... especially economically' and that '[h]igher rents boost landlords' profits while making it more difficult for tenants to build up (intergenerational) wealth' (Kadelke, 2023: 67).

Thus far we have focused on the distribution of income and wealth between landlord households and tenant households. There is, however, another channel through which income and wealth inequality is linked to rental housing. Although it has not, to my knowledge, been studied, rental income also flows to wealthier households via their ownership of shares in property companies, REITs or via pensions. As Kholodilin & Kohl (2023: 181) argue:

> Funded privatized pensions have increasingly found residential real estate a promising investment in times of decreasing yields on government bonds and other safe assets. In extreme cases, such as Switzerland with its more than 60% tenancy rate in the population and one of the most privatized pension systems, tenants essentially provide a private pension to the nation's retirees by paying large shares of their income in rent.

This is one way in which the financialization of rental housing plays a part in the wider story of rental housing and economic inequality.

Summarizing the relationship between economic inequality and the PRS, we can say that rental affordability makes it harder for lower-income households to access quality housing, thus driving the unequal distribution of housing itself, but is also part of a much wider story of economic inequality. The growth of the PRS, and the trend towards higher rents and declining affordability, is a driver of both wealth and income inequality. In terms of income inequality, high rents reduce renters' ability to afford their non-housing expenses, thus potentially driving them into poverty. But it is also crucial to recognize that high rents also increase the concentration of income among wealthier households, because the rental spending of renters is a form of income for landlords. Indeed, 'by extracting higher rents, multiple property owners enhance their own wealth accumulation while increasing rent burdens among tenants – thus obstructing them from accumulating wealth' (Hochstenbach, 2022). In other words, landlords are getting richer *because* tenants are getting poorer. Christophers (2020: 373) argues that '[n]othing today bespeaks exclusion from the "common wealth" more than being locked out of homeownership – and nothing bespeaks the likelihood of remaining excluded more than paying half of one's income in rent'. As Kadelke (2023: 67) notes, in the context of rental housing we must 'emphasize the relational character of inequalities. Landlords and tenants are not just members of different social groups who wave to each other in a friendly way.' Or as Rosenthal and Vilchis (2024: 21) put it, somewhat more polemically:

> The third of Americans who rent their housing make these payments to a handful of corporations and the mere 6.7 percent of the population who own that housing. This is a transfer of wealth from over 100 million tenants to just over 11 million landlords. The poorest Americans are overwhelmingly tenants; the richest own real estate. Rent is an engine of inequality. If you've played the board game Monopoly, you understand the idea: a roll of the dice and a purchase allow you to extract rents until everyone else is bankrupt.

From this perspective, the problem of rental unaffordability should be reframed as one of income redistribution.

Understanding housing inequality through the lens of class, tenure and generation

In the above discussion of housing inequality, we have seen that there are three key variables that seem to be central to the relationship between inequality and the revival of the PRS: age/generation, class and tenure. The relationship between, and relative salience of, these three ways of thinking about inequality has been subject to academic debate. This section attempts to untangle this debate and to conceptually clarify the relationship between age, class and tenure in understanding contemporary housing inequality.

I begin by arguing for the importance of tenure, and the PRS in particular, as a way of thinking about housing inequality. The utility of tenure has recently been subject to question within academic debate. Zavisca and Gerber (2016: 348), for example, critique the tendency to limit analysis of housing to 'dichotomous measures of housing tenure', i.e. the dichotomy between owning and renting. Zhang (2023: 3) argues that 'the enormous role of housing tenure in structuring our knowledge about housing can hardly be overstated'. He also notes that:

> Notwithstanding the importance of housing tenure, the concept per se has rarely been made clear. Rather, it has largely been taken-for-granted as discretely categorized, consisting of owning, renting, and a myriad of other tenures. These tenure categories are perceived as self-evidently the most logical way of grouping housing for the purpose of analysis. (Zhang, 2023)

Zhang (2023) further argues that this dichotomy obscures our understanding of housing today, and proposes a reformulation of how we think about systematic differences in housing experiences in terms of the relationship of households to the financial system, thus drawing on the recent literature on financialization. For example, heavily indebted homeowners can be just as precarious as renters, and experience the associated impacts of housing insecurity and eviction, as demonstrated in the wake of the Global Financial Crisis (García-Lamarca & Kaika, 2016). This re-conceptualization, it is argued, 'stretches across the owning-renting divide to enable a holistic perspective' (Zhang, 2023). In this sense, Zhang builds on recent literature examining the 'edges of homeownership', which looks at the ways in which heavily indebted homeowners and renters have much in common, and both can be understood as sharing a common experience of precarity (Haffner et al., 2017). Conceptualizing inequality through the lens of the PRS has also been seen as somewhat 'Anglo-centric', as it reflects the homeownership norm and weakly regulated rental sectors of the English-speaking world (Howard, 2025; Howard et al., 2024).

These debates should give us pause for thought when considering the value of a tenure-based, and specifically PRS-centred, approach to analysing contemporary housing inequality. However, there is a robust rationale for a tenure-based focus, although, in light of the above noted objections, it does require clarification of the conceptual grounding of such a focus. There are three core arguments that support a focus on private renting as a tenure in the analysis of housing inequality. First, many core aspects of housing which are of value to households, and as such impact wellbeing, are unequally distributed between tenures, especially between private renters and homeowners. It is true, as argued in much of the literature, that housing outcomes are not neatly divided between tenures, and that heavily indebted and low-income homeowners can experience much the same issues as low-income renters. There are thus grounds for some caution here, but nevertheless the consistency of the findings, across many jurisdictions, that private tenants have worse affordability outcomes, higher levels of housing insecurity, and lower levels of wellbeing and residential satisfaction, makes its own argument for the importance of tenure distinctions. These negative outcomes are not confined to private renters, and the lines between the tenures are blurred, but at the very least they suggest that tenure should continue to be one of the lenses through which we analyse housing inequality.

Second, and more importantly, many of the issues that are disproportionately experienced by renters are directly related to the PRS *as a tenure*. Wealth inequality is the most obvious of these, as the difference between renting and purchasing an asset, which is the core difference between the two tenures, is central to understanding this form of inequality. Security of tenure, discussed in much greater detail in the next chapter, is another such issue, as in all countries that I am aware of, property owners have stronger security of tenure than renters. Even in those countries that have the most well-regulated rental sectors, security of tenure is weaker for renters, and in some cases is becoming more insecure (e.g. the Netherlands, see Huisman, 2020), and there are particular renter cohorts that experience acute insecurity, such as those on very low incomes (e.g. Germany, see Soederberg, 2018) and those living in informal and subletting segments of the PRS (e.g. Sweden, see Listerborn, 2023). Moreover, at least in some of these countries, ownership is increasingly polarized in terms of age and class (Howard et al., 2024).

This brings us to our third and final argument. The PRS as a tenure is not only a valid, but a crucial analytical lens through which to understand the political economy of housing because it is characterized by *distinctive property and market relationships*, and these are central to the forms of housing inequality experienced by renters. The PRS differs from the other major tenures (homeownership and social housing) in that marketized relationships of housing provision are central to the experience of renting privately (McKee & Harris, 2025). For homeowners, marketized relationships pertain primarily at the point of purchase. However, once householders have entered homeownership, they obtain a degree of autonomy from market relationships. Saunders, discussing the work of Rose in his classic article on housing and class, points out that the relative autonomy from marketized relationships provided by homeownership helps to explain why homeownership aspirations are often prevalent among the working class:

> [W]orking-class aspirations for home ownership were in the nineteenth century a response to the increasing erosion of personal autonomy by capitalist social relations, and thus represented an attempt to reestablish a 'separate sphere in the sense of seeking out, and through the fabric of everyday life, a distinct cultural space for gaining as much control as possible over the purpose and direction of our lives'. (Saunders, quoting Rose, 222)

There is no doubt that the financialization of homeownership has transformed the tenure, especially for those on low incomes or who are exposed to interest-rate shocks and property busts. However, within homeownership this is a relatively unusual outcome. Similarly, while social or public housing tenants experience power relations vis-à-vis their landlord, and can have some of the worst housing outcomes, their relationship with their landlord is not marketized. In contrast, within private rental, tenants depend on a fully marketized relationship for the provision of their home every day of their tenancy and for every issue that arises, from needing a washing machine replaced to having repairs done. Private renting, then, represents what we might call the 'total subsumption' of housing within marketized social relationships.

This 'total subsumption' stems from the fact that, as argued in Chapter 2, the landlord retains ownership of the property within which the tenants create their home. A number of crucial things follow from this. It means that the tension between the housing provider and the resident is 'baked in' to the experience of private renting, frequently leading to disputes, housing insecurity and stress for the tenant (Byrne & McArdle, 2022; McKee & Harris, 2025). It also means that there is a core, unequal and objective property relationship at the heart of the experience of private renting that is distinct from other tenures. Tenants thus share a collective, and objective, relationship to residential property, as well as a collective, although of course varying, subjective experience of that relationship. Note here a key point: whereas the critiques of a tenure-based focus typically focus on housing *outcomes* to argue that tenure is over-emphasized, I am emphasizing here housing and property *relationships* as a way of conceptually clarifying the analytical utility of a tenure-based perspective.

Moving on from the issue of tenure, another aspect of how contemporary housing inequality is understood which has been subject to much critical commentary is the issue of age and generation, and especially the term 'generation rent' (Christophers, 2017; Howard, 2025; Howard et al., 2024). These criticisms have focused on a number of ways in which a generational focus can obscure the most important political economic processes associated with the decline of homeownership and housing inequality today (Howard, 2025). Firstly, it has been argued that differences within generational cohorts (intra-cohort inequalities) can be more salient than those between generations (inter-cohort differences) (Forrest & Hirayama, 2015; Howard, 2025). This is especially the case for class differences, as income levels are very closely related to ability to access homeownership (McKee, 2012). Thus, while there may appear to be a general exclusion from homeownership for younger households, in the medium term better-off households will in fact be able to transition to homeownership. From this point of view, it can be argued that a generational lens, which has proved particularly popular in the media, not only obscures the crucial class dynamics that drive housing inequality, but also risks depoliticizing housing inequality by reducing it to a 'boomers Vs millennials/Gen Z' framing (Howard, 2025; Stringer, 2025).

Secondly, treating generations as separate and competing cohorts obscures the obvious fact that they are, quite literally, related to one another (Christophers, 2017). This is salient because older generations can transfer savings and property assets to younger generations, thus enabling *some* younger households to acquire residential property (Ronald, 2018). Moreover, the so-called 'bank of mum and dad' phenomenon can allow those who benefit from it to acquire residential property with lower levels of debt than would otherwise be possible, which of course has a decisive impact on their housing costs and household wealth (Köppe, 2020). Thus, inter-generational relationships do not only take the form of competition, and can reproduce housing inequality along class lines. Inter-generational and intra-generational inequalities are, in other words, related, and reproduce one another (Ronald, 2018).

Thirdly, and finally, from a more analytical perspective, while housing can be distributed very unevenly across ages, it does not follow that age is the salient variable, or that age-related processes are the salient processes, from the point of view of understanding how this inequality arises and is sustained. Christophers (2017) points out that many economic resources are concentrated among older households. For example, CEOs tend to be much older than entry-level employees. Yet we do not see inequalities of capital, income or occupational position as *arising from age* or taking the form of generational inequality. Instead, we use concepts like class because these capture the salient political economy processes that help us to understand these forms of inequality, such as ownership of capital or possession of valuable knowledge, skills and qualifications.

All of these arguments offer a crucial corrective to the simplistic media narrative around 'generation rent', and can help us to formulate a clear conceptualization of the forms of housing inequality discussed in the chapter. However, if we focus solely on class, we will also omit something crucial from our analysis. Specifically, generation and age must remain important parts of our analysis, and not only because the distribution of ownership of residential property (both homeownership and landlordism) is skewed towards older households. Generation is salient to understanding housing inequality because the transition between historical housing regimes is a crucial political economy process, an understanding of which is required if we are to explain the nature of housing inequality today. In other words, the shift from relatively inclusive housing regimes, as discussed in Chapter 3, characterized by wider access to social housing and homeownership, towards more exclusive housing regimes, characterized by reduced access to social housing and a greater concentration of ownership of residential property, is central to the phenomenon we are trying to understand. Moreover, because housing has been, especially in the latter part of the twentieth century, an incredibly well-performing investment asset in most advanced economies (Adkins et al., 2021), the fact that housing accumulates value *over time* goes a long way to explaining why generational cohorts who acquired housing under favourable conditions *in the past* have higher levels of wealth today. Thus, there is a *temporal dimension* to housing inequality which manifests as generational differences. Adkins et al. (2021) argue that '[w]hat distinguishes successive generations, then, is less a difference in absolute wealth holdings than a difference in modes of access to wealth'. We can rephrase this to highlight the significance of *different modes of access to housing across time*. One way of thinking about this is to argue that the term 'generation rent' is unhelpful if it is used to manufacture a simplistic competition between generations, imagined as homogeneous and separate entities, but does have political and analytical utility if it is used to name the transition to a fundamentally more unequal housing system.

The generational perspective, then, is not an obstacle to understanding that must be overcome, but something that must be incorporated within, and related to, analyses of class and tenure. Thus, I argue that generational housing inequality should be understood as interacting with, and being mediated by, class, inter-generational transfers and tenure (Ronald, 2018). Thus, I propose a framework for understanding contemporary housing inequality based on the three dimensions of class, tenure and generation. Generation captures the historical modes of access to housing (i.e. the modes that pertain to different housing regimes as they change over time). Class determines, within a given housing regime, ability to access housing, and the conditions under which it is accessed (e.g. the level of debt taken on, or the proportion of income spent on housing, the value, quality and location of the house that is accessed). It also shapes the crucial inter-generational transfers that allow access to housing to be reproduced along class lines through different generations (Ronald, 2018), and therefore across different historical housing regimes. Tenure captures the distinctive property and social relations that constitute the different tenures, and the PRS in particular, and thus shape the nature, affordability and experience of home.

This chapter has examined housing inequality today and its relationship to the PRS, demonstrating the far-reaching impacts of the recent growth of the PRS in terms of the economic dimension of housing inequality. The evidence discussed throughout this chapter presents a compelling case that the growth of the PRS, in combination with increasing affordability issues is a significant driver of inequality today and, as such, should be a major concern for policy-makers and activists. This further highlights the distinction set out in Chapter 3 between the inclusive housing regimes associated with the mid and late twentieth century, and the exclusive regimes associated with the era of neoliberalism and financialization. In spite of criticisms of tenure and generational analyses of housing inequality, the chapter has argued that these lenses, along with class, are fundamental to understanding contemporary housing inequality, especially as it relates to the PRS.

6

Making a Home in Someone Else's Asset

Insecurity and Power in Lightly Regulated Rental Markets

The previous chapter focused on the economic dimensions of housing inequality, such as affordability and wealth and income inequality. There is, however, more to housing and home than this economic dimension. Such a narrow, and somewhat economistic, focus can obscure some of the most important forms of housing inequality (DeLuca & Rosen, 2022: 344), and risks reproducing the general tendency to 'reduce housing to a commodity' (Zavisca & Gerber, 2016: 348). To go beyond this, we need to broaden our understanding, moving beyond a focus on the distribution and cost of housing understood as a simple 'good' or economic asset, to an emphasis on what I call 'inequality of home', i.e. the unequal distribution of the complex set of resources associated with home, and crucial to human wellbeing.

The importance of home to the analysis of the PRS has already been discussed in Chapter 2, but this issue requires further scrutiny because it opens up an important, and under-appreciated, dimension of the political economy of housing: the complex dynamics of home making and unmaking when living in 'someone else's asset'. Home, in this sense, has been neglected within political economy perspectives. With the notable exception of Madden & Marcuse (2016), most political economy accounts focus on the production and distribution of housing, and typically emphasize the role this plays within the capitalist system as a whole, or within macroeconomic processes. Researchers in housing studies, on the other hand, have recently produced a large body of insightful research into renters' everyday experiences of 'making home'. This research sheds light on the subjective, affective and relational dimension of the nature of private rental housing, and deepens our understanding of what it means to be a tenant. The contention of this chapter is that the subjective, relational and experiential dimensions of home and home making should not be peripheral or secondary to the political economy of housing. On the contrary, these issues should be central to our analysis, for a number of reasons.

The first of these reasons relates to the peculiar nature of housing as a commodity. In Chapter 2 we saw that housing is a unique commodity, due to the issues of spatial fixity, scarcity, and the uniqueness of each dwelling. But beyond this, what is most striking about rental housing is the fact that it involves an individual or family making their lives in an asset owned by someone else. When it comes to the PRS, households are quite literally living 'inside' a commodity. The economic form of PRS housing and the nature of 'home' in the sector are thus two sides of the same coin.

The second relates to the landlord–tenant relationship. Private rental housing is characterized, or even constituted, by the social relationship between the landlord and the tenant. In Chapter 2, we discussed this from the perspective of understanding the nature of rent. But it is also relevant in the sense that this economic relationship is arguably the most decisive and impactful factor when it comes to tenants' experiences of housing (Morris et al., 2021). As noted in the previous chapter, a characteristic feature of the PRS, and one that distinguishes it from other tenures, is that it is permeated by the marketized social relationship between landlord and tenant (McKee & Harris, 2025), and therefore it is a form of housing which is totally subsumed within the market. The 'commodity form' of rental housing and the market relationship that permeates all dimensions of the lives of private tenants are thus crucial to understanding this form of housing. To put it succinctly, the experience of home matters if we are interested in a comprehensive understanding of the political economic relationships that are central to the PRS. Just as an analysis of class cannot ignore the experience of work and of employee–employer relationships, so too the analysis of private rental housing, and the commodity form and property relations at its core, must explore the terrain of subjective experience.

The third reason why tenants' subjective experience of home matters is because it is at the heart of the politics of private rental housing. Tenant politics, as will be discussed in much more depth in the next chapter, has always been an expression of the frustration tenants feel and, especially, the tension between landlord and tenant. How the landlord–tenant relationship is experienced by tenants shapes the conflict between them, as well as how issues in rental housing become politicized and, in some instances, subject to collective organizing on the part of tenants. Here we should recall that homes are not merely 'consumed' by tenants, they are also produced by tenants. As we will see throughout this chapter, the ways in which the production of home is constrained and undermined by the commodity form of PRS housing and the landlord–tenant relationship is of crucial importance. Thus, what we might call the 'terrain of home making' is a vital site of conflict and antagonism within the PRS. In this sense, this chapter recalls one of the original, if today sometimes less appreciated, *raison d' être's* of political economy: to understand the forms of conflict and antagonism that are generated by material economic relations (including those of property ownership) and how these relate to the possibility and potential for social change.

The focus on the experiential dimension in this chapter offers a fourth and final insight in relation to the political economy of the PRS, and of housing in general. Analysing this dimension requires qualitative and interpretative research that can capture the experiential, subjective, affective and relational aspects of PRS housing. Qualitative research, especially interview and ethnographic data, rarely forms of part of political economy analyses of housing but, without this type of data, private rental housing remains a kind of 'black box', and our focus is limited to how such housing is produced and allocated, thus limiting our understanding of what type of economic phenomena is at stake, and what types of relationships are at play (McKee & Harris, 2025). This chapter thus brings together much of the recent qualitative/ethnographic research on the PRS, with a view to incorporating an analysis of home and home making, more typically associated with the field of housing studies, into a political economy perspective.

Centring home and home making within our analysis returns us to the concepts of social reproduction and especially ontological security, already introduced in Chapter 2. These concepts help to capture core dimensions of home, particularly security, control, autonomy and agency. Analysis of how these aspects of home can be undermined or constrained within the PRS, and how this is an important aspect of contemporary inequality, has often relied on the concept of housing precarity. Although the concept remains somewhat ill-defined, Clair et al. (2019, quoted in Waldron, 2023) define housing precarity as a 'state of uncertainty which increases a person's real or perceived likelihood of experiencing an adverse event, caused (in part) by their relationship with their housing provider, the physical qualities, affordability, security of their home and access to essential services'. Waldron, for example, notes that:

> [H]ousing precarity ... is capable of incorporating a range of concerns regarding insecurity, quality and access, as well as affordability. Indeed, such an approach recognizes that housing is far from a solely economic concern for renters and is a complex resource that is linked to individuals' sense of ontological security, their personal relationships and social reproduction. (Waldron, 2023)

It is these latter issues (ontological security, relationships and social reproduction) which are the focus of this chapter, and there is a growing qualitative literature that draws on the concept of housing precarity to capture the pervasive sense of insecurity experienced by PRS tenants (Bobek et al., 2021; Bone, 2014; Byrne & Sassi, 2023; Huisman, 2020; Listerborn, 2023; McArdle & Byrne, 2022; Soaita & McKee, 2019, 2020; Waldron, 2022, 2023, 2024). As Vicki Spratt (2022: 37) notes in her book *Tenants*: 'If you do not own your own home in Britain, precarity is a fact of life. It comes in the form of rent rises, eviction notices and knowing that, ultimately, your landlord has control over the one place in the world you should feel safe' (quoted in McKee & Harris, 2025: 3).

The concept of precarity is most useful in that it allows us to capture the outcome of the multiple and intersecting forms of 'inequality of home' experienced by private renters. Precarity describes how these forms of inequality generate a *distinctive experience of home* which is characteristic of private rental housing. Of course this differs in different contexts, with a small number of European countries notably offering significant protections for tenants. But, even within these countries, the tendency is towards greater housing insecurity and economic inequality for renters (e.g. in the Netherlands see Huisman, 2020; for Germany see Soederberg, 2018; for Austria see Kadi, 2015). The concept of precarity, moreover, links the experience of housing insecurity with the wider class dynamics associated with contemporary capitalism. As noted in Chapter 3, the revival of the PRS can be situated within the wider erosion of the post-war class compromise and the attendant processes of financialization and neoliberalization. Most of the literature on precarity focuses on the labour market dynamics at play here, specifically the prevalence of short-term, non-standard and insecure forms of employment (Standing, 2011). Understanding the forms of inequality experienced by today's renters places housing in dialogue with these labour market dynamics and a class analysis of contemporary political economy. While more research is needed in this area, the link between housing and labour market precarity has already been demonstrated empirically (Arundel & Lennartz, 2020).

As indicated in the title of this chapter, the focus is on lightly regulated rental sectors, and especially English-speaking countries. This is in large part because literature on the subjective experience of private rental is overwhelmingly focused on the lightly regulated English-speaking countries (Soaita & McKee, 2020), and the discussion here reflects this and is thus limited by a degree of Anglo-centrism. It is also because these contexts are most fruitful in terms of exploring the challenges associated with home making in the PRS and the power asymmetries between landlord and tenant which will be central to the arguments in this chapter. Some might argue that the discussion in this chapter, indeed, is not relevant to rental sectors with stronger tenant protections, such as security of tenure. More research is needed in this regard, but my intuition is that the core economic nature of the PRS means that the same issues are relevant across different jurisdictions, but they are mediated in different ways by policy (and culture). This is not to minimize the importance of policy, as the ways in which the market power of landlords, for example, are mediated can certainly prove crucial to tenants' experience of security and home. It is to highlight, nevertheless, that policy to protect tenants is implemented in those jurisdictions precisely because it is required in order to allow tenants to enjoy security or create meaningful homes, despite the landlord's ownership of the property they make their home in. Notwithstanding this, there is certainly a need for more research on the nature of home and home making in these more tightly regulated markets to gain a fuller understanding of the extent to which issues like insecurity should be understood as more or less inherent in the economic form of PRS housing, or rather as a function of policy regimes, which vary across contexts.

The chapter begins by examining the 'meaning of home' and practices of home making in private rental housing, exploring issues such as security, privacy and control. The following section explores these questions within the marketized social relationship between landlord and tenant, before discussing how different types of landlord are of relevance to understanding tenants' experiences.

Making home in the private rental sector

Chapter 2 has already touched on the conceptualization of home, emphasizing the concept of ontological security, which is widely used within the housing studies research. As argued, ontological security 'identifies the interrelationships between the physical dimensions of housing (such as basic safety and security) and the psycho-social dimensions of home such as privacy, emotional security and identity' (Hulse & Milligan, 2014). Home is thus understood as a 'safe haven in which individuals can be themselves and from which they can derive an enhanced sense of emotional security' (Walshaw quoted in Easthope, 2004: 581). Ontological security can be further broken down into a number of core dimensions, in particular security (the stability and predictability of a home over time and into the future), privacy (the ability to control who accesses a dwelling, and freedom from surveillance and scrutiny within the dwelling), and control (the ability to exert agency and autonomy with regard to the physical and aesthetic organization of the dwelling, and to express one's identity through the dwelling). The sense of autonomy and agency that residents experience within and over their dwelling are cross-cutting themes that relate all three of these dimensions. These features all share two important attributes that are of direct relevance to the present chapter's focus on home making.

Firstly, all three dimensions of home relate to the subjective experience of home. They thus take shape via practices of meaning-making through which residents construct their home and their affective relation to it (Clapham, 2011; Soaita & McKee, 2019).

Secondly, and relatedly, they are not permanent qualities of a physical dwelling, but rather produced and reproduced through everyday activities through which home is 'assembled' (Soaita & McKee, 2019). This is something we are all familiar with because the first thing we do upon moving into a new dwelling is to make it 'feel like' home by making changes to the dwelling, organizing belongings in a particular fashion, and establishing routines of social reproduction, care, recreation and rest. Ontological security does not emerge spontaneously from the fact of inhabiting a dwelling, but rather through *practices* through which we construct a sense of ownership, control, stability, privacy and safety (Soaita & McKee, 2019).

The PRS, however, is 'the least conducive tenure for establishing feelings of home within Anglophone societies' (Bate, 2018: 10), and this relates precisely to how these practices of home making can be constrained, undermined, interrupted and destabilized by private rental housing (McArdle & Byrne, 2022). The lack of security often associated with the PRS is captured in qualitative evidence that highlights renters' feelings of transience and impermanence (McKee & Harris, 2025; Rolfe et al., 2023; Soaita & McKee, 2020). One study, for example, highlights how the subjective sense of insecurity feeds into a sense of renting as a temporary or transitory form of accommodation: 'You don't ever get fully invested to get all your stuff out. You'll still be kind of living in suitcases or boxes' (tenant quoted in Soaita & McKee, 2019). The feeling of impermanence, described by a participant in one study as the fact that 'you've always got that fear that you're not going to have a roof over your head at some stage' (tenant quoted in Morris et al., 2017), is often expressed by tenants in terms of this sense of 'living out of a box'. These issues can be particularly acute for renters with children:

> In the back of your mind you're always wondering, when will I have to move? ... And the main thing is for the children ... Like growing up you'd like to be able to have them in one home, you know, especially the little one ... She doesn't understand the concept of renting. She thinks it's her house and she sees the [for sale] sign going up and ... she just asks a million questions ... Not having security is the biggest thing. (Tenant quoted in Morris et al., 2017: 662)

In relation to privacy, a variety of issues, including being forced to share with strangers, or landlords entering the property without permission (Hoolachan et al., 2017: 70), can undermine tenants' experience of home. Landlords conducting property inspections and engaging in 'intrusive behaviour' (Verstraete & Moris, 2019) likewise undermine ontological security by eroding feelings of control and privacy (Lister, 2005).

The ways in which tenants' control over their dwellings is undermined is also widely documented in the research. This typically takes the form of tenants' inability to furnish, paint or otherwise alter dwellings (Bate, 2018, 2021; Easthope, 2014). For example, McKee and Harris's (2023) research found that the unwillingness of landlords to carry out repairs and lack of choice over furnishings undermined tenants' feeling of control with regard to their home (see also Rolfe et al., 2023). It also relates to constraints on tenants' ability to make decisions about what happens within the dwelling, including having guests over or smoking. The inability to have pets is frequently referenced in the literature as a key example of how tenants' experience of home can be undermined, almost in an infantilizing fashion (Power, 2017). Control over home is also related to questions of identity expression, autonomy and agency (Madden & Marcuse, 2016; Soaita & McKee, 2019).

At an extreme, the lack of control and privacy can manifest as abuse and harassment on the part of landlords (Cowan et al., 2000), even in the form of sexual exploitation, as seen in recent 'sex for rent' controversies in a number of countries (DeLuca & Rosen, 2022; Tester, 2008). Some cohorts can be particularly vulnerable to abuse and harassment, for example older renters (Izuhara & Heywood, 2003).

The challenges and complexities tenants face in making a home in the PRS are exemplified in research I conducted with Juliana Sassi (Byrne and Sassi, 2023), examining the experiences of tenants during the Covid-19 pandemic in Ireland. In common with many jurisdictions, Ireland introduced an eviction moratorium which was in effect during much of the period between March 2020 and April 2021. Thus, for a brief period, Irish tenants enjoyed almost complete *de jure* security of tenure. This changed tenants' experience of security and their relationship with their landlord. One research participant, when asked if the eviction moratorium impacted how confident she would feel to negotiate with her landlord, replied: 'I'd feel like the law was on my side, that nothing is going to change while the ban is there.' Likewise, one migrant tenant described gaining confidence from the eviction ban: 'It was a factor for me, because when you have this security it is easier for you to assert yourself ... I was confident to talk [to the landlord] because I knew that the government was taking steps to ensure that no one would become homeless during the pandemic.' The eviction moratorium gave a degree of additional security, enabling tenants to feel some certainty about the immediate future, and thus provided greater confidence and autonomy.

The temporary nature of the eviction ban, however, meant that insecurity was deferred, rather than eliminated, and thus tenants continued to experience a sense of long-term insecurity. This also related to the landlord–tenant relationship. One loneparent tenant, for example, noted the difficulty tenants face in challenging an eviction, even if the law is on their side:

> If they [the landlord] did want you out and you were digging your heels, they could make life very, very difficult for you if they wanted to, you know. I am not really sure if that [eviction ban] offers anyone any real assurance ... I'm sure there would be landlords there who would make your life miserable.

Thus, while the eviction ban offered *de jure* protections from the physical loss of home, it did not offer protections against the erosion of home, the unmaking of home, via a deterioration of the landlord–tenant relationship. Interestingly, moreover, the experience of making and unmaking home was different for different cohorts of tenants. For example, for lone parents, long-term security was paramount. Lone parents tended to spend more time talking about a pervasive sense of insecurity and about long-term security, especially focused on their children's experience of home: 'It is, it's our home ... it bothers me that I can't have security here. It bothers me more for [my daughter].' In contrast, Brazilian tenants were far more likely to move home during the pandemic. Interestingly, this was not due to evictions. Residential mobility among this cohort was instead driven by research participants' attempts to find more suitable homes, and frequent moves were common among this cohort, even before the pandemic. These tenants moved to escape poor housing conditions and also to find more amenable people with whom to house share. Although these moves were 'voluntary', in the sense that they were not landlord-initiated, they still implied a level of housing insecurity which took its toll. One Brazilian tenant said: 'psychologically it's very hard. We felt afraid. We are in the middle of a pandemic and have to move to a new place, with people we do not know. So mentally it was very difficult.' Displacement was thus deployed as a form of home *unmaking* by Brazilian research participants in order to respond and adapt to new forms of housing stress generated by the intersection of structural problems in the PRS (e.g. overcrowding) and the Covid-19 pandemic. Home *unmaking* emerges here as part of how dwellers exercise agency to respond to inadequate housing conditions and create opportunities for home *making* in the pandemic context.

We see throughout these examples a kind dialectic of stabilization and destabilization of home, within which tenants endeavour to create elements of security, privacy and control, and these are challenged or negated (McArdle & Byrne, 2022). According to much of the research, this is perhaps the defining characteristic of what it means to live in a private rental home. It is a way of experiencing home that is not entirely unique to the PRS, but does relate to one of the distinct features of the sector, namely making a life and a home in an 'asset' that belongs to another. The following quote from a tenant, which appears in Hoolachan et al. (2017: 70), expresses a sentiment which will be familiar to many renters and anyone who has conducted qualitative research with tenants: 'it's somebody else's home and we are just living there!' At the heart of the struggle over home making in the PRS, then, is the social relationship between landlord and tenant, to which we now turn.

The landlord–tenant relationship

The challenges of home making discussed above are all, to one extent or another, associated with the relationship between the tenant and the landlord (Byrne & McArdle, 2022; Rolfe et al., 2023). The insecurity that tenants experience in terms of *de jure* security of tenure arise not just from the absence of protections for tenants, but also from their necessary corollary: the powers granted to landlords to terminate tenancies or evict tenants (Madden & Marcuse, 2016). Similarly, the limiting of tenants' rights to alter dwellings emerges from the primacy of property rights over tenancy rights. Landlords, as already mentioned, can also abstain from undertaking repairs or dealing with issues relating to standards and maintenance. Landlords can even influence the design, aesthetics and lifestyle choices; for example, banning pets or smoking, or preventing tenants from painting the walls of the dwelling or hanging pictures (McKee & Harris, 2025; Power, 2017). By entering a dwelling without permission landlords can exercise power over tenants' experience of privacy. Indeed, from the very outset 'landlord selectivity', the power of landlords to choose or reject tenants, sets up a power dynamic. Even when a tenancy is terminated, landlords can continue to exercise power over tenants by withholding a deposit, leading to a potential risk of homelessness, or refusing to provide a reference, which can also jeopardize a tenant's ability to obtain future accommodation. Thus, the ways in which ontological security is undermined for tenants in most cases arise directly out of the action of landlords and are a function of the power imbalance, or asymmetry, between landlords and tenants.

One of the clearest ways in which the impact of this social and power relationship over tenants' experience of home is demonstrated is the example of 'revenge evictions', and other forms of 'retaliatory conduct', i.e. actions taken in order to penalize tenants for challenging their landlord or complaining to a third party, such as a local authority. Retaliatory conduct often takes the form of 'revenge evictions', and links together the issues of security of tenure, the power of landlords and the agency of tenants. This is a significant issue that impacts on tenants' ability to advocate for themselves (Allen, 2011; Byrne & McArdle, 2022). In the UK context, for example, Shelter's (2014) *Safe and Decent Homes* report found that over 300,000 tenants experience some form of landlord retaliation each year.

Retaliatory actions are a key example of the ways in which tenants' agency may be constrained. This is particularly the case for tenants with 'few alternative housing options', who 'devise strategies to ensure they do not upset their landlord. These include not complaining about poor living conditions and prioritizing rent payments over basic needs' (Hoolachan et al., 2017). Hoolachan and McKee (2019: 20) interviewed nineteen actors in housing at the local and national scale in Scotland. One of these interviewees described the situation: 'The tenants feel they're in this really weird situation where they're being ripped off. They're living in a house that's built for half as many of them as there are but they're terrified about asking for repairs or anything because they feel that the landlords have all the cards ... ' (see also Hulse et al., 2019).

Certain cohorts of tenants are particularly vulnerable to revenge evictions. For example, there is evidence that low-income tenants face 'revenge evictions' more frequently (Hoolachan et al., 2017). Desmond's (2016) ethnographic research also finds that there are divergences in how tenants relate to landlords and manage conflict when it arises, and their risk of being subject to revenge evictions. Many low-income black women, he notes, 'preferred to avoid direct confrontation with their landlord'. His research also suggests they had good reason to do so, as women who reported their landlord due to minimum standards violations 'greatly increased their risk of eviction, for there are few things landlords detested more than a clipboard-in-hand building inspector scrutinizing their property for fine-grained code violations' (Desmond, 2016: 115).

Despite the power asymmetry between landlords and tenants, nevertheless, it is a mistake to assume tenants are merely passive victims within this dynamic. The agency of tenants on a political level is the subject of Chapter 7. However, here it is important to acknowledge the 'everyday agency' and forms of resistance that tenants employ in making home and negotiating their relationship with their landlord (McArdle & Byrne, 2022). Research has documented the various strategies tenants employ to navigate the precarity of the PRS (Waldron, 2024) and their relationship with their landlord (Lister, 2004). Strategies adopted by tenants can take an economic form (withholding rent until repairs are carried out), or a social one (not offering the landlord tea to signal they are not welcome) (Lister, 2004, see also McCardle & Byrne, 2022). McCardle and Byrne (2022) also show the everyday practices through which tenants resist housing insecurity and the ways in which their home making is undermined, for example keeping careful records that relate to conflicts with their landlord for use in a potential formal dispute at a later point. Similar informal 'counter-strategies' are deployed by tenants who navigate the discriminatory practices of landlords by omitting information or adopting the traits that are deemed desirable by landlords (Verstraete & Moris, 2019), and the agency of tenants is also evident in how they find creative ways to 'make home' despite the constraints that can characterize private renting (Rolfe et al., 2023).

Most of what we know about private tenants' experiences of home comes from the housing studies literature. While economic factors are discussed within that literature, for example how the 'tightness' of rental markets reduces tenants' agency and empowers landlords, in general the insecurity experienced by tenants and the power imbalance within the landlord–tenant relationship are treated as policy problems, which is unsurprising given that policy does play an important role in establishing the respective rights and obligations of landlords and tenants and the relevant enforcement regime, for example. What is often missing, however, is an appreciation of the fact that policy in the rental sector must also be conceptualized as mediating a relationship which is fundamentally economic, i.e. that it is at its core a market transaction. As argued in Chapter 2, the nature of 'home making' in the PRS, and hence tenants' subjective experiences of home, as well as the nature of the landlord–tenant relationship, derive first and foremost from how rental housing functions as a commodity. At its core, rental housing is characterized by the fact of making a home in someone else's asset. The subjective experience of home in the sector reflects this fundamental material and political economic reality. The tensions that tenants experience in relation to home, and the everyday conflicts and antagonisms that pervade the power relation between landlord and tenant, are thus not incidental features of the sector, but rather directly related to the core economic form upon which it is predicated. In short, where tenants lack security in their home, it is *because* their landlord enjoys security with regard to their asset and where tenants lack control over their home, it is *because* their landlord holds control of their asset.

Here we should note again that it is paramount to avoid approaching the PRS as solely or primarily consisting of tenants' access to a particular form of housing (i.e. an approach which frames the PRS in terms of the residents' relationship with a dwelling), instead opting for an approach which frames the PRS as a political economic *relationship* between landlord and tenant. This relationship, once again, is a property-based relationship and the power asymmetry between landlords and tenants is thus also based on property relationships. By delving into the qualitative literature and tenants' subjective experience of home, we can draw out how this power/property relationship has decisive impacts on the nature of 'home making' in the PRS, or at the very least in the lightly regulated markets that are the focus of this chapter. However, it is also important to recognize that this experience can vary hugely, just as economic relationships in any sphere of the economy will vary depending on the characteristics of the specific actors (in our case landlords and tenants) involved. One of the most important ways in which it can vary in the PRS relates to the varieties of landlords that are found in the sector, an issue I turn to in the next section.

Varieties of landlord–tenant relationship

This chapter has so far focused on the everyday power asymmetry between landlords and tenants as it pertains to the issue of home making and the subjective experiences of tenants. In doing so, it has drawn primarily on literature from the relatively lightly regulated English-speaking countries, and has focused on the situations in which the landlord is an individual (rather than a company or institution) and deals directly with the tenant. Nevertheless, it is of course the case that the nature and form the relationship between landlord and tenant takes vary widely, within and between countries.

The most obvious way that the relationship between landlord and tenant varies is in different housing regimes. As was noted in the last chapter, several authors have advanced criticisms of what can sometimes be an Anglo-centric account of the PRS which focuses on chronic housing insecurity (see for example Howard et al., 2024). Indeed, it is important to point out that in many Western European countries, especially the Nordic and continental cases, there are far more protections for tenants, and landlords enjoy less power. This indicates that the chronic levels of insecurity discussed here are not inevitable or natural, nor are they inherent to the PRS, but are rather the outcome of a number of different forces, in particular housing policy and housing markets. With regard to the former, the relative rights and obligations of tenants and landlords are of course set out in different national legislation, and vary quite widely. With regard to the latter, the nature of investment in the PRS also shapes tenants' experiences and issues of home making. For example, an insightful comparative article by Kemp and Kofner (2010) demonstrates that German landlords are often more focused on long-term capital gains, as opposed to rental income, and this has implications for how they approach their role as landlords, as well as their ability to operate within a context of strong security of tenure protections for tenants. There are, however, a number of caveats. Firstly, and most importantly, what these examples show are that the unequal property relations between landlord and tenant are *mediated* by policy and market contexts. The relationship thus takes different forms, and the power asymmetry can be more or less intense. Secondly, viewed from a global perspective, countries which provide very strong protections for tenants are the exception, rather than the rule. In the vast majority of countries globally, and even within the advanced economies, tenants' rights are typically weak, including the English-speaking world, most of the Mediterranean, and virtually all of Central and Eastern Europe.

Another example of the varying forms the relationship between landlord and tenant can take is the difference between individual landlords, sometimes called 'mom and pop' landlords, and larger or institutional landlords. Smaller landlords are typically unprofessional (in the sense that they have, or had, another form of employment and source of income), and have a small number of properties. In many countries, indeed, the majority of landlords have one or two properties. While it is sometimes thought that in countries with well-regulated sectors, such as Germany, most rental housing is provided by institutions, in fact small landlords dominate in the vast majority of countries. Nevertheless, since the Global Financial Crisis, and as discussed in detail in Chapter 4, financial institutions have expanded investment in the PRS, and this has given rise to a new body of research examining their impacts on housing and tenants' experiences. Of particular interest to the present chapter is the question of what the difference between larger or smaller landlords might mean in terms of the nature of the landlord–tenant relationship. Henry Gomory's (2021) research is instructive here. He focuses on how eviction practices vary between small and large landlords. His research finds that large landlords file for eviction 186 per cent more than small landlords, and also do so over smaller amounts of rent arrears. Gomory (2022: 1778) suggests that the contrasting nature of the social relationships at play helps to explain this divergence:

> Small landlords' relationships with tenants and flexible decision-making styles might make them draw on extra-economic considerations, such as subjective impressions of tenants, when making eviction decisions. In contrast, large landlords' formal relationships with tenants might frame the transaction as purely economic, and their bureaucratized management practices might ignore extra-economic considerations.

Similarly, Shiffer-Sebba (2020) investigates the divergent practices of what she calls deliberate and 'circumstantial' landlords, the latter referring to landlords who obtain a rental property without explicitly investing in the sector, for example through inheritance. She finds that deliberate landlords are more likely to base their decisions and actions on the logic of profit maximization, while circumstantial landlords often value 'social closeness' in terms of their relationship with their tenants. This is also evident in McKee and Harris's (2025) research, which found instances of landlord 'flexibility' in relation to negotiating reduced rents for a period (see also Desmond, 2016). Research has also examined the differences between rent setting and other landlord practices in poor versus non-poor neighbourhoods, showing that the different market contexts lead to different investment strategies and consequently different types of relationship between landlord and tenant (Desmond & Wilmers, 2019).

These insights draw our attention to the fact that landlord–tenant relationships are sometimes characterized by inter-personal dynamics and informal rules (McKee & Harris, 2025). A contrasting case is where rental properties, while owned by individual, small-scale landlords, are managed by estate agents or property managers. On the one hand, estate agents can depersonalize the relationship between a tenant and individual landlord, and can manage properties in a more professional way (Verstraete & Moris, 2019). This can be of benefit to a tenant because, in some instances, when dealing directly with a landlord, tenant–landlord conflicts become unduly personal, emotional and distressing for tenants (Byrne & McArdle, 2022). On the other hand, there is evidence that real-estate agents can also discriminate against certain categories of tenants, and if they control access to a significant segment of rental housing this can of course create systematic barriers to entry (Bonnet & Pollard, 2021; Furst & Evans, 2017; Verstraete & Moris, 2019).

It is also, of course, the case that not all relationships between landlords and tenants are negative. Interestingly, however, McKee and Harris's qualitative research with tenants in the UK found that:

> [E]ven when participants praised their landlord it was evidently still a highly unequal relationship. Despite initially reporting being satisfied, further probing highlighted some tenants simply did not want to raise legitimate issues or escalate complaints where issues remained unaddressed because of the risk of damaging the relationship with their landlord ... this fear existed even when the relationship between the landlord and their tenant was described positively. (McKee & Harris, 2025: 7)

Thus, while flexibility and informality can sometimes be of benefit to tenants, they ultimately raise challenges for the PRS:

> This is ultimately a market relationship and not based on a social contract like social housing tenants enjoy. Yet, it is also a marketized relationship centred on informality, moral judgements, unwritten codes, and imperfect communication. (McKee & Harris, 2025: 15)

Conclusion: The subjective composition of the PRS and the landlord–tenant relationship

While the nature and form of landlord–tenant relationships vary to a considerable degree, the power asymmetry between landlords and tenants is at the heart of the experience of home in lightly regulated PRS markets. This asymmetry is at the same time between landlords and tenants, and between the related power of property/assets over home (Madden & Marcuse, 2016). The dimension of power, it is important to highlight, operates or manifests precisely on the terrain of home-making. The power of the landlord arises first and foremost from their ability to bring to an end the tenant's access to their home, and more generally to determine the condition within which the tenant's home making can occur. From this perspective, it can be argued that 'insecurity of tenure is not necessarily about transient living but about proprietors' power *over* tenants, coded into legislative and regulatory frameworks ...' (Soaita and McKee, 2019: 154, *authors' emphasis*). Similarly, Brickell (2012: 229) notes that 'the intimate and personal spaces of home – and their loss – are closely bound up with, rather than separate from, wider power relations'. Desmond's (2016: 117) ethnographic account of renting in Milwaukee captures this aspect of life in the rental sector powerfully:

> [E]victions are not simply the consequence of tenants' 'misbehaviour' or landlords' financial accounting, nor are they governed strictly by formal or deterministic rules. Evictions also are the outcome of interactions among people occupying different positions in social hierarchies and possessing different dispositions and interactional styles, conditioned by these positions.

Moreover, is worth highlighting that the fact that PRS tenants have systematically fewer protections than homeowners and (typically at least) social renters is no coincidence, and it does not merely arise from cultural norms that favour homeownership. It also arises from the fact that limiting the rights and protections of tenants, and conversely empowering landlords, is a pre-requisite for having a PRS in the first place. Without the ability to set rents and recover properties, landlordism is not economically viable. Indeed, in those countries that introduced extremely strong tenant protections in the inter-war and post-war eras, the PRS more or less disappeared over time (Harloe, 2021). This is not to say that tenancy rights and rent regulation cannot be strengthened, but there is an obvious, and inherent, tension between tenants' rights and investment in rental housing. Thus, the form of property relationships which are characteristic of rental housing have consequences both for tenant subjective experience of home and for policy-making and regulation.

The concept of precarity is one way of conceptualizing the pervasive sense of insecurity, instability, inability to shape their dwelling and their future that come up so frequently within the qualitative research. Precarity relates to both policy and market factors, and is also linked to wider forms of socio-economic inequality such as class, race, migration status, etc. It helps to position housing precarity as something which can be specific to housing experiences, but is also linked to a wider set of transformations in labour markets and welfare states associated with contemporary political economy.

In the case of private rental housing, as argued throughout this chapter, precarity is first and foremost associated with the power asymmetry between landlords and tenants, the analysis of which is of crucial importance to a political economy perspective, because it is here, in the struggle over 'home', that the unequal property relations at the heart of the PRS manifest most clearly (Byrne, 2020c). If we limit our analysis to 'housing', in the sense of the production and allocation of dwellings, we miss the fact that the more intangible everyday practices of home making are the core terrain of tension and conflict (McKee & Harris, 2025).

'Bringing home back in', so to speak, is an important antidote to its under-emphasis within the political economy literature. This neglect is related to the assumption, discussed in the introduction to this book and Chapter 2, that the experience of residents is merely one of 'consumption'. While relations of production have always been central to political economy, housing consumption has typically been dismissed as a secondary concern (Gray, 2018a). From this point of view, the relationship between a tenant and their landlord is of no more importance than that between a customer and a shopkeeper. However, as argued in Chapter 2, this is an erroneous perspective because tenants do not merely consume home, they also produce it through everyday practices of home making. The value of home arises out of the tenant's everyday activities, and the conflict between landlords and tenants, as documented throughout this chapter, manifest precisely in relation to the practices of home making and unmaking through which this value is produced and destroyed. To ignore the production of home in the context of the PRS is thus to miss what should be at the very core of any political economy analysis. Furthermore, this chapter has also shown that understanding the social relations associated with the production of home in the PRS requires a qualitative and interpretive lens because it centres the forms of everyday meaning making through which home is constituted (see Byrne, 2020c). Capturing the subjective, experiential, relational and affective dimension of private rental homes is therefore not merely a nice 'add on' that can enrich a political economy of housing, rather it is the very beating heart of such an approach.

It is important to add that, while my analysis emphasizes the importance of the production of home, it in no way follows that the relationship between tenant and landlord is somehow analogous to that between worker and capitalists (for a discussion of this issue which remains relevant, see Saunders, 1984). The social dynamics of production are central to how the capitalist system functions and how almost all commodities are produced and circulated, as well as to profit making, the accumulation of capital and economic growth. In other words, they relate to structural and systemic issues within capitalism as a political economic system. The social relations of the PRS do not play anything like this role in the wider economic system. The purpose of the analysis advanced here is not to make claims about the relationship between the PRS and the capitalist system, however, it is to shed insights into the nature of this particular form of housing and the experiences of tenants, and reveals how the social relations of PRS housing relate to the nature of conflict and antagonism within the sector, and consequently the possibilities and potential for politicization.

Finally, the material basis for tenant politics is, at least in part, the tension and antagonism that is embedded in the processes of home making described in this chapter. It is here that the frustrations of tenants germinate, and the sense of unfairness that is so often expressed takes shape. The politicization of the PRS, to which we turn in the next chapter, in other words, depends upon the subjective experience of the objective property relations associated with private rental homes.

7
The Politics of the Private Rental Sector
Tenants as Agents

In a recent article, Gil and Palomera (2024) note the remarkable politicization of the PRS which has occurred across numerous jurisdictions over the last few years (see also Stringer, 2025). In June 2021, the then Swedish government collapsed amid an attempt to deregulate rents for newly constructed housing. In the same year, a referendum in Berlin passed a motion calling for the expropriation of PRS housing owned by the city's institutional landlords. A year beforehand the Berlin government had already introduced an unprecedented rent freeze (although this was later repealed when it was found to be unconstitutional). Meanwhile, in March 2023 the Irish media was convulsed for two weeks by a debate about the withdrawal of a ban on evictions, a measure originally introduced as a temporary response during the Covid-19-era public health restrictions. In July 2024, Joe Biden, before he withdrew his candidacy, promised to introduce rent controls for corporate landlords as part of his campaign for the US presidential election.

These vignettes are noteworthy because they reflect the return of the PRS, and the challenges faced by tenants, as a major political issue and a focus of policy change. The decline of the PRS throughout the latter part of the twentieth century was accompanied by a decline in the sector's political salience. In recent decades, in the majority of advanced economies, it is safe to say that homeownership and social housing have been the dominant issues in housing politics. This trend is clearly in reverse today, as indicated by the above examples. But these examples are part of a wider tendency, evident in the wave of increased rent regulation across numerous jurisdictions since about 2015. Since that year, the following European countries have either introduced or strengthened rent regulation: Germany, Ireland, Scotland, Spain, France and the Netherlands. Indeed, some have argued that we are currently moving away from the neoliberal housing policies that have dominated in recent decades, to a form of 'post-neoliberalism' (Byrne, 2024; Kadi et al., 2021; Schipper, 2015) or 'regulated marketization' (Hochstenbach, 2023; Hochstenbach & Ronald, 2020), i.e. towards greater regulation of the PRS and stronger protections for tenants.

The trend towards the politicization of the PRS is also evident in the recent revival of tenant organizations responding to issues like unaffordability, insecurity and poor standards (Stringer, 2025). This return of tenant politics highlights what will be the key focus of this chapter: tenants, and residents in general, do not simply 'consume' housing. Nor are they simply passive victims of housing policy and markets. They are also active agents capable of working collectively to transform housing conditions at an individual level and at the level of the housing system (Madden & Marcuse, 2016). Here it is important to recall that, as argued in Chapter 1 of this book, the social relationship between landlord and tenant, the core social relationship that constitutes private rental housing, is one of inequality (of property), power and conflict. Throughout this book we have seen how this relationship is situated within, interacts with and is reinforced by a host of wider social, political and economic structures and forms of stratification. Within this context it is easy to mis-represent tenants as helpless, disempowered victims, and to miss the fact that the agency of tenants is one of the most powerful forces that shape the nature of the PRS, as well as the everyday experiences of tenants. Residents-as-agents are capable of acting individually and collectively, formally and informally, to shape their housing reality. Thus, in advancing a political economy framework for the PRS which emphasizes its inherently political nature, because of the dimensions of power and conflict, we must remember that this does not consist solely of the power landlords can exercise over tenants, but also in the ability of tenants to contest this power, and even, in some circumstances, to transform the political economic structures that are the basis of landlord power.

I have already touched on the agency of tenants in one sense in Chapter 2, where I emphasized that tenants actively produce home through everyday practices of meaning making and social reproduction. But the capacity of tenants (and residents in general) to exercise agency goes far beyond this, from organizing rent strikes locally to establishing national organizations with the explicit aim of overcoming issues of inequality and power asymmetry within the PRS (Card, 2024). The Swedish Tenants' Union, established in 1927 and today the largest civil society organization in Sweden, being one example of the latter. Indeed, as I will argue below, tenants have played an important role in shaping the development of housing systems in the twentieth century. Consequently, any attempt to understand the PRS, and explain its evolution over time, must engage with tenant politics (Köppe & Byrne, 2024). Too often, accounts of how housing systems develop focus solely on market developments or the intentions and actions of policy-makers. Even in today's literature, tenants are typically presented as victims of the issues in the PRS, and the potential for policy interventions to overcome these issues is normally discussed without any reference to how tenants can themselves be involved in the process of policy change. Thus, when it comes to private rental housing, we need put the 'political' back in political economy. As Madden and Marcuse put it, '[n]o account of the housing system is complete without an understanding of the collective power of inhabitants' (2016: 83, see also Card, 2024).

In this chapter, I delve into how tenants organize politically to transform their own housing realities and the wider housing system. The chapter begins with a discussion of historical examples of tenant organizing, focusing on the rent strikes of the early twentieth century, before providing an overview of the organizational forms, strategies and politics of contemporary tenant activists. The Chapter then provides a more in-depth account of two important examples of contemporary tenant politics. The first example is that of the Barcelona and Madrid Tenants' Unions, established in 2017 but already influential organizations in the development of housing politics in Spain and Catalunya. I discuss the organizational form of these unions and their campaigns, focusing on how they have developed 'tenant power' from the bottom up, enabling them to advocate for, and often achieve, significant policy change at the regional and national level. As tenant activists Rosenthal and Vilchis (2024: 11) argue in the US context, 'we organize tenants as political subjects whose task it is to beat back the power of real estate and change the world'. The second example is the *Deutsche Wohnen und Co. Entiegen* campaign in Berlin, which, as mentioned, called for and won a referendum in support of the mass expropriation of PRS housing in the city, offering a compelling example of how tenant activism can sometimes radically expand the horizons of housing policy and politics.

History of tenant activism: A brief history of rent strikes

Before diving into contemporary tenant politics, it is interesting to consider its history. Of course, there is a long history of tenant movements in an agrarian context, but the politics of private rental housing begins with industrialization, urbanization and the commodification of housing markets in the nineteenth century. These developments, of course, gave rise to the issues of unaffordable housing, chronic overcrowding and appalling housing conditions, famously documented by Engels (1993). With the vast majority of urban households renting privately, it is unsurprising that the early twentieth century is the most well-known and well-documented period of tenant activism. Perhaps the most well-known example here is the Glasgow rent strikes of 1915. The context for the strikes was a population increase in Glasgow, combined with slum housing conditions for many workers, war profiteering by landlords and frequent evictions (Damer, 1980). Before the First World War, Scottish tenants had been agitating and organizing around housing for some time. In 1913, the Scottish Federation of Tenants' Associations was established, and in 1914 the Glasgow Women's Housing Association was formed (Gray, 2018a). In 1915 a major rent strike began which culminated in an estimated 25,000 tenants participating. The rent strikes were articulated, in the context of the First World War, as a response to the profiteering of unpatriotic landlords who were exploiting, and often evicting, families, including those left behind by soldiers. During the strike, eviction attempts were sometimes blocked, including by groups of women (Gray, 2018a). Women were also involved in and led mass protests. Ultimately, on November 28, the Rents and Mortgage Interest (War Restrictions) Bill, which established rent controls, was introduced at the House of Commons, and became law in December of 1915. Rent controls would remain in place until their repeal under Thatcher in 1988.

Lawson's work on the history of the rent strike in New York City provides another important historic example. Similarly to the Glasgow case, the background to New York's wave of early twentieth-century rent strikes was appalling housing conditions in the city's rapidly expanding rental sector. However, it was a series of sharp rent increases at the outset of the twentieth century, triggered by inadequate housing supply, partially related to tenement demolition and the influx of migrants, that provided the immediate catalyst. In 1904, against this backdrop, a rent strike began 'involving 800 tenement houses and the threat of eviction for 2,000 tenants in the Jewish quarter of the Lower East Side, marked the first use of this strategy in New York City' (Lawson, 1984: 235). A second wave of rent strikes occurred in the winter of 1907/09, and once again during the First World War. Just like in Glasgow, housing shortages and associated rent increases caused by the the war led to hardship for tenants. This gave rise to a long period of rent strikes between 1917 and 1920 across working-class neighbourhoods in Brooklyn, Manhattan and the Bronx. By 1920, the scale of organized tenant resistance to rent increases was such that the state legislature introduced rent controls.

In Buenos Aires, in the Autumn of 1907, residents of some of the city's tenements responded to rent increases of almost fifty per cent by refusing to pay, which spread quickly to other buildings thus kicking off a prolonged mass rent strike. One of the tenants in the first building to go on strike, one Antonio Rinaldi, set up a tenant committee and wrote to the landlord to ask for a reduction in rent:

> Dear Sir; In light of the high cost of rent in this building ... and the tenants being in general agreement ... we have decided to ask you, as the owner, to reduce the rent by 30 percent. That is our only request. We await a prompt and favourable response to our request. Yours truly, The Tenants. (quoted in Baer, 1993: 355)

By the end of the year, tenants of more than 2,000 buildings were participating in the strike, ultimately incorporating approximately 120,000 tenants, around ten per cent of the city's inhabitants, as well as spreading to other cities. Tenants established neighbourhood and building committees to collectively organize the rent strike and coordinate with the trade union movement (Baer, 1993).

The above are just three examples of some of the most well-known instances of mass tenant mobilization during the early twentieth century. Guzmán and Ill-Raga (2022) point out that there were many more mass rent strikes throughout the period, including Barcelona in 1904, Milan in 1909, Vienna in 1910, Leeds in 1914, Paris in 1919, Stockholm in 1920, and Barcelona again in 1931.

We can learn much about the political significance of rent strikes, and tenant organization more generally, from the historical examples discussed here. The role of women, and the 'sphere of social reproduction', in housing struggles emerges as one important theme (Guzmán & Ill-Raga, 2022). Mary Barbour, one of the most famous leaders of the Glasgow 1915 rent strikes, organized tenant committees and anti-eviction actions during the strikes, with her supporters becoming known as 'Mrs. Barbour's Army'. More generally, Gray (2018b: 61) argues that during the Glasgow rent strikes:

> [I]t was women who led the rent strikes, and women who forced the housing issue on the ground, even if strong support from other organizations was crucial ... the vast majority of rent strikers were working-class 'housewives', whose actions were based on their everyday experience of tenement life ...

The role of women was also notable in the Buenos Aires case. Juana Rouco Buela, for example, was a leading figure in the 1907 rent strikes in Buenos Aires (Munoz et al., 2024). Women also played leadership roles both in street protest and in civil disobedience and direct action. The demonstrations of the time are often remembered as 'broom parades' because women protesters marched with brooms to represent the sweeping away of landlords and high rents. Baer (1993: 359) also gives the example of a prominent eviction attempt which was blocked by women activists:

> Some of the women in the *conventillo* (tenement) tried to bar the police from entering. They closed all the doors and windows to the street. Those officers who did get into the building were attacked by broom-wielding women and had boiling water poured onto them from the patio above.

In Buenos Aires, Baer (1993: 358) notes that for many women the home was the site of both social reproductive and productive, paid, labour:

> Women responded to the issue of housing because it affected their lives in many ways. Some worked at home, taking in ironing or washing, which often was prohibited by the owner's rules ... They were defending more than their homes; they were protecting their families and preserving their means of employment.

The politics of the PRS, then, are also the politics of home and therefore relate directly to questions of social reproduction and production in the domestic sphere (Guzmán & Ill-Raga, 2022; Madden, 2025).

Furthermore, the relationship between tenant politics and labour movements is also evident through the examples discussed here. In Glasgow, labour organizations, including both socialist political parties and trade unions, were instrumental in commencing and supporting the rent strikes, and the threat of industrial actions was one the factors that led to the introduction of rent controls (Gray, 2018b). In both Buenos Aires (Baer, 1993) and New York (Lawson, 1984), the tenants who organized the rent strikes had experience as part of the organized labour movement and borrowed tactics and strategies from this experience. In relation to the former, Baer (1993: 367) argues that '[t]enant organization had many points of contact with organized labour and drew on that experience', and highlights the fact that the tenants were supported by both of the main national trade unions. Interestingly, Lawson notes that the rent strike, as the name suggests, was a tactic influenced by tenants' engagement in workplace politics. The very name 'tenant unions' for the organizations created by the rent strikers indicates the parallels between tenant activism and working-class political organization. To quote Lawson (1984: 239): '[c]alling the organizations they formed in their buildings "tenant unions", they collectively withheld their rents from their landlords just as they had frequently withheld their labour from their employers'. As discussed further below, what is most noteworthy is how discourses, practices and organizational forms from workplace-based politics readily map on to conflicts between landlords and tenants.

Finally, as Gray (2018a: xix) argues in a point relevant for all the historical cases discussed here, 'probably the greatest lesson from the 1915 rent strikes is that the threat and practice of collective tenant organization and direct action is a prerequisite condition for radical housing transformation'. While the Buenos Aires rent strike was ultimately unsuccessful, in the New York and Glasgow cases, collective action by tenants, including rent strikes and the creation of formal tenants' organizations, had enormous impacts in terms of legislation, policy and the development of the housing system. In the NYC case, the intensity of rent strikes and the scale of working class participation 'were interpreted as posing political threats in addition to pressing the landlords individually' (Lawson, 1984: 238). This fear triggered a dramatic policy reaction: 'So great was the fear generated ... that in 1920 the state legislature enacted rent controls in order to defuse the issue' (Lawson, 1984: 238). Similarly, in Glasgow, in response to the tenant-led rent strikes the British government established rent controls for the first time. As noted by Gray (Gray, 2018a: xviii):

> Rent Strikes in Glasgow are now widely acknowledged as a decisive event in a wider national struggle that shaped both the British tenants' movement and British housing policy dramatically, with few historical events exhibiting such a close causal link between urban struggle and state intervention.

It is important here to underline that the introduction of rent controls, a policy which eventually spread to almost every country in the Western world as the twentieth century progressed, was enormously significant on a number of levels (Harloe, 2021). First, it was a policy of aggressive state intervention introduced despite the *lassiez-faire* economic ideology which dominated at the time. The introduction of rent controls made strong regulation of the PRS part of mainstream policy, and normalized what was in effect both a form of housing decommodification and a mechanism to mitigate the power asymmetry between landlords and tenants. In this sense, forcing the introduction of rent controls represented a remarkable political achievement. Second, as rent controls at the time usually involved freezing rents, they dramatically curtailed the impact of housing commodification on tenants and hence had an immediate impact on the material conditions of urban workers

(Gray, 2018a). Third, the introduction of rent controls had huge knock-on consequences for the development of housing systems. This first generation of 'hard' rent controls played a role in the decline of PRS housing, which in turn was one of the crucial factors which pushed states towards direct public provision of rental housing, i.e. social or public housing (Harloe, 2021). Recent cross-national comparative research has shown that the development of social housing is closely related to the introduction of rent controls and the decline of PRS investment in the early and mid twentieth century (Kholodilin et al., 2022). It is important to note that there are a variety of factors at stake here. Issues such as the world wars, related periods of rampant inflation, changing labour market conditions, etc., and ideological support for homeownership all played into the politics of the PRS in the twentieth century in general and specific measures such as rent controls. Nevertheless, the above discussion of historical tenant organizations demonstrates the agency of tenants and tenant activism have had a wide-reaching impact on the development of housing systems (Köppe & Byrne, 2024).

The new tenant politics: Tenant organizing after the Global Finance Crisis

As the size of the PRS dwindled across advanced economies throughout the twentieth century, so the prevalence of tenant organizations and politics also declined. Several countries with large rental sectors, notably Sweden and Germany, retained large tenant unions.[4] These became more professionalized, formal organizations whose main focus was legal support for tenants and advocating for tenants in relation to legislation and policy. The Swedish Union of Tenants, for example, is the largest civil society organization in the country, and plays a formal role in the negotiation of rent increases at a national level. However, in most countries tenant organization declined quite radically in the latter part of the twentieth century. The breakdown of the post-war housing pact and the decline of

---

4 The Swedish National Tenants' Union has more than 500,000 members (Gustafsson et al., 2019), while Germany's largest tenant union, Berliner Mieterverein, has over 160,000 members (Guzmán, 2024).

the 'promise of homeownership', however, means that some of the conditions which gave rise to earlier periods of tenant agitation are relevant once again. As discussed in earlier chapters, the dynamics of contemporary housing are associated with greater inequality, the concentration of ownership of residential property, new forms of housing-based stratification, and growing precarity. The latter, in particular, is linked to both rent increases and evictions, and makes for experiences of housing shared by many tenants and likely to trigger dissatisfaction and contention.

Against this backdrop, we are currently seeing a resurgence of tenant activism across many countries (Byrne, 2017, 2018; Card, 2024; Gil & Palomera, 2024, Stringer, 2025), including Scotland, England, the US, Spain, Portugal, Greece, Poland, Germany and Ireland. In Sweden, where tenant politics has traditionally been dominated by the national tenants' union, a new generation of grassroots tenant organizations has emerged to resist the financialization of the PRS, renoviction and displacement (Gustafsson et al., 2019; Polanska & Richard, 2021). Similarly, local movements in Germany over the last decades have arisen in response to the same set of issues, ultimately culminating in the Deutsche Wohnen und Co (DWE) referendum campaign, discussed in detail below (Vollmer & Gutiérrez, 2022). In Poland, where tenants have been completely marginalized within the political system and housing policy, tenant organizations have recently emerged and, in collaboration with squatter movements, attempted to put the issues faced by tenants on the political agenda (Jezierska & Polanska, 2018). Michener and SoRelle (2023) provide an account of a plethora of tenant organizations that have sprung up across the US over the last few years, most of which have a grassroots structure and focus on creating 'tenant power' through legal support, case work and campaigning (see also Card, 2024).

In Ireland, CATU, formed in 2017, is a union-based organization (i.e. with fee-paying members) that, at the time of writing, has around 2,000 members, and organizes via the provision of support to individual tenants, including legal support and direct action, such as campaigns against individual landlords and anti-eviction actions (Gavin & O'Callaghan, 2024). The London based organization Digs, based in the Hackney area, focused on similar activities. It was not a union but went on to be part of forming the London Tenants' Union, another membership-based organization (Stringer, 2025;

Wilde, 2022). Other recently created unions focusing on a mix of support for individuals, direct actions and campaigning include the Scottish organization Living Rent (Saunders et al., 2018), and the Barcelona and Madrid Tenants' Unions (Gil & Palomera, 2024; Guzmán, 2025).

These organizations are developing new ways of responding to the growing conflict between tenants and landlords and between housing as a right and housing as a financialized asset (Stringer, 2025). They typically aim to organize tenants *en masse* and to change the structural conditions and policies which condemn tenants to a life of high rents, frequent evictions and low-quality housing. One common feature of the new breed of tenant organizations is that they provide a collective response to, and support for, issues faced by individual tenants, for example where a particular household is experiencing the threat of a rent increases or eviction, or is living in poor housing (Byrne, 2018; Guzmán, 2025; Munoz et al., 2024; Saunders et al., 2018). The support provided by tenant organizations involves providing information about tenants' rights, negotiating with landlords, media campaigns targeting specific landlords, and taking legal cases. It also includes direct action and civil disobedience to protect tenants, for example by blocking attempted evictions or intimidation by landlords (Byrne, 2017; Guzmán, 2025). For tenant organizers, this form of activism is about transforming individual housing crises into collective experiences, in order to de-individualize the experience of renting (Rosenthal & Vilchis, 2024; Stringer, 2025), but also to politicize that experience by showing that by working together tenants can overcome the power imbalance between landlords and tenants (Byrne, 2018; Gil & Palomera, 2024; Guzmán, 2025).

Collectively organizing tenants in response to the power asymmetries between landlord and tenants is (Racu, 2024), of course, reminiscent of the core logic of labour movements and of trade unions, as already noted in the discussion of historic rent strikes. In this respect, it is noteworthy that the most prominent form of tenant activism and organization is the tenants' union. Indeed, as Martin Luther King once argued, '[t]o produce change people must be organized to work together in units of power, [including] economic units such as groups of tenants who join forces to form a tenant union or to organize a rent strike' (quoted in Rosenthal & Vilchis, 2024: 99). It is no coincidence that the union form translates so well from the workplace context to that of rental housing. The relationship between landlord and tenant, as argued in Chapter 2, is characterized by a power imbalance with its roots in unequal property relationships, and generates opposing material interests. The control a landlord may exercise over a tenant's home, moreover, lends them power over the latter's conditions for social reproduction. In all of this, the nature of the landlord–tenant relationship closely resembles that between employer and employee. It is for this reason that the union form of organizing has played such a prominent role for both workers and tenants (Stringer, 2025).

In addition to supporting individual tenants, tenant organizations campaign for changes at the level of policy and legislation (Card, 2024; Kallin et al, 2024). While supporting individual tenants can be seen as 'reactive', in the sense that it responds to problems generated by the nature of the housing system, campaigning for policy change is proactive, in that it seeks to transform the housing system itself. In the next sections of this chapter, two in-depth case studies are provided of tenant-led campaigning. But at an international level, these types of campaign are proliferating. Living Rent emerged as a national tenants' organization in response to a consultation opened by the Scottish government in relation to security of tenure. In common with England and Wales, Scotland had one of the weakest forms of security of tenure in Western Europe. The inclusion of 'no-fault evictions' meant that tenants enjoyed effectively zero security. Living Rent used this opportunity to engage with tenants, to shape debate and discourse around tenants' rights and to impact on policy change. Their main tactic was to set up street stalls and go 'door-knocking' to engage tenants, inform them of the consultation and encourage them to make a submission to the consultation process. They designed a postcard addressed to the relevant government department which tenants could fill out. At the end of the process, Living Rent delivered sacks full of the postcard submissions to the consultation process. The focus of their campaign was on security of tenure including ending 'no-fault evictions' and securing long-term security for tenants. However, rent levels, which were not originally included within the consultation process, were also raised by Living Rent and their campaign included a demand for rent controls. The campaign met with a surprising level of success. Policy reforms were introduced that far exceeded what might have been anticipated. Indefinite security of tenure was introduced and regulation of rent increases, which was initially not even included in the policy agenda, was also introduced (Kallin et al., 2024).

Card's (2024) work on Los Angeles provides insight into the policy impact of the new generation of US tenant organizations. In LA, a city where 95 per cent of residents identified homelessness as the greatest threat to the city and 81 per cent said 'protecting tenants' is 'extremely important' or a 'major' priority for the city, the tenant movement has been growing since 2009, when a statewide coalition called Tenants Together was founded, with new tenant unions springing up such as the Los Angeles Tenants' Union (Card, 2024). In 2018, many of these organizations came together to campaign for legislative change (known as Proposition 10) that would repeal legislation that prohibited local government from effectively regulating rents. While the 'Prop 10' campaign failed, it helped to put rent regulation on the political map and eventually contributed to the passing of some forms of rent regulation in California a few years later, in 2020 (Card, 2024).

Tenant organizations have furthermore developed a language to speak from a tenants' perspective, reflecting the experiences of tenants but also articulating 'tenants' as a social actor and as a collective (Byrne, 2018). In this sense, they strive to develop the political identity of tenants (Gavin & O'Callaghan, 2024; Stringer, 2025). Creating this sense of solidarity is a specific goal of tenant organizations to counter the individualizing nature of renting (Wilde, 2022). Part of this process involves collecting and sharing individual experiences of renters. For example in both Ireland (Byrne, 2018) and England (Wilde, 2022; Stringer, 2025), tenant organizations have collected tenants' experiences of poor conditions, high rents and abusive behaviour by landlords. This type of intervention is significant because in many countries the issues experienced by tenants are more or less invisible in terms of the media or the political debate, which tends to overwhelmingly focus on homeowners (Byrne, 2018). This also relates to interventions in the media and in social media, both of which have been used to great effect by the new generation of tenant organizations (cf. Polanska & Richard, 2021). In many countries there were no organizations directly representing tenants in the PRS until recent years. Groups like the Barcelona Tenants' Union and Living Rent have thus filled a gap within the media ecosystem by representing tenants' perspectives and challenging landlord narratives.

In the remainder of the Chapter, I take a more detailed look at two of the most significant examples of what I call 'the new tenant politics', the first examining tenant unions in Catalunya and Spain, and the second looking at the recent campaign to expropriate corporate landlords in Berlin.

Tenant unions in Barcelona and Madrid: From everyday activism to legislative change

Among the most impactful of the new generation of tenant organizations are the tenant unions established in 2017 in Barcelona and Madrid (Guzmán, 2025). The Spanish housing system, as we have already seen in Chapter 3, has been characterized by enormous volatility over the last number of decades, with its extraordinary homeownership boom and bust, followed by a wave of repossessions and subsequent resurgence of private renting. The Spanish case certainly conforms to Fuller's (2019) argument that there is a relationship between the intensity of housing financialization in a given country and the extent to which housing is politicized in that country. More specifically, there are two contextual features that are important to understand before we delve into a discussion of these two remarkable new tenant unions. First, over the same period in which the proportion of households who rent privately doubled, the policy and market context for renters deteriorated. In terms of policy, the Urban Letting Act of 2013 shortened the tenancy period from five to three years and deregulated rents, as well as reducing the period after which a tenant can be evicted due to rent arrears. In terms of the market dynamics, average rents have increased markedly, especially in larger cities. In Barcelona, for example, average rents increased by fifty per cent in the decade to 2023 (Guzmán, 2025). These changes have made the precarious nature of housing for those locked out of homeownership one of the most salient challenges in terms of housing in Spain, and have brought renewed attention to the rental sector. It is also important to note that over this period there has also been a significant financialization of the PRS, with a growing role for institutional landlords such as REITs, private equity firms, etc.

The second important aspect of the context relates to the politics of housing in Spain and especially of social movements. The post-GFC context, with its unprecedented wave of mass evictions and home repossessions, gave rise to a social movement known as the *Plataforma de Afectados por la Hipoteca* (PAH), which could be translated as the Mortgage Victims Platform (Colau & Alemany, 2014; Di Feliciantonio, 2017). This organization, which originally emerged in Barcelona in 2009, grew to become a national grassroots movement and became central to housing politics in Spain. The PAH's main objective was to stop repossessions and evictions of homeowners. Moreover, under Spanish law homeowners whose property has been repossessed but are in negative equity are required to pay the outstanding loan amount. Given the extent of the property price collapse, this meant that households could find themselves evicted, homeless and subject to an enormous debt burden. The PAH thus also campaigned for this outstanding debt burden to be cancelled where a property had been repossessed by the mortgage provider.

What is most relevant about the PAH for the purposes of the present chapter is its approach to housing politics and to the political organization of residents. It is worth discussing this in some detail as it has been enormously influential on the Barcelona and Madrid Tenants' Unions (Guzmán, 2025), as well as more broadly in Europe and even further afield. The core foundation of their work was what we might call 'direct action case work'. This involved collectively supporting an individual household who is in danger of repossession and eviction. This support included legal support, such as supporting a household in negotiations with their mortgage provider. But it also involved undertaking direct action such as occupying the offices of the mortgage provider and, perhaps most importantly, using protests to physically block evictions (Gil & Palomera, 2024). This latter action, known as '*stop desauhcios*' (stop evictions), was not only effective in preventing households from becoming homeless, it also created leverage for the household over their mortgage provider and highlighted in the media and public debate the plight of 'mortgage victims' (Colau and Alemany, 2014). On the basis of these actions and tactics, the PAH spread across Spain and grew into the largest housing movement in Europe for quite some time.

Most importantly in terms of understanding tenant politics, the legacy of the PAH has been to introduce a number of important ideas into what social movement theorists call the 'repertoire' of housing movements (Guzmán, 2025). First, the PAH showed that while housing issues are usually experienced by households in an isolated and individualizing fashion, because these issues are typically faced by many households they can be transformed into collective issues and solidarity can be created between households facing similar challenges, as well as with society at large. As Guzman (2025: 850) notes, the PAH's 'success relied on transforming individual feelings of loneliness, guilt, and injustice stemming from financially broken homeowners into a shared political struggle to contest the lucrative debt driven housing regime actively promoted by the state and the financial sector'. In this sense, the PAH showed that political action can transform residents into agents within the political system. Moreover, the PAH showed that legal support, advice, accompaniment and advocacy for individual households could be combined with direct action and civil disobedience in a way that both produced concrete results for the individual households and politicized housing. With regard to the latter, it did so by challenging the power of dominant actors within the housing system, in this case banks and mortgage providers, and by highlighting how political institutions often supported these dominant players more than they did households whose right to housing was being undermined.

While the PAH continues to exist, in more recent years, as noted, attention has moved from the repossession crisis created by the financial crash, to issues of rent affordability and eviction experienced by private tenants. Guzman's (2025) research on Barcelona's tenant union provides an account of one individual, Rosa, whose trajectory captures this transition. She fell into mortgage arrears in 2013 after becoming unemployed during the GFC. Rosa attended the PAH's assemblies, and the PAH negotiated on her behalf with the eventual result that her outstanding mortgage debt was cancelled in 2015. She was able to remain in her house as a tenant paying a 'social' (i.e. affordable) rent to her former mortgage provider, Caixabank. However, when her four-year tenancy came to an end, Caixabank increased her rent by fifty per cent to bring it into line with market rents. Having experienced the sharp edge of heavily indebted homeownership, Rosa now found herself experiencing the challenges increasingly plaguing private tenants. In response to this rent hike, Rosa joined the Tenants' Union of Barcelona (TUB) and became a tenant activist.

The core political logic of the Spanish tenant unions is that individually tenants are weak but collectively they are strong (Racu, 2024), and thus that collectively organizing tenants, and especially a union model, can reverse the power asymmetry between landlords and tenants, empowering tenants with regard to their individual housing issues but also at a societal level. The TUB (at the time of writing) has almost 3,000 members (Guzmán, 2025). At a practical level, the TUB organizes regular 'tenant assemblies' where tenants can come to receive support in relation to a rent hike, eviction or similar issue. The tenant assembly is also where support for that individual is organized. This may include, just like the PAH, legal support and advice as well as direct action and campaigns (Guzmán, 2025). The need for direct action arises from the fact that the legislative and policy framework in Spain is often unable to protect tenants from an eviction and rent increase, thus requiring an extralegal response. At a discursive level, the TUB highlights the conflict between tenants and landlords in order to construct a political identity among tenants (Guzmán, 2025).

In recent years, one of the most common issues experienced by tenants attending one of the TUB's tenant assemblies is what the Union calls 'invisible evictions'. This refers to when a tenant, at the end of a tenancy cycle, receives a rent increase from their landlord. If they don't accept this increase their tenancy will not be renewed, and they will be evicted, something that is provided for under Spanish rental legislation. This led to the establishment of the *Ens Quedem* (Stay Put) campaign in 2018, which encourages tenants to reject rent increases and remain in their homes. Tenants taking part continue to pay their original rent, despite the tenancy having legally expired. The TUB communicates with the landlord on the tenant's behalf in an effort to enter negotiations. If they are unsuccessful in this, the Union will typically attempt to exert pressure on the landlord through activities such as occupying the landlord's offices or those of the relevant real-estate agent, organizing demonstrations, and canvassing door-to-door to generate support for the tenant (Guzmán, 2025). The ultimate goal is to pressure the landlord into agreeing to a 'fair rent' and renewing the tenancy.

In some cases, the tactics of the Stay Put campaign are applied by multiple tenants against a shared landlord, one such example being the case of 121 households in the small city of Sant Joan Despí, near Barcelona. The apartments of these households had been acquired by Goldman Sachs, which attempted to introduce rent increases of seventy per cent on average:

> [M]any households started to organize by holding meetings in several carparks and by joining the Union. They created banners during collective workshops and hung them from their balconies ... They held a general assembly during which they unanimously decided to stay put once their contracts expired, to keep paying the same rent, and to keep fighting until they achieved a fair negotiation. Finally, after what many claim to be the biggest demonstration in Sant Joan Despí since the 1970s, Goldman Sachs accepted a collective bargaining process, with the Union representing all of its tenant members – this was the first such process in Spanish history to take place between a corporate landlord and a tenant organization ... In the end, Goldman Sachs and the building residents reached an agreement on a moderate rent increase. (Gil & Palomera, 2024: 640)

In cases where evictions do occur, the tenant unions are also involved in organizing anti-eviction actions. The wider political logic of the Stay Put campaign is that '[b]y achieving individual tenant victories, such as signing new leases with affordable rents, the Stay Put campaign aims at accruing tenant bargaining power' (Guzmán, 2025: 861).

The Stay Put campaign is an example of how direct action case work can generate 'tenant power' vis-à-vis their landlord in response to specific housing issues experienced by tenants. The other dimension of the work of the Barcelona and Madrid Tenants' Unions is their role in generating political leverage for tenants in terms of regional and national politics: 'Since the unions were created, their priority has been to influence and directly draft housing policies through a bottom-up approach, in order to change the rules of the housing system' (Gil & Palomera, 2024). For example, the TUB successfully campaigned for the introduction of rent regulation by the Catalan Regional Parliament in 2020, and they worked with the Madrid Tenants' Union to campaign for national legislation, eventually enacted in 2023. Gill and Palomera (2024: 630) describe this aspect of tenant activism in terms of 'counter-hegemonic legislative strategies' which seek to 'create, maintain, and escalate active social conflicts around housing ... through organized offensive tenant disobedience, by creating new narratives and by challenging the prevailing common sense around housing'. In terms of the context for the Union's campaigning work, from 2020 Spain has been governed by a coalition between the centre left and the more radical left. The inclusion of Podemos, a radical left party established after the GFC, was particularly significant due to its close relationship with the PAH movement, mentioned above. A similar coalition also existed in Catalunya, which is where the tenant unions had their first major victory in terms of legislative change.

During the summer of 2020, the TUB engaged in negotiations with the Catalan government as well as various other parliamentary parties, with the aim of achieving an ambitious form of rent regulation. It is worth noting that, prior to the formation of the tenant unions, rent regulation was not on the political agenda either in Catalunya or in Spain (Gil & Palomera, 2024). The union employed protest actions, media work and negotiation with political parties to win over a parliamentary majority for their rent control proposals. The legislation was approved in September 2020, and was viewed by the union as a major victory (Gil & Palomera, 2024). The legislation introduced rent regulation in Barcelona and 61 other cities, and even included measures of rent reduction in some instances. Two years later, it was found to be unconstitutional on the grounds that the Catalan parliament did not have the power to regulate rents, and that this could only be done on a national level. However, the tenants unions then worked at a national level using the same tactics of campaigning, protesting and negotiating with political parties to influence the formulation of new housing legislation that provided for rent regulation at a national level, which was introduced in 2023 (Kallin et al, 2024). While this legislation was somewhat watered down, it was viewed by the unions as a qualified success. The role of tenant unions in legislative change should not be overstated, as there were a variety of political forces that favoured such change, in particular the prominent role of far-left 'challenger' parties (e.g. Podemos) and successive socialist-led governments. However, the above certainly demonstrates the agency of tenant unions, and therefor of tenants themselves, in shaping the development of housing systems at the national level

(Gil & Palomera, 2024). The prominent role of the Barcelona and Madrid organizations in street-level mobilization, media discourses and direct negotiation with government at municipal and national scales indicates the renewed significance and influence of tenant activism in shaping political and policy developments in the sector.

Deutsche Wohnen und Co.: Enteignen: expanding the horizon of housing politics

In 2021, German politics was rocked by the passing of a radical referendum in Berlin calling for the expropriation of hundreds of thousands of rental housing units owned by corporate landlords. This seismic event in housing politics raises the possibility of the radical redistribution of housing as one way to respond to the housing crises that are increasingly common across Europe. The Berlin context is notable because Germany is one of the very few advanced economies where homeowners are the minority, and in the city itself 85 per cent of households are renters (Dancygier & Wiedemann, 2024). Berliners have also faced challenges in terms of housing supply and affordability in recent years. Research suggests that 25 per cent of German households are somewhat or strongly concerned about being forced out of their neighbourhood because of rising rents, a figure which rises to 33 per cent for cities with more than 100,000 inhabitants (Dancygier & Wiedemann, 2024). Until the mid 2000s, Berlin enjoyed moderate rent levels. But between 2009 and 2018, average rents increased by forty per cent, and asking rents by more than eighty per cent. Moreover, forty per cent of Berliners pay more than thirty per cent of their income on housing costs, a common indicator of unaffordability (Vollmer & Gutiérrez, 2022).

In response to these challenges, housing and tenant organizations had been growing in Berlin for many years (Card, 2024). Vollmer and Gutiérrez (2022) highlight the examples of Kotti and Co. and the Otto-Surh tenant organization. The former started in 2011 and organized in opposition to rent increases for tenants living in former social housing. Otto-Suhr, similarly, organized tenants of former social housing. Around 2016, the two organizations worked together across common issues faced by tenants of Deutsche Wohnen, Berlin's largest landlord and a symbol of the financialization of the city's PRS. The issues faced by such tenants included rent increases and lack of maintenance and repair. Vollmer and Gutiérrez (2022: 65) thus argue that 'the campaign to expropriate Deutsche Wohnen & Co emerged from a strong and diverse tenant movement capable of aligning lower and middle-income interests and shaping the public discourse on a housing state of emergency'.

In terms of the campaign itself, *Deutsche Wohnen & Co Enteignen* (Expropriate Deutsche Wohnen & others, henceforth DWE) is a movement striving to transfer formerly public housing that is now in the hands of these corporate landlords back into public ownership. The referendum proposed the 'expropriation' of private rental housing owned by institutional investors with more than 3,000 units, which would result in the expropriation of approximately 240,000 units, representing one of the most radical forms, if not the most radical form, of housing redistribution ever undertaken in a capitalist society (Dancygier & Wiedemann, 2024). The campaign aims to tackle the housing crisis (rising rents, limited supply) at its roots, reshape property relations within the city, and to give people an avenue through which to shape housing politics, instead of just reacting to them. Dancygier and Wiedemann (2024: 15) sum up the politics of the campaign as follows:

> Their growing market share, the campaign argued, allowed these companies to push up rents citywide, while their quest for profits translated into cost-cutting leading to poor maintenance and service. The appropriate policy solution was therefore expropriation. In the view of the campaign, public ownership would not only ensure that citizens no longer finance the profits of investors but also improve affordability directly: the public entity owning and managing expropriated apartments would set rents directly; preserve affordable housing and prevent displacement in areas facing gentrification; increase housing supply by densifying properties; and allocate housing based on need, not income.

At some points during the campaign, up to several hundred activists joined the organizational assembly of the campaign, and Vollmer and Gutiérrez (2022: 59) argue that approximately 2,000 people were active at some point or another.

Even though the expropriation of these rental properties may sound like a radical measure (certainly more radical than a rent cap, a measure that was deemed unconstitutional in 2021 after the Berlin senate tried to implement it), it has a legal basis in the German constitution. Article 15 of the German constitution states:

> For the purpose of socialization, land, natural resources, and means of production can be transferred to public ownership or other forms of public economy through a law that regulates the type and extent of compensation.[5]

Additionally, Article 14(3) of the constitution clarifies that:

> Expropriation is only permitted for the public good. It may only be made by law or based on a law that regulates the type and extent of compensation. The compensation must be determined by fairly weighing up the interests of the general public and those involved.

Therefore, the German constitution permits the socialization of land, means of production and natural resources for the common good if the previous owners are adequately compensated. Although this article of the constitution has never been used before, and therefore has no legal precedent, the DWE campaign argues that it provides legal ground for the expropriation of properties from corporate landlords. Moreover, it must also be noted that much of the stock which the campaign focuses on originated in the social/public sector and was privatized over recent decades. Thus, the call for expropriation is in a sense a call to turn the clock back and to restore the more non-market-oriented housing regime of the post-war era (Dancygier & Wiedemann, 2024).

DWE has drawn up several models detailing how corporate landlords could be compensated in case of expropriation. These models consider factors such as the number of properties being expropriated and the balancing of interests between corporate landlords and the broader societal good. While they each differ, the underlying logic is that compensation should be at below-market prices. One model proposes the funding of compensation, management expenses and upkeep exclusively through the (lower) rents paid by tenants in the collectively managed properties. This would naturally result in relatively low compensation payouts to corporate landlords. Other models focus solely on compensating

---

5 Thanks to Cora Saxenberger for her research assistance on the DWE campaign. The two articles of the German constitution were also translated by her.

for the land value of the properties, while some also factor in elements such as inflation. There is even a suggestion that a symbolic compensation as low as €1 could be feasible.

Should the expropriation be successful, what happens with the properties is determined by two crucial principles of the DWE campaign: *Vergesellschaftlichung* and *Gemeinswirtschaft*, describing, respectively, the form of ownership/management and the purpose of the properties. *Vergesellschaftlichung* (socialization/transferral into common ownership) refers to the public or common ownership and management of the properties. The expropriated properties are to be commonly owned and democratically managed. DWE aims to create a system that combines direct citizen participation with representation through district councils. While DWE plans on creating a new, independent institutional body tasked with monitoring the management of the properties, it also emphasizes a grassroots or 'bottom-up' structure focused on empowering residents. Additionally, citizens of Berlin who do not live in expropriated properties will also have the opportunity to participate in decision making in some form.

Gemeinswirtschaft (common administration) refers to the purpose according to which these properties will be managed. DWE argues that the management of the expropriated properties should be on the basis of the common or social good of the residents and the city of Berlin. Within the properties, they aim to create affordable living spaces for current and future generations, encourage housing construction, give space to small businesses, and create child and youth centres, art spaces and spaces for political groups. They also plan to offer accommodation for refugees, create domestic violence shelters, foster accessible living, and prevent evictions and homelessness. They emphasize ecological and environmentally sustainable politics, the use of renewable energy, a focus on social politics, the elimination of discrimination, and the blockage of future privatization. The rationale behind these management and ownership values is described as an economy of solidarity, an opportunity to envision more equitable and sustainable ways of distributing housing and political power in the city of Berlin.

So far, the initiative has received widespread support from various stakeholders, such as trade unions and tenant unions, ecological movements and local businesses, as well as from the majority of Berlin citizens, according to DWE's surveys. DWE brought their idea before the Berlin senate in the form of a referendum which, after several rounds of collecting signatures, and repeatedly surpassing the required numbers, was held in September 2021. 59.1 per cent of votes supported the expropriation campaign. Therefore, the Berlin Senate was required to take measures to facilitate expropriation and socialization, a task which was assigned to an expert commission. Danciger and Wiedman (2024) analyse political support for the DWE movement, and find that it is mainly driven by the values Berliners hold in relation to housing which emphasize its importance as a social right as opposed to a financial asset.

The DWE campaign has been very critical of the new centre-right Berlin government, which they accuse of failing to fulfil its democratic duty (since the Senate made little effort to facilitate the expropriation of corporately owned rental properties). Therefore, in September 2023, DWE decided to re-start the process for a referendum, this time with an already formulated legislative proposal which the citizens will decide on directly. However, as of May 2024, they are still working on the formulation of said proposal and no specific date for a new referendum has been presented yet.

Stepping back, the DWE case demonstrates a number of important points in terms of contemporary housing and tenant politics. It shows that the dynamics of the political economy of the PRS in a particular context will shape the forms of protest and conflict that are likely to arise. In this case, the central role played by financialization in the development of the German PRS over the last two decades, and the specific form it has taken with regard to the mass privatization of former social and public rental housing, is central to understanding how and why a form of tenant politics focused on mass expropriation emerged. Second, it provides further evidence, already clear from other cases such as Ireland (Byrne, 2019) and Spain (Gil & Palomera, 2024), that corporate and financial landlords are more likely to be the subject of contentious politics and a political 'backlash' (Dancygier & Wiedemann, 2024). As Dancygier and Wiedemann (2024: 11) argue: 'the uncoordinated transactions of ordinary citizens and small landlords are less likely to evoke worries about unchecked market power associated with large-scale investor purchase ... citizens may perceive institutional investors not simply as neutral market actors, but as powerful operators representing an excessive form of capitalism that harms citizen welfare'.

Beyond this, what is most interesting about the DWE example is how tenant activism can radically expand the horizon in terms of housing politics. Their demands, whatever one's view might be on how feasible or indeed desirable they might be, involve an extremely ambitious form of mass decommodification or PRS housing, as well as the mass redistribution of housing from the PRS to the public sector. The campaign thus aims at an extensive 'deconcentration' of ownership of residential housing, thus directly addressing the core political economy dynamic analysed in Chapters 3 and 4. Although it has yet to be achieved, the DWE example shows how the contentious and antagonistic nature of the PRS can give rise to its politicization in ways which can make possible its radical transformation.

Conclusion

Tenant politics has a long history, but in recent years we have seen a revival of activism in the PRS. This relates to the transformations at the level of political economy discussed throughout this book. The processes of financialization and neoliberalization that have characterized housing systems in many advanced economies in recent decades have given rise to issues of affordability, insecurity, increased difficulty in accessing homeownership, and greater housing inequality. They have gone hand in hand with the concentration of ownership of residential housing among better-off households and financial institutions. These developments, combined with an understanding of the inherently antagonistic relationship between landlord and tenant, help us to understand the dramatic politicization of the PRS, discussed throughout this chapter, and the associated resurgence of tenant activism. As Gil and Palomera (2024: 629) argue, '[t]he post-2008 crisis resolution regime has thus fostered new struggles and the emergence of various forms of tenants' organizations worldwide'. The new generation of tenant organizations is focused on three core areas of activity: forms of collective support for tenants facing specific housing challenges; the creation of tenant unions, i.e. membership-based structures; and campaigning for legislative change.

Through these activities, tenant organizations seek to create what we might call a 'tenant identity', or the emergence of tenants as a collective political subject, i.e. a self-conscious political actor. From everyday forms of activism at the neighbourhood level, to communication in traditional and social media, activists strive to create discourses and narratives that provide a shared sense of housing experiences among tenants, and to politicize these experiences by highlighting how they arise from political choices, and how these choices typically favour landlords over tenants. The task of tenant organizations is thus to make the latent antagonism between landlords and tenants explicit (Gil & Palomera, 2024), and 'to organize tenants within relationships of mutual aid, where tenants' struggles not only produce individual outcomes but also contribute to the formation of new political identities and subjectivation processes' (Gil & Palomera, 2024). Most importantly, tenant organizations are about 'reversing the asymmetrical power relation between tenants and landlords' (Polanska & Richard, 2021: 202) at an everyday level, but also at a wider social and political economic level, and thus to transform both everyday housing realities and the housing regime at the level of institutions and legislation. Forms of tenant activism can thus 'catalyse a conflict to reconfigure the power relations between tenant and landlord' (Guzmán & Ill-Raga, 2022), and aim to 'organize tenants as political subjects whose task it is to beat back the power of real estate and change the world' (Rosenthal & Vilchis, 2024: 11).

Conclusion
Transforming the Social Relations of Home

Tenant politics, as discussed in the previous chapter, directly targets the power imbalance between landlord and tenant and the unequal property relations associated with the PRS, and aims to transform the housing experience of renters. Many tenant organizations, unsurprisingly, also advocate for specific demands for the reform of the rental sector, including enhanced security of tenure and the regulation of rents, as well as more radical demands such as that of expropriation discussed in the context of the DWE campaign in the previous chapter. These issues raise the possibility of the transformation of the PRS and questions about how we should think about that transformation. This chapter builds on the framework set out throughout the previous chapters to address these questions.

The analysis and critique developed throughout this book highlight the power dynamics and inequalities that characterize the PRS. The perspective adopted is one that seeks to denaturalize the PRS, i.e. to avoid assuming *a priori* that private rental housing is natural, normal or inevitable. The critical orientation of such a perspective can contribute to how we think about the PRS, its role within the housing system and its future as a housing tenure. In this concluding chapter, however, I do not wish to discuss, and even less so recommend, specific forms of policy intervention to address the issues that are all too common in rental housing across many contexts. The reason is simple: the appropriateness of policy responses is entirely dependent on specific contexts and must be discussed in relation to the structures of a given housing system, the specific nature of the problems at play, the dynamics of supply and demand and other relevant public policy objectives. Policy objectives must also reflect democratic wishes within a given society. Generally speaking, I take the approach that there is not one set of 'correct' policies, but rather each society must decide what it is they want out of their housing system and then consider which policy interventions are most likely to deliver that. Rather than delving into the evidence around particular forms of rent controls or the appropriate structure of rent subsidies, it is more useful and appropriate, in the context of a book such as this, to set out an *orientation* to how we think about PRS transformation.

Drawing on the arguments threaded throughout this book, I think about this orientation in terms of addressing the major obstacles associated with PRS housing from the point of view of *equality* and of the *experience of home*. Throughout this book we have seen two major recurring concerns. On the one hand, the micro-political economic dynamics that manifest in the landlord–tenant relationship, and in particular the form of 'market tyranny' landlords may exercise over tenants' homes. On the other hand, the wider macro-political economy structures and processes associated with the distribution of housing and hence with the concertation of ownership of residential housing, be it in the hands of multiple-property-owning households or institutional landlords. Consequently, in thinking about the transformation of the PRS, these are, I argue, the two core dynamics that we should seek to overcome or at least ameliorate. While there are potentially many ways that these can be addressed, here I discuss two broad categories of intervention: *housing demarketization* and *housing deconcentration*. The first of these is focused on transforming the micro-political relations between landlord and tenant and therefore the experience of home and of inequality within that experience. The second of these is focused on the distribution of housing and therefore on the macropolitical-economic-structural inequalities associated with property relationships at a social level.

Demarketization: Security and affordability

The concept of de-marketization I develop here does not refer to the replacement of private rental housing with public housing, or some other form of decommodified housing, but rather to mitigating, eroding and transforming market-based social relationships and practices within the context of the PRS. For the most part, this means eroding the power of landlords over tenants and therefore enhancing the latter's experience of security, safety, control and privacy within the home. The two most important avenues through which this takes place relate to security of tenure, on the one hand, and rents and affordability on the other. Starting with the issue of security, this goes beyond the ability of renters to remain in their home for a long time period. It also includes the predictability of the tenancy, i.e. the extent to which the tenancy can only end under defined circumstances and subject to some sort of regulatory oversight; what these defined circumstances are and how predictable they are; and to what extent these circumstances are under the control of the tenant. Moreover, the nature and quality of the relationship between tenant and landlord is crucial. In Chapter 6 I argued that privacy and control are central to tenants' experience of home. With regard to the former, the ability to determine who comes and goes from the dwelling, including the landlord, is usually the principal issue. With regard to the latter, what is most important is the tenant's ability to shape the nature and design of a dwelling, as well as the types of practices that can take place, such as hanging pictures, painting and decorating, having guests stay over, smoking or having pets.

All of these issues can be dealt with via legislation that protects tenants, guarantees security of tenure and sets out the rights and obligations of landlords, as of course happens in many countries. All such forms of legislation involve enhancing the rights of home over the rights of property, and therefore enhancing the rights of residents over owners. A central aim of the transformation of the PRS should be to move the sector in this direction, to the extent that it is possible within a given set of circumstances.

Rent regulation, similarly, can restrict the power of landlords to set rents or to introduce rent increases. The issue of rent relates to both the experience of home and to wider socio-economic inequality, as discussed in Chapter 5. In terms of home, the power to increase rents undermines security of tenure in general, and can also be used as a retaliatory mechanism by landlords, and thus contributes substantially to the power asymmetry between landlords and tenants (Kallin et al., 2024). But high rents of course also contribute to the affordability problems that disproportionately affect private tenants and involve the transfer of income from generally lower-income tenants to generally higher-income landlords, which fosters both income inequality and wealth inequality. The regulation of rents can take many different forms, and there is a large debate on their design and impact (Marsh et al., 2023); the point here is that any system that moves rent setting away from market dynamics and reduces the power and autonomy of landlords with regard to rents can play a role in demarketizing the PRS, challenging the unequal power dynamic between landlord and tenant, and consequently enhance tenants' experience of home (Kallin et al., 2024). Another mechanism which can address some of the issues with high rents and affordability is rent subsidies, i.e. income subsidies that cover a portion, or all, of a tenant's rent costs. Such policies exist in many countries in one form or another. Rent subsidies, by reducing affordability issues for low-income tenants, can mitigate inequality and support wider access to the market. They can also play a redistributive role just as any social transfer can. Rent subsidies, however, like any policy, can raise some challenges or unintended consequences, in particular inflating rent prices by subsidizing demand, and they also involve the transfer of public money to landlords, which may reinforce socio-economic inequality.

Finally, all of the above measures are only meaningful in the context of robust enforcement. Non-compliance with legislation and 'black markets' are commonplace across many jurisdictions, which speaks to the fact that the PRS is generally challenging to regulate. There are many reasons for this, including the economics of the sector (Harloe, 2021), but chief among them are that the vast majority of rental sectors are dominated by relatively small-scale, non-professional landlords (i.e. landlords with another source of income), and the fact that the landlord–tenant power asymmetry undermines tenants' ability to advocate for themselves effectively, e.g. for fear of retaliatory evictions (Byrne & McArdle, 2022). Strengthening security of tenure and limiting landlord control over rents can support effective enforcement because they empower tenants to advocate for themselves. The existence of large and effective tenant unions, as in Sweden and Germany, can also play an important role in supporting tenants to challenge abusive or non-compliant practices, thus facilitating enforcement within the sector and generally fomenting a more professional culture.

Deconcentrating the ownership of residential property

The above discussion of demarketization addresses a suite of forms of intervention in the PRS which are, of course, already widely implemented in many (though not enough) jurisdictions. The second set of interventions are perhaps less often explicitly discussed with regard to addressing issues in the PRS. This is because they don't relate directly to the PRS itself, but rather to the transformation of the housing system more widely, and specifically the distribution of the housing stock and the consequent degree of concentration of ownership (Baxter-Clow et al., 2022). Here I (briefly) discuss three forms of intervention which all relate to *deconcentrating ownership of residential property*: growing alternative tenures; disincentivizing or restricting PRS investment; and the direct redistribution of residential property.

Growing alternative tenures

The development of alternative tenures, historically, played a central role in the decline of the PRS and the development of more equitable housing systems, as discussed in Chapter 3. Although not typically discussed in these terms, this involved altering the distribution of ownership of residential property such that ownership was more widely distributed. This can happen primarily via the development of social or public housing or via the expansion of homeownership. The expansion of homeownership, mainly via direct supports of one form or another to low- and middle-income households, led to housing systems in which property ownership became much more dispersed among all households. Of course, there were and continue to be households and social groups excluded from this process (Florida & Feldman, 1988), but overall the deconcentration of residential property ownership involved was fundamental to the development of more egalitarian housing systems, in terms of experience of housing (e.g. greater security of tenure, less dependence on landlordism) and also in terms of wealth inequality. This does not mean that promoting homeownership is not without its downsides, or that it is necessarily the appropriate response today. Rather, it is simply to note that more equal housing systems can be developed in this fashion. Of course, in many countries the 'ideology of homeownership' remains dominant, such that it is the preferred or aspirational tenure for the majority of households, which makes pro-homeownership policies often politically palatable.

There are, of course, other ways to redistribute ownership in a more egalitarian fashion besides supporting individual ownership of residential property. These include forms of ownership such as cooperative housing and community land trusts, or what scholars sometimes refer to as 'housing commons' (Ferreri & Vidal, 2022; Huron, 2015). Cooperatives, in particular, make possible forms of non-individual ownership of residential property, enhancing security, autonomy, control and affordability, and supporting a more equitable distribution of housing (Balmer & Gerber, 2018; Ganapati, 2010). Both cooperatives and community land trusts continue to be popular approaches with housing activists all over the world, and although they typically face challenges due to lack of government support and the dominance of private property models, in principle there is no reason why they could not become major housing tenures.

It should also be noted that while supporting individual property ownership may be seen as a form of neoliberal housing policy, as in the famous cases of Thatcher's 'property-owning democracy' and Bush Senior's 'ownership society' (Béland, 2007), there have also been cases of more radical approaches. The Singapore housing system is a case in point. Singapore has an extremely high homeownership rate (as high as 90% by some estimates), but this is overwhelmingly provided through state intervention (Phang, 1996). Most interestingly, from the point of view of redistribution, its housing system was built on mass state compulsory acquisition of land as part of the post-independence process of decolonization (Haila, 1990, 2015). The post-colonial redistribution of land also played a role in the development of mass homeownership in Ireland (Norris, 2016).

The other main avenue through which ownership can be deconcentrated is through widening public ownership of housing in the form of social or public housing. This is of course different from the more equitable distribution of ownership among individual households. Instead, it works by reducing the proportion of the housing stock which is privately owned, and supporting collective ownership through the state. Of course, this also has the benefit of demarketizing or decommodifying housing, supporting security of tenure, and enhancing affordability. Once again, there are a variety of ownership models through which this can take place. In many countries, such as Austria, social housing originally developed via associational bodies, including workers' associations, which led to the development of what today are known as limited-profit housing associations, which own roughly twenty per cent of the housing stock, and as much as forty per cent in Vienna (Marcuse, 1986). In Denmark, following agitation from social tenants in the 1970s, ownership of social housing associations was democratized such that tenants collectively own the housing estate or development that their home is part of and elect the board of management (Høghøj, 2023). In many other countries, of course, social housing was developed, owned and managed by local authorities, municipalities or other state institutions. Social housing is not, of course, a panacea. It is important to be cognizant that when private ownership is replaced by public ownership the power relationships between landlord and tenant do not disappear, but are rather reconfigured, typically taking on a form which is characterized by bureaucratic power rather than market power and the profit motive. Nevertheless, it is clear that non-market rental housing has been much more effective at delivering secure, higher-quality and more affordable homes in many countries.

Both social housing and homeownership, it has been argued, can lack the flexibility and low barriers to entry of private rental. The PRS has typically served households seeking quick access to housing or transitory housing, such as recently arrived immigrants, students, or individuals experiencing family breakdown. Social housing, which usually involves waiting lists, and homeownership, with its high transaction costs and the legal complexity of purchasing, are less suitable here. In this regard, it is worth considering the potential of non-market forms of flexible, transitory or short-term housing that may be able to offer similar flexibility and low barriers to entry but in a more egalitarian, secure and affordable fashion. One such example is publicly supported student housing in Denmark, which is provided via a similar model to the country's social housing sector, but is set up to meet the short-term housing needs of university students. This model has helped ensure that Denmark has one of the lowest levels of young adults remaining in the family home of any European country, at just 3.6 per cent (compared with Mediterranean countries, for example, where the figure is more than 45%) (Disch & Slaymaker, 2023). It is not impossible to imagine some form of public tenure that could meet these kinds of needs, especially of recent migrants, who in many countries face the worst housing conditions in the PRS.

Disincentivizing and restricting PRS investment

The measures discussed above involve a re-balancing of housing systems via the expansion of tenures other than the PRS. However, the concentration of ownership of residential property can also be reduced via the direct disincentivizing or even restriction of PRS investment. Such measures can, once again, take many forms. Indeed, strengthening tenant protections such as enhanced security of tenure or rent controls can potentially play this role, although our understanding of the direct relationship between PRS regulation and investment is limited (Marsh et al., 2023). An example is the Danish 'Blackstone law', discussed in Chapter 4, which disincentivized the acquisition and renovation of PRS stock by institutional investors by tightening rent regulation. Other measures can include taxation, either targeting rental property in general or specific cohorts. In Ireland, for example, stamp duty on the purchase of residential property was recently increased in the case of institutional landlords purchasing ten or more single-family dwellings (Byrne, 2024). The planning system can also be used, as has recently been done to restrict buy-to-let investment in the Netherlands (Hochstenbach, 2023). Financial regulation is another avenue through which investment in rental property can be restricted or discouraged, for example by regulating buy-to-let mortgages (as in the case of Ireland). Finally, regulation of short-term lets in response to Airbnb and similar platforms looks likely to be another important area where the concentration of property ownership can be challenged.

While there has recently been an increase in these types of measures in some countries, in response to the resurgence of private renting, throughout the twentieth century many such measures played a role in creating property markets which disadvantaged landlordism and therefore tipped the balance in favour of either homeownership or collective ownership, or both (Harloe,

2021). The extent to which such policies will be appropriate or wise, of course, depends on specific contextual factors. These include the extent of demand for PRS housing and the availability of alternative tenures. Nevertheless, influencing investment dynamics has been, and is likely to continue to be, one of the crucial ways in which the nature of the housing system, and in particular ownership of residential property, can be reshaped.

The direct redistribution of residential property

The role of land redistribution in the development of the Singaporean (and Irish) housing system during the twentieth century was already alluded to above. It is, however, also possible to directly redistribute existing residential property. This can take place (and indeed has taken place) via three main avenues. First, the acquisition of existing private rental stock via local authorities or other public/collective bodies. When local authorities acquire such property it is sometimes referred to as 'municipalization', and there are some contemporary examples of this in places such as the UK (Baxter & Elliott, 2024; Diner, 2023). This process obviously involves the transfer of housing stock from private to public hands, thus deconcentrating ownership, and can support the provision of secure and affordable non-market housing.

Second, PRS housing can be expropriated by public institutions. This is obviously a more radical measure which will likely run into issues of constitutionality in many countries. Nevertheless, as we have seen in the case of the DWE referendum campaign in Berlin, expropriation may sometimes be legally possible. Moreover, in more 'revolutionary' contexts, such as the Soviet Union, mass housing nationalization was a central housing policy (Kalyukin & Kohl, 2020). The nature, extent and politics of expropriation can thus vary considerably, and I appreciate that readers will have different views about the legitimacy of such measures. Nevertheless, expropriation is the most immediate way to undermine the concentration of ownership of property and radically redistribute it.

Third, squatting is an important if much-neglected feature of many housing systems. Housing scholars focusing on the Global South have long emphasized the crucial role squatting, mainly in the form of informal settlements, plays in processes of urbanization and in the housing systems of cities like Nairobi and Lagos (Vasudevan, 2015). In the advanced economies, however, squatting also played an important role in certain contexts, notably in the wake of the Second World War (Watson, 2019) and in counter-cultural movements in Europe in the 1970s (Katsiaficas, 2014). But the practice still continues today. It was an important tactic of the Spanish PAH (*Plataforma de Afectados por la Hipoteca*) movement, discussed in Chapter 7, in the wake of the post-GFC foreclosure crisis, for example (Di Feliciantonio, 2017a).

Squatting is an interesting example of deconcentration because it does not redistribute housing among private households or from public to private, but rather from private ownership (and sometimes public ownership) into a form of 'non-ownership'. For this reason, it has been argued that squatting can represent a form of 'commoning' of housing and urban space and the pre-figurative construction of new forms of providing and managing housing (Di Feliciantonio, 2017b; Polanska & Weldon, 2020; Ruiz Cayuela & García-Lamarca, 2023). Of course, squatting is illegal and is often subject to police action, which means it is not always a sustainable form of housing, and readers will, once again, have different views on its legitimacy, but it remains one of the most direct ways in which the ownership of residential property is politicized, contested and re-imagined.

The direct redistribution of residential property from the PRS may strike some readers as radical, and indeed some of the above measures are radical. But here it should be borne in mind that the redistribution of housing in the opposite direction, i.e. from public to private hands, has been a normal and mainstream form of housing policy intervention for many decades. It has taken place through the mass sale of council housing to individual tenants in countries like Ireland and the UK, through the mass sale of former social housing to private equity funds in countries like Germany and Sweden, and through the wholesale privatization of the housing system in the case of former Soviet Union countries in Central and Eastern Europe. This raises the very interesting question as to why it is seen as politically feasible to redistribute property from public to private and to foment the concentration of ownership of residential property, but not the reverse? This is thus an avenue of housing system transformation that merits further consideration and indeed research.

Conclusion

The arguments put forward above are not meant as 'policy fixes' to solve the specific issues manifesting in different specific contexts. They instead try to sketch out a normative and political orientation that can underpin how we think about the transformation of the PRS at a general level. As evident throughout this book, this view is predicated on a commitment to housing justice in the sense of creating a more equal housing system and in the sense of reducing the unnecessary suffering of tenants by extending access to secure, affordable homes. More specifically, the ideas of demarketization and deconcentration as what we might call guiding principles of PRS, and wider housing system, reform take aim at the power asymmetry between landlord and tenant that manifests at the everyday level of the social relation of home, on the one hand, and the structural transformation of property relations at a society-wide level, on the other.

As mentioned above, the extent to which any specific measure is appropriate or wise in a given context is, however, always going to be context-dependent. Rent regulation, for example, can be effective and has a strong normative rationale, but it may not be the smart move in contexts which are particularly supply constrained or where affordability is not the most pressing issue. A further caveat is warranted here. As evident from the above, I don't discuss the supply of PRS property. In a sense the supply issue is addressed above in that I argue for the expansion of alternative tenures, but beyond this I don't have much to say on it. This may strike the reader as odd because in much public debate supply seems to be the only thing talked about. The supply of housing is indeed crucial, and has a direct bearing on the issues discussed throughout this book. Where supply is limited and few rental properties are available, landlords are further empowered vis-à-vis tenants. Conversely, in markets with ample supply tenants will have much more power in negotiating with their landlord, as demonstrated in recent research on the Romanian PRS (Soaita, 2024). However, the availability of PRS housing is tied up with a much wider set of factors relating to the supply and demand of housing, from population growth, levels of household formation and household sizes (on the demand side), to the planning system, finance and the availability of skilled workers (on the supply side).

These issues go far beyond the matters that directly relate to the policy and politics of the PRS, and indeed they go beyond my own areas of expertise, and so I will leave those questions to others.

The future of the PRS and its relationship to inequality remains to be decided. Whether the sector will continue to play a prominent role in advanced economies, or go into another phase of decline, remains to be seen. Certainly, there is plenty of indication that economic incentives will drive the further accumulation of residential property by wealthier households and large institutions. But there are already clear signs that policy-makers are pushing back against this tendency (Byrne, 2024; Hochstenbach, 2023). While market factors and policy trends will play an important role, tenants themselves, and the wider politicization of housing, are likely to be just as important in shaping the future transformation of the housing system. Hopefully this book has set out a political economy perspective which can illuminate what is at stake in all of this and a framework for analysing the PRS, the experiences of tenants, and the sector's relationship with inequality, politics and housing justice.

References

Aalbers, M. B. (2016). *The Financialization of Housing: A Political Economy Approach*. Routledge.
Aalbers, M. B. (2017). The variegated financialization of housing. *International Journal of Urban and Regional Research*, 41(4), 542–54.
Aalbers, M. B. (2019). Introduction to the forum: From third to fifth-wave gentrification. *Tijdschrift Voor Economische En Sociale Geografie*, 110(1), 1–11.
Aalbers, M. B. & Christophers, B. (2014). Centring housing in political economy. *Housing, Theory and Society*, 31(4), 373–94.
Aalbers, M. & Ward, C. (2016). 'The shitty rent business': What's the point of land rent theory? *Urban Studies*, 53(9), 1760–83.
Acharya, B., Bhatta, D. & Dhakal, C. (2022). The risk of eviction and the mental health outcomes among the US adults. *Preventive Medicine Reports*, 29, 101981.
Adkins, L., Cooper, M. & Konings, M. (2021). Class in the 21st century: Asset inflation and the new logic of inequality. *Environment and Planning A: Economy and Space*, 53(3), 548–72.
Aglietta, M. (2000). *A Theory of Capitalist Regulation: The US Experience* (Vol. 28). Verso.
Ahmari, S. (2023). *Tyranny, Inc. How Private Power Crushed American Liberty – and What to Do About It*. Penguin, Random House.
Allen, K. (2011). *Submission on Retaliatory Eviction to the DECC Green Deal Consent Barriers and Retaliatory Eviction Working Group*. National Private Tenants Organization.
An, B. Y. (2023). The influence of institutional single-family rental investors on homeownership: Who gets targeted and pushed out of the local market? *Journal of Planning Education, and Research*, 44(4), 2231–50.
Andersen, M. J., Williamson, A. B., Fernando, P., Wright, D. & Redman, S. (2018). Housing conditions of urban households with Aboriginal children in NSW Australia: Tenure type matters. *BMC Public Health*, 18, 1–13.
Andreucci, D., García-Lamarca, M., Wedeking, J. & Swyngedouw, E. (2017). 'Value grabbing': A political ecology of rent. *Capitalism Nature Socialism*, 28(3), 28–47.
Ansell, B. (2023). Aspiration Nation? Who believes in a 'British Dream'? Not the young, that's for sure. *Political Calculus*. https://benansell.substack.com/p/aspiration-nation

Arundel, R. (2017). Equity inequity: Housing wealth inequality, inter and intragenerational divergence, and the rise of private landlordism. *Housing, Theory and Society*, 34(2), 176–200.

Arundel, R. & Doling, J. (2017). The end of mass-homeownership? Changes in labour markets and housing tenure opportunities across Europe. *Journal of Housing and the Built Environment*, 34(2), 176–200.

Arundel, R. & Lennartz, C. (2020). Housing market dualization: Linking insider–outsider divides in employment and housing outcomes. *Housing Studies*, 35(8), 1390–414.

Arundel, R., Li, A., Baker, E. & Bentley, R. (2022). Housing unaffordability and mental health: Dynamics across age and tenure. *International Journal of Housing Policy*, 24(1), 44–74.

Arundel, R. & Ronald, R. (2021). The false promise of homeownership: Homeowner societies in an era of declining access and rising inequality. *Urban Studies*, 58(6), 1120–40.

August, M. & Walks, A. (2018). Gentrification, suburban decline, and the financialization of multi-family rental housing: The case of Toronto. *Geoforum*, 89, 124–36.

Baer, J. A. (1993). Tenant mobilization and the 1907 rent strike in Buenos Aires. *The Americas*, 49(3), 343–68.

Baker, E., Mason, K., Bentley, R. & Mallett, S. (2014). Exploring the bi-directional relationship between health and housing in Australia. *Urban Policy and Research*, 32(1), 71–84.

Baker, J., Lynch, K., Cantillon, S. & Walsh, J. (2016). *Equality: From Theory to Action*. Springer.

Ball, M. (1983). *Housing Policy and Economic Power: The Political Economy of Owner Occupation*. Methuen Inc.

Ball, M. (2011). *Investing in Private Renting: Landlord Returns, Taxation and the Future of the Private Rented Sector*. Residential Landlords Association.

Ball, M. (2024). Are landlords so different? Comments on residential accumulation: A political economy framework. *Housing, Theory and Society*, 41(1), 27–31.

Balmer, I. & Gerber, J. (2018). Why are housing cooperatives successful? Insights from Swiss affordable housing policy. *Housing Studies*, 33(3), 361–85.

Bartels, C. & Schröder, C. (2020). *The Role of Rental Income, Real Estate and Rents for Inequality in Germany*. Forum for a New Economy, Berlin.

Bate, B. (2018). Understanding the influence tenure has on meanings of home and homemaking practices. *Geography Compass*, 12(1).

Bate, B. (2021). Making a home in the private rental sector. *International Journal of Housing Policy*, 21(3), 372–400.

Baxter, D. & Elliott, J. (2024). *Bringing Private Homes into Social Ownership can Rewire the Housing System*. Joseph Rowntree Foundation.

Baxter, D., Elliott, J. & Earwaker, R. (2022). *Making a House a Home: Why Policy Must Focus on the Ownership and Distribution of Housing*. Joseph Rowntree Foundation.

Béland, D. (2007). Neo-liberalism and social policy: The politics of ownership. *Policy Studies*, 28(2), 91–107.

Bentley, R., Baker, E., Mason, K., Subramanian, S. V. & Kavanagh, A. M. (2011).

Association between housing affordability and mental health: A longitudinal analysis of a nationally representative household survey in Australia. *American Journal of Epidemiology*, 174(7), 753–60.

Beswick, J., Alexandri, G., Byrne, M., Vives-Miró, S., Fields, D., Hodkinson, S. & Janoschka, M. (2016). Speculating on London's housing future: The rise of global corporate landlords in 'post-crisis' urban landscapes. *City*, 20(2), 321–41.

Blunt, A. & Dowling, R. (2006). *Home*. Routledge.

Bobek, A., Pembroke, S. & Wickham, J. (2021). Living in precarious housing: Non-standard employment and housing careers of young professionals in Ireland. *Housing Studies*, 36(9), 1364–87.

Bone, J. (2014). Neoliberal nomads: Housing insecurity and the revival of private renting in the UK. *Sociological Research Online*, 19(4), 1–14.

Bonnet, F. & Pollard, J. (2021). Tenant selection in the private rental sector of Paris and Geneva. *Housing Studies*, 36(9), 1427–45.

Brenner, N., Peck, J. & Theodore, N. (2010). Variegated neoliberalization: Geographies, modalities, pathways. *Global Networks*, 10(2), 182–222.

Brenner, R. (2006). *The Economics of Global Turbulence*. Verso.

Brill, F. (2022). Governing investors and developers: Analysing the role of risk allocation in urban development. *Urban Studies*, 59(7), 1499–517.

Brown, P., Newton, D., Armitage, R. & Monchuk, L. (2020). *Lockdown. Rundown. Breakdown: The COVID-19 Lockdown and the Impact of Poor-Quality Housing on Occupants in the North of England*. The Northern Housing Consortium.

Byrne, M. (2016a). 'Asset price urbanism' and financialization after the crisis: Ireland's National Asset Management Agency. *International Journal of Urban and Regional Research*, 40(1), 31–45.

Byrne, M. (2016b). Bad banks and the urban political economy of financialization. *City*, 20(5), 685–99.

Byrne, M. (2016c). Entrepreneurial urbanism after the crisis: Ireland's 'Bad Bank' and the redevelopment of Dublin's docklands. *Antipode*, 48(4), 899–918.

Byrne, M. (2017). Organizing tenants in the rentier society. *RoarMag*. https://roarmag.org/essays/organizing-tenants-rentier-society/

Byrne, M. (2018). Tenant self-organization after the Irish crisis: The Dublin Tenants' Association. In N. Gray (ed.), *Rent and its Discontents: A Century of Housing Struggle* (pp. 85–101). Rowman & Littlefield.

Byrne, M. (2019). The political economy of the 'residential rent relation': Antagonism and tenant organizing in the Irish rental sector. *Radical Housing Journal*, 1(2), 9–26.

Byrne, M. (2020a). Generation rent and the financialization of housing: A comparative exploration of the growth of the private rental sector in Ireland, the UK and Spain. *Housing Studies*, 35(4), 743–65.

Byrne, M. (2020b). Stay home: Reflections on the meaning of home and the Covid-19 pandemic. *Irish Journal of Sociology*, 28(3), 351–5.

Byrne, M. (2020c). Towards a political economy of the private rental sector. *Critical Housing Analysis*, 7(1), 103–13.

Byrne, M. (2021a). *Institutional Investment in the Private Rental Sector in the Wake of the Covid-19 Pandemic: A Review of International 'Grey Literature' and Reflections on the Irish context*. Geary Institute for Public Policy.

Byrne, M. (2021b). *The Impact of COVID-19 on the Private Rental Sector: Emerging International Evidence.* Geary Public Policy Institute. https://publicpolicy.ie/perspectives/the-impact-of-covid-19-on-the-private-rental-sector-emerging-international-evidence/

Byrne, M. (2024). Post-neoliberalization and the Irish private rental sector. *Housing Studies*, 39(7), 1658–77.

Byrne, M. & McArdle, R. (2020). *Security and Agency in the Irish Private Rental Sector.* Threshold.

Byrne, M. & McArdle, R. (2022). Secure occupancy, power and the landlord–tenant relation: A qualitative exploration of the Irish private rental sector. *Housing Studies*, 37(1), 124–42.

Byrne, M. & Norris, M. (2022). Housing market financialization, neoliberalism and everyday retrenchment of social housing. *Environment and Planning A: Economy and Space*, 54(1), 182–98.

Byrne, M. & Sassi, J. (2023). Making and unmaking home in the COVID-19 pandemic: A qualitative research study of the experience of private rental tenants in Ireland. *International Journal of Housing Policy*, 23(3), 523–42.

Card, K. (2024). From the streets to the statehouse: How tenant movements affect housing policy in Los Angeles and Berlin. *Housing Studies*, 39(6), 1395–421.

Carvalho, R., Liu, T., Zhang, F., Yu, R. & Oh, E. (2023). Key themes of build-to-rent: Developing a conceptual framework for achieving successful developments through a systematic literature review. *Buildings*, 13(8), 1926.

Casier, C. (2024). The coliving market as an emergent financialized niche real estate sector: A view from Brussels. *Housing Studies*, 39(9), 2355–76.

Castells, M. (1983). *The City and the Grassroots: A Cross-Cultural Theory of Urban Social Movements.* University of California Press.

Chisholm, E., Howden-Chapman, P. & Fougere, G. (2020). Tenants' responses to substandard housing: Hidden and invisible power and the failure of rental housing regulation. *Housing, Theory and Society*, 37(2), 139–61.

(2010). On voodoo economics: Theorising relations of property, value and contemporary capitalism. *Transactions of the Institute of British Geographers*, 35(1), 94–108.

Christophers, B. (2015). The limits to financialization. *Dialogues in Human Geography*, 5(2), 183–200.

Christophers, B. (2017). Intergenerational inequality? Labour, capital and housing through the ages. *Antipode*, 50(1), 101–21.

Christophers, B. (2020). *Rentier Capitalism: Who Owns the Economy, and Who Pays for It?* Verso Books.

Christophers, B. (2022a). Mind the rent gap: Blackstone, housing investment and the reordering of urban rent surfaces. *Urban Studies*, 59(4), 698–716.

Christophers, B. (2022b). The role of the state in the transfer of value from Main Street to Wall Street: US Single-Family Housing after the Financial Crisis. *Antipode*, 54(1), 130–52.

Christophers, B. (2023). The rentierization of the United Kingdom economy. *Environment and Planning A: Economy and Space*, 55(6), 1438–70.

Clair, A., Reeves, A., McKee, M. & Stuckler, D. (2019). Constructing a housing precariousness measure for Europe. *Journal of European Social Policy*, 29(1), 13–28.

Clapham, D. (2011). The embodied use of the material home: An affordance approach. *Housing, Theory and Society*, 28(4), 360–76.

Colau, A. & Alemany, Adriá. (2014). Mortgaged lives: From the housing bubble to the right to housing. *Mortgaged Lives: From the Housing Bubble to the Right to Housing*. Journal of Aesthetics & Protest Press.

Colburn, G., Walter, R. J. & Pfeiffer, D. (2021). Capitalizing on collapse: An analysis of institutional single-family rental investors. *Urban Affairs Review*, 57(6), 1590–625.

Cowan, D. S., Marsh, A., Niner, P., Kennett, P. & Forrest, R. (2000). *Harassment and Unlawful Eviction of Private Rented Sector Tenants and Park Home Residents*. UK Department of Environment, Transport and the Regions.

Crouch, C. (2009). Privatised Keynesianism: An unacknowledged policy regime. *British Journal of Politics and International Relations*, 11(3), 382–99.

Daly, P. (2022). Barrington Lecture 2022/23: Institutional investment in housing: Financialisation 2.0 in the case of Ireland. *Journal of the Statistical & Social Inquiry Society of Ireland*, 52.

Damer, S. (1980). State, class and housing: Glasgow 1885–1919. In *Housing, Social Policy and the State*. Croom Helm.

Dancygier, R. & Wiedemann, A. (2024). The financialization of housing and its political consequences. *American Journal of Political Science* (early online).

De Mille, A. G. (1944). Henry George: The fight for Irish freedom. *American Journal of Economics and Sociology*, 3(2), 251–72.

DeLuca, S. & Rosen, E. (2022). Housing insecurity among the poor today. *Annual Review of Sociology*, 48, 343–71.

Desmond, M. (2016). *Evicted: Poverty and Profit in the American City*. Broadway Books.

Desmond, M. & Wilmers, N. (2019). Do the poor pay more for housing? Exploitation, profit, and risk in rental markets. *American Journal of Sociology*, 124(4), 1090–124.

Dewilde, C. (2018). Explaining the declined affordability of housing for low-income private renters across Western Europe. *Urban Studies*, 55(12), 2618–39.

Dewilde, C. & Lancee, B. (2013). Income inequality and access to housing in Europe. *European Sociological Review*, 29(6), 1189–200.

Di Feliciantonio, C. (2017a). Social movements and alternative housing models: Practicing the 'politics of possibilities' in Spain. *Housing, Theory and Society*, 34(1), 38–56.

Di Feliciantonio, C. (2017b). Spaces of the expelled as spaces of the urban commons? Analysing the re-emergence of squatting initiatives in Rome. *International Journal of Urban and Regional Research*, 41(5), 708–25.

Diner, A. (2023). *Beyond New Build: Repurposing Private Rented Housing to Deliver a New Generation of Social Homes for England*. New Economics Foundation.

Disch, W. & Slaymaker, R. (2023). *Housing affordability: Ireland in a cross-country context*. ESRI.

Dong, H. (2018). The impact of income inequality on rental affordability: An empirical study in large American metropolitan areas. *Urban Studies*, 55(10), 2106–22.

Downey, D. (2014). The financialization of Irish homeownership and the impact of the global financial crisis. In A. MacLaran & S. Kelly (eds), *Neoliberal Urban Policy and the Transformation of the City: Reshaping Dublin*. Palgrave.

Dustmann, C., Fitzenberger, B. & Zimmermann, M. (2022). Housing expenditure and income inequality. *Economic Journal*, 132(645), 1709–36.

Dyck, I., Kontos, P., Angus, J. & McKeever, P. (2005). The home as a site for long-term care: Meanings and management of bodies and spaces. *Health & Place*, 11(2), 173–85.

Easthope, H. (2004). A place called home. *Housing, Theory and Society*, 21(3), 128–38.

Easthope, H. (2014). Making a rental property home. *Housing Studies*, 29(5), 579–96.

Ecker, J. (2016). Queer, young, and homeless: A review of the literature. *Child & Youth Services*, 37(4), 325–61.

Edel, M. (1982). Home ownership and working-class unity. *International Journal of Urban and Regional Research*, 6(2), 205–22.

Engels, F. (1993). *The Condition of the Working Class in England*. Oxford University Press.

Ferreri, M. & Vidal, L. (2022). Public-cooperative policy mechanisms for housing commons. *International Journal of Housing Policy*, 22(2), 149–73.

Fields, D. (2017). Unwilling subjects of financialization. *International Journal of Urban and Regional Research*, 41(4), 588–603.

Fields, D. (2017). Urban struggles with financialization. *Geography Compass*, 11(11), e12334.

Fields, D. (2018). Constructing a new asset class: Property-led financial accumulation after the crisis. *Economic Geography*, 94(2), 118–40.

Fields, D. & Uffer, S. (2016). The financialization of rental housing: A comparative analysis of New York City and Berlin. *Urban Studies*, 53(7), 1486–502.

Fields, D. & Vergerio, M. (2022). *Corporate Landlords and Market Power: What Does the Single-Family Rental Boom Mean for our Housing Future?* UC Berkeley.

Florida, R. L. & Feldman, M. M. A. (1988). Housing in US Fordism. *International Journal of Urban and Regional Research*, 12(2), 187–210.

Forrest, R. & Hirayama, Y. (2015). The financialization of the social project: Embedded liberalism, neoliberalism and homeownership. *Urban Studies*, 52(2), 233–44.

Forrest, R. & Hirayama, Y. (2018). Late homeownership and social re-stratification. *Economy and Society*, 47(2), 257–79.

Foye, C., Clapham, D. & Gabrieli, T. (2018). Home-ownership as a social norm and positional good: Subjective wellbeing evidence from panel data. *Urban Studies*, 55(6), 1290–312.

Fuller, G. W. (2019). *The Political Economy of Housing Financialization*. Agenda Publishing.

Fuller, G. W. (2021). The financialization of rented homes: Continuity and change in housing financialization. *Review of Evolutionary Political Economy*, 2, 551–70.

Fullilove, M. T. (2001). Root shock: The consequences of African American dispossession. *Journal of Urban Health*, 78(1), 72–80.

Furst, R. T. & Evans, D. N. (2017). Renting apartments to felons: Variations in real estate agent decisions due to stigma. *Deviant Behavior*, 38(6), 698–708.

Gabor, D. & Kohl, S. (2022). *My Home is an Asset Class*. The Greens/EFA.
Ganapati, S. (2010). Enabling housing cooperatives: policy lessons from Sweden, India and the United States. *International Journal of Urban and Regional Research*, 34(2), 365–80.
García-Lamarca, M. (2021). Real estate crisis resolution regimes and residential REITs: Emerging socio-spatial impacts in Barcelona. *Housing Studies*, 36(9), 1407–26.
García-Lamarca, M. (2022). *Non-Performing Loans, Non-Performing People: Life and Struggle with Mortgage Debt in Spain*. University of Georgia Press.
García-Lamarca, M. & Kaika, M. (2016). 'Mortgaged lives': The biopolitics of debt and housing financialisation. *Transactions of the Institute of British Geographers*, 41(3), 313–27.
Gavin, T. & O'Callaghan, C. (2024). Resisting root shock in the collapsed city: Constructing community and the fight to stay put through tenant organizing in Dublin. *Built Environment*, 50(2), 360–72.
George, H. (1879). *Progress and Poverty: An Inquiry Into the Cause of Industrial Depressions, and of Increase of Want with Increase of Wealth. The Remedy*. D. Appleton.
George, H. (1881). *The Irish Land Question* (Vol. 3, Issue 8). D. Appleton.
Gil García, J. & Martínez López, M. A. (2023). State-led actions reigniting the financialization of housing in Spain. *Housing, Theory and Society*, 40(1), 1–21.
Gil, J. & Palomera, J. (2024). Can tenant unions challenge neoliberal housing governance? The emergence of a new movement in Spain and its impact on post-neoliberal housing policy. *Housing, Theory and Society*, 41(5), 628–56.
Gilderbloom, J. I. & Appelbaum, R. P. (1987). Toward a sociology of rent: Are rental housing markets competitive? *Social Problems*, 34(3), 261–76.
Gold, S. (2020). Is housing hardship associated with increased adolescent delinquent behaviors? *Children and Youth Services Review*, 116.
Gomory, H. (2022). The social and institutional contexts underlying landlords' eviction practices. *Social Forces*, 100(4), 1774–805.
Goodyear, R. (2017). A place to call home? Declining home-ownership rates for Māori and Pacific peoples in New Zealand. *New Zealand Population Review*, 43, 3–34.
Gotham, K. F. (2009). Creating liquidity out of spatial fixity: The secondary circuit of capital and the subprime mortgage crisis. *International Journal of Urban and Regional Research*, 33(2), 355–71.
Grander, M. & Westerdahl, S. (2024). Financialization the Swedish way: How Vonovia became the largest owner of former public housing estates through transaction pathways and calculative practices. *Journal of Urban Affairs*, 1–19 (early online).
Gray, N. (2018a). Rent unrest: From the 1915 rent strikes to contemporary housing struggles. In N. Gray (ed.), *Rent and its Discontents: A century of housing struggle*. Rowman & Littlefield.
Gray, N. (2018b). Spatial composition and the urbanization of capital: The 1915 Glasgow rent strikes and the housing question reconsidered. In N. Gray (ed.), *Rent and its Discontents*. Rowman & Littlefield.
Guironnet, A., Bono, P. H. & Kireche, N. (2024). The French touch to the

financialisation of housing. Institutional investment into the Paris city-region (2008–2021). *Housing Studies*, 39(12), 2985–3006.

Gurney, C. M. (2023). Dangerous liaisons? Applying the social harm perspective to the social inequality, housing and health trifecta during the Covid-19 pandemic. *International Journal of Housing Policy*, 23(2), 232–59.

Gustafsson, J. (2024). Renovations as an investment strategy: Circumscribing the right to housing in Sweden. *Housing Studies*, 39(6), 1555–76.

Gustafsson, J., Hellström, E., Richard, Å. & Springfeldt, S. (2019). The right to stay put: Resistance and organizing in the wake of changing housing policies in Sweden. *Radical Housing Journal*, 1(2), 191–200.

Guzmán, J. (2023). The housing/financial complex in Spain: After the 2008 global financial crisis. *International Journal of Urban and Regional Research*, 47(6), 900–16.

Guzmán, J. (2025). The repertoire of housing contention: The birth of the Stay Put campaign in Barcelona. *Housing Studies*, 40(4), 845–66.

Guzmán, J. & Ill-Raga, M. (2022). Rent strikes: Revolutions at Point Zero. *Socialism and Democracy*, 36(1–2), 193–215.

Haffner, M. (2023). Private renting in the Netherlands: Set to grow? In P. Kemp (ed.), *Private renting in the advanced economies*. Policy Press.

Haffner, M. E., Ong, R., Smith, S. J. & Wood, G. A. (2017). The edges of home ownership – the borders of sustainability. *International Journal of Housing Policy*, 17(2), 169–76.

Haila, A. (1990). The theory of land rent at the crossroads. *Environment and Planning D: Society and Space*, 8(3), 275–96.

Haila, A. (2015). *Urban Land Rent: Singapore as a Property State*. John Wiley.

Harloe, M. (2021). *Private Rented Housing in the United States and Europe*. Routledge.

Harris, E. & Nowicki, M. (2020). 'Get smaller?' Emerging geographies of micro-living. *Area*, 52(3), 591–9.

Harvey, D. (1974). Class-monopoly rent, finance capital and the urban revolution. *Regional Studies*, 8(3–4), 239–55.

Harvey, D. (1976). Labor, capital, and class struggle around the built environment in advanced capitalist societies. *Politics & Society*, 6(3), 265–95.

Harvey, D. (2012). *Rebel Cities: From the Right to the City to the Urban Revolution*. Verso.

Harvey, D. (2018). *The Limits to Capital*. Verso.

Hearne, R. (2020). *Housing Shock: The Irish Housing Crisis and How to Solve it*. Policy Press.

Hick, R., Pomati, M. & Stephens, M. (2022). *Housing and Poverty in Europe: Examining the Interconnections in the Face of Rising House Prices*. Cardiff University.

Hick, R., Pomati, M. & Stephens, M. (2024). Housing affordability and poverty in Europe: On the deteriorating position of market renters. *Journal of Social Policy*, 1–24 (early online).

Hiscock, R., Kearns, A., MacIntyre, S. & Ellaway, A. (2001). Ontological security and psycho-social benefits from the home: Qualitative evidence on issues of tenure. *Housing, Theory and Society*, 18(1–2), 50–66.

Hochstenbach, C. (2022). Landlord elites on the Dutch housing market: Private landlordism, class, and social inequality. *Economic Geography*, 98(4), 327–54.

Hochstenbach, C. (2023). Balancing accumulation and affordability: How Dutch housing politics moved from private-rental liberalization to regulation. *Housing, Theory and Society*, 40(4), 503–29.

Hochstenbach, C. & Ronald, R. (2020). The unlikely revival of private renting in Amsterdam: Re-regulating a regulated housing market. *Environment and Planning A: Economy and Space*, 52(8), 1622–42.

Høghøj, M. (2023). Negotiating citizenship on the urban periphery: Mass housing, resident democracy, and acts of citizenship in 1970s' Denmark. In H. Droste (ed.) *Urban Life in Nordic Countries*. Routledge.

Holm, A., Alexandri, G., Bernt, M., Belotti, E., Bortolotti, A., Watt, P., Hodkinson, S. & Aaudycka, B. (2023). Housing policy under the conditions of financialisation. Sciences Po.

Hoolachan, J. & McKee, K. (2019). Inter-generational housing inequalities: 'Baby boomers versus millenials'. *Urban Studies*, 56(1), 210–25.

Hoolachan, J., McKee, K., Moore, T. & Soaita, A. (2017). Generation rent and the ability to 'settle down': Economic and geographical variation in young people's housing transitions. *Journal of Youth Studies*, 20(1), 63–78.

Horton, A. (2021). Liquid home? Financialisation of the built environment in the UK's 'hotel-style' care homes. *Transactions of the Institute of British Geographers*, 46(1), 179–92.

Housing Europe (2022). *The State of Housing in Europe 2022 Mid-Term Update*. Housing Europe.

Howard, A. (2025). Seven propositions about 'generation rent'. *Housing, Theory and Society*, 42(1), 1–22.

Howard, A., Hochstenbach, C. & Ronald, R. (2024). Understanding generational housing inequalities beyond tenure, class and context. *Economy and Society*, 53(1), 135–62.

Hubbard, P. (2009). Geographies of studentification and purpose-built student accommodation: Leading separate lives? *Environment and Planning A*, 41(8), 1903–23.

Huisman, C. J. (2020). *Insecure Tenure: The Precarisation of Rental Housing in the Netherlands*. Unpublished PhD thesis.

Hulse, K. (2023). Growth and change: Private renting in Australia in the 21st century. In P. Kemp (ed.), *Private Renting in the Advanced Economies* (pp. 18–41). Policy Press.

Hulse, K. & Milligan, V. (2014). Secure occupancy: A new framework for analysing security in rental housing. *Housing Studies*, 29(5), 638–56.

Hulse, K., Morris, A. & Pawson, H. (2019). Private renting in a home-owning society: Disaster, diversity or deviance? *Housing, Theory and Society*, 36(2), 167–88.

Hulse, K. & Yates, J. (2017). A private rental sector paradox: Unpacking the effects of urban restructuring on housing market dynamics. *Housing Studies*, 32(3), 253–70.

Huron, A. (2015). Working with strangers in saturated space: Reclaiming and maintaining the urban commons. *Antipode*, 47(4), 963–79.

Immergluck, D. & Law, J. (2014). Investing in crisis: The methods, strategies, and expectations of investors in single-family foreclosed homes in distressed neighborhoods. *Housing Policy Debate*, 24(3), 568–93.

Inchauste, G., Karver, J., Kim, Y. S. & Abdel Jelil, M. (2018). *Living and Leaving: Housing, Mobility and Welfare in the European Union*. World Bank.

Izuhara, M. & Heywood, F. (2003). A life-time of inequality: A structural analysis of housing careers and issues facing older private tenants. *Ageing & Society*, 23(2), 207–24.

Jacobs, K., Atkinson, R. & Warr, D. (2024). Political economy perspectives and their relevance for contemporary housing studies. *Housing Studies*, 39(4), 962–79.

James, L., Daniel, L., Bentley, R. & Baker, E. (2024). Housing inequality: A systematic scoping review. *Housing Studies*, 39(5), 1264–85.

Janoschka, M., Alexandri, G., Ramos, H. O. & Vives-Miró, S. (2020). Tracing the socio-spatial logics of transnational landlords' real estate investment: Blackstone in Madrid. *European Urban and Regional Studies*, 27(2), 125–41.

JCHS (2024). *America's Rental Housing 2024*. Joint Centre for Housing Studies, Harvard University.

Jezierska, K. & Polanska, D. V. (2018). Social movements seen as radical political actors: The case of the Polish tenants' movement. *VOLUNTAS: International Journal of Voluntary and Nonprofit Organizations*, 29(4), 683–96.

Jordà, Ò., Schularick, M. & Taylor, A. M. (2016). The great mortgaging: Housing finance, crises and business cycles. *Economic Policy*, 31(85), 107–52.

Kadelke, P. (2023). Landlords vs tenants= top vs bottom? Class positions in rental housing in Germany. *Critical Housing Analysis*, 10(1), 66–76.

Kadi, J. (2015). Recommodifying housing in formerly 'Red' Vienna? *Housing, Theory and Society*, 32(3), 247–65.

Kadi, J., Vollmer, L. & Stein, S. (2021). Post-neoliberal housing policy? *European Urban and Regional Studies*, 1–22.

Kallin, H., Gray, N., Beswick, J., Dempsey, D., Donnachie, S., Gavin, T., Gustafsson, J., Kholodilin, K., Mironova, O., Monroy, E., Palomera, J., Persdotter, M., Vilenica, A. & Wallstam, M. (2024). Rent controls in comparative perspective: Reflections on an international symposium. *Radical Housing Journal*, 6(2), 237–51.

Kalyukin, A. & Kohl, S. (2020). Continuities and discontinuities of Russian urban housing: The Soviet housing experiment in historical long-term perspective. *Urban Studies*, 57(8), 1768–85.

Katsiaficas, G. (2014). *City is Ours: Squatting and Autonomous Movements in Europe from the 1970s to the Present*. PM Press.

Kearns, A., Hiscock, R., Ellaway, A. & Macintyre, S. (2000). 'Beyond four walls'. The psycho-social benefits of home: Evidence from West Central Scotland. *Housing Studies*, 15(3), 387–410.

Kelly, S. (2014). Light-touch regulation: The rise and fall of the Irish banking sector. In A. MacLaren and S. Kelly (eds/) *Neoliberal Urban Policy and the Transformation of the City: Reshaping Dublin*. Palgrave.

Kemp, P. (2015). Private renting after the global financial crisis. *Housing Studies*, 30(4), 601–20.

Kemp, P. (2023). New trajectories in private rental housing. In *Private Renting in the Advanced Economies*. Policy Press.

Kemp, P. A. & Kofner, S. (2010). Contrasting varieties of private renting: England and Germany. *International Journal of Housing Policy*, 10(4), 379–98.

Kholodilin, K. A. & Kohl, S. (2023). Rent price control – yet another great equalizer of economic inequalities? Evidence from a century of historical data. *Journal of European Social Policy*, 33(2), 169–84.

Kindermann, F. & Kohls, S. (2018). Rental markets and wealth inequality in the Euro-area. University of Regensburg.

Kitchin, R., O'Callaghan, C., Boyle, M., Gleeson, J. & Keaveney, K. (2012). Placing neoliberalism: The rise and fall of Ireland's Celtic Tiger. *Environment and Planning A*, 44(6), 1302–26.

Köppe, S. (2020). Passing it on: Inheritance, coresidence and the influence of parental support on homeownership and housing pathways. In R. Ronald and C. Lennartz (eds) *Housing Careers, Intergenerational Support and Family Relations* (pp. 78–100). Routledge.

Köppe, S. & Byrne, M. (2024). The political economy of housing. In B. Greve & A. Moreira (eds), *Handbook on the Political Economy of Social Policy*. Edward Elgar Publishing.

Kuenzel, R. & Bjornbak, B. (2008). The UK housing market: Anatomy of a house price boom. *ECOFIN Country Focus*, 5(11).

Lawson, R. (1984). The rent strike in New York City, 1904–80: The evolution of a social movement strategy. *Journal of Urban History*, 10(3), 235–58.

Le Grand, J. & Robinson, R. V. (2016). *Economics of Social Problems*. Springer.

Lee, K. Y. (2000). *From Third World to First: The Singapore Story*. Harper Collins.

Lenhard, J., Coulomb, L. & Miranda-Nieto, A. (2022). Home making without a home: Dwelling practices and routines among people experiencing homelessness. *Housing Studies*, 37(2), 183–8.

Lima, V. (2020). The financialization of rental housing: Evictions and rent regulation. *Cities*, 105, 102787.

Lima, V., Hearne, R. & Murphy, M. P. (2023). Housing financialisation and the creation of homelessness in Ireland. *Housing Studies*, 38(9), 1695–718.

Lister, D. (2004). Young people's strategies for managing tenancy relationships in the private rented sector. *Journal of Youth Studies*, 7(3), 315–30.

Lister, D. (2005). Controlling letting arrangements? Landlords and surveillance in the private rented sector? *Surveillance and Society*, 2(4), 513–28.

Listerborn, C. (2023). The new housing precariat: Experiences of precarious housing in Malmö, Sweden. *Housing Studies*, 38(7), 1304–22.

Logan, J. R. & Molotch, H. (2007). *Urban Fortunes: The Political Economy of Place*; with a new preface. University of California Press.

López, I. & Rodríguez, E. (2011). The Spanish model. *New Left Review*, 69(3), 5–29.

Loughran, K. (2015). The Philadelphia negro and the canon of classical urban theory. *Du Bois Review: Social Science Research on Race*, 12(2), 249–67.

Lukes, S., de Noronha, N. & Finney, N. (2019). Slippery discrimination: A review of the drivers of migrant and minority housing disadvantage. *Journal of Ethnic and Migration Studies*, 45(17), 3188–206.

Lyons, R. (2024). *Institutional Investment and the Private Rental Sector in Ireland*. Irish Institutional Property.

Madden, D. (2024). Beyond the limits of rentability: Revalorizing urban space in late neoliberalism. *Environment and Planning F*, 26349825241241316.

Madden, D. (2025). Social reproduction and the housing question. *Antipode*, 57(2), 578–98.

Madden, D. & Marcuse, P. (2016). *In Defense of Housing: The Politics of Crisis*. Verso Books.

Madigan, R. & Munro, M. (1991). Gender, house and 'home': Social meanings and domestic architecture in Britain. *Journal of Architectural and Planning Research*, 8(2), 116–32.

Marcuse, P. (1986). A useful instalment of socialist work: Housing in Red Vienna in the 1920s. In R. Bratt, C. Hartmann & A. Meyerson (eds), *Critical Perspectives on Housing*. Temple University Press.

Marsh, A., Gibb, K., Harrington, N. & Smith, B. (2023). *The Impact of Regulatory Reform on the Private Rented Sector*. UK Collaborative Centre for Housing Evidence.

Marsh, A., Gibb, K. & Soaita, A. M. (2023). Rent regulation: Unpacking the debates. *International Journal of Housing Policy*, 23(4), 734–57.

Martin, C., Hulse, K. & Pawson, H. (2018). *The Changing Institutions of Private Rental Housing: An International Review*. Australian Housing and Urban Research Institute.

Mason, K. E., Baker, E., Blakely, T. & Bentley, R. J. (2013). Housing affordability and mental health: Does the relationship differ for renters and home purchasers? *Social Science & Medicine*, 94, 91–7.

McArdle, R. & Byrne, M. (2022). Rootlessness: How the Irish private rental sector prevents tenants feeling secure in their homes and tenant's resistance against this. *Geoforum*, 136, 211–18.

McBride, T. (2006). John Ferguson, Michael Davitt & Henry George – Land for the people. *Irish Studies Review*, 14(4), 421–30.

McCabe, C. (2011). *Sins of the Father: Tracing the Decisions that Shaped the Irish Economy*. History Press Ireland.

McCarthy, B. (2024). *Institutional Investment and Residential Rental Market Dynamics*. Central Bank of Ireland.

McKee, K. (2012). Young people, homeownership and future welfare. *Housing Studies*, 27(6), 853–62.

McKee, K. (2019). Generation rent is a myth – Housing prospects for millennials are determined by class. *The Conversation*. 3 January. https://theconversation.com/generation-rent-is-a-myth-housing-prospects-for-millennials-are-determined-by-class-108996

McKee, K. & Harris, J. (2025). The role of landlords in shaping private renters' uneven experiences of home: Towards a relational approach. *International Journal of Housing Policy*, 25(1), 101–18.

McKee, K., Soaita, A. M. & Hoolachan, J. (2020). 'Generation rent' and the emotions of private renting: Self-worth, status and insecurity amongst low-income renters. *Housing Studies*, 35(8), 1468–87.

Mee, K. (2009). A space to care, a space of care: public housing, belonging, and care in Inner Newcastle, Australia. *Environment and Planning A: Economy and Space*, 41(4), 842–58.

Meen, G. & Whitehead, C. (2020). *Understanding Affordability: The Economics of Housing Markets*. Policy Press.

Michener, J. & SoRelle, M. (2023). Politics, power, and precarity: How tenant organizations transform local political life. In S. Anzia (ed.), *Interest Groups in U.S. Local Politics* (pp. 31–58). Springer Nature Switzerland.

Moreno, M. (2024). *Inequality and Wealth Distribution among Irish Households: Introducing new Distributional Wealth Accounts*. Central Bank of Ireland.

Moreno Zacarés, J. (2024). Residential accumulation: A political economy framework. *Housing, Theory and Society*, 41(1), 4–26.

Morris, A., Hulse, K. & Pawson, H. (2017). Long-term private renters: Perceptions of security and insecurity. *Journal of Sociology*, 53(3), 653–69.

Morris, A., Hulse, K. & Pawson, H. (2021). Finding a rental property and feeling at home. In A. Morris, K. Hulse & H. Pawson (eds), *The Private Rental Sector in Australia: Living with Uncertainty* (pp. 87–105). Springer.

Munoz, G., Vilenica, A. & Quiroz, M. (2024). Dismantling rentier logic: Tenant struggles in Argentina. *Radical Housing Journal*, 6(1), 125–37.

Nethercote, M. (2020). Build-to-rent and the financialization of rental housing: Future research directions. *Housing Studies*, 35(5), 839–74.

Nethercote, M. (2024). Monopoly dynamics and the rise of UK single-family rental. *Geoforum*, 148, 103907.

Norris, M. (2016). *Property, Family and the Irish Welfare State*. Palgrave Macmillan.

Norris, M. & Byrne, M. (2015). Asset price Keynesianism, regional imbalances and the Irish and Spanish housing booms and busts. *Built Environment*, 41(2).

Norris, M. & Coates, D. (2014). How housing killed the Celtic Tiger: Anatomy and consequences of Ireland's housing boom and bust. *Journal of Housing and the Built Environment*, 29(2), 299–314.

Ó Broin, E. (2019). *Home: Why Public Housing is the Answer*. Merrion Press.

Ó Riain, S. (2013). The rise and fall of Ireland's Celtic Tiger: Liberalism, boom and bust. In *The Rise and Fall of Ireland's Celtic Tiger: Liberalism, Boom and Bust*. Cambridge University Press.

Obeng-Odoom, F. (2022). Georgist political economy. In F. Stilwell, D. Primrose & T. B. Thornton (eds) *Handbook of Alternative Theories of Political Economy*. Edward Elgar Publishing.

OECD (2021). *Building for a Better Tomorrow: Policies to Make Housing More Affordable*. In *Employment, Labour and Social Affairs Policy Briefs*. Organization for Economic Cooperation and Development.

Palomera, J. (2014). How did finance capital infiltrate the world of the urban poor? Homeownership and social fragmentation in a Spanish neighborhood. *International Journal of Urban and Regional Research*, 38(1), 218–35.

Pareja-Eastaway, M. & Sánchez-Martínez, T. (2011). El alquiler: Una asignatura pendiente de la Política de Vivienda en España. *Ciudad y Territorio Estudios Territoriales*, 167: 53–70.

Pareja-Eastaway, M. & Sámchez-Martínez, T. (2017). Social housing in Spain: What roles does the private market play? *Journal of Housing and the Built Environment*, 32(2), 377–95.

Pfeiffer, D., Schafran, A. & Wegmann, J. (2021). Vulnerability and opportunity: Making sense of the rise in single-family rentals in US neighbourhoods. *Housing Studies*, 36(7), 1026–46.

Phang, S.-Y. (1996). Economic development and the distribution of land rents in Singapore: A Georgist implementation. *American Journal of Economics and Sociology*, 55(4), 489–501.

Pittini, A. (2012). *Housing Affordability in the EU: Current Situation and Recent Trends.* CECODHAS, European Social Housing Observatory.

Polanska, D. V. & Richard, Å. (2021). Resisting renovictions: Tenants organizing against housing companies: Renewal practices in Sweden. *Radical Housing Journal*, 3(1), 187–205.

Polanska, D. V. & Weldon, T. (2020). In search of urban commons through squatting: The role of knowledge sharing in the creation and organization of everyday utopian spaces in Sweden. *Partecipazione e Conflitto*, 13(3).

Polanyi, K. (2001). *The Great Transformation: The Political and Economic Origins of Our Time*. Beacon Press.

Power, E. (2017). Renting with pets: A pathway to housing insecurity? *Housing Studies*, 32(3), 336–60.

Power, E. & Mee, K. (2020). Housing: An infrastructure of care. *Housing Studies*, 35(3), 484–505.

Power, E. R. (2019). Assembling the capacity to care: Caring-with precarious housing. *Transactions of the Institute of British Geographers*, 44(4), 763–77.

Racu, V. (2024). The Madrid Tenants' Union. In *The Rent Strike (Podcast)*. https://podcasts.apple.com/us/podcast/the-madrid-tenants-union-part-1/id1741048740?i=1000659087798

Raymond, E. L., Duckworth, R., Miller, B., Lucas, M. & Pokharel, S. (2018). From foreclosure to eviction: Housing insecurity in corporate-owned single-family rentals. *Cityscape*, 20(3), 159–88.

Raymond, E. L., Miller, B., McKinney, M. & Braun, J. (2021). Gentrifying Atlanta: Investor purchases of rental housing, evictions, and the displacement of Black residents. *Housing Policy Debate*, 31(3–5), 818–34.

Residential Tenancies Board. (2024). *Profile of the Register – Private Registration Statistics Q2 2023 – Q1 2024*. Residential Tenancies Board.

Revington, N. & August, M. (2020). Making a market for itself: The emergent financialization of student housing in Canada. *Environment and Planning A: Economy and Space*, 52(5), 856–77.

Rex, J. & Moore, R. (1967). *Race, Community and Conflict: A Study of Sparkbrook*. Oxford University Press.

Rolfe, S., McKee, K., Feather, J., Simcock, T. & Hoolachan, J. (2023). The role of private landlords in making a rented house a home. *International Journal of Housing Policy*, 23(1), 113–37.

Rolnik, R. (2019). *Urban Warfare: Housing under the Empire of Finance*. Verso Books.

Ronald, R. (2008). *The Ideology of Homeownership: Homeowner Societies and the Role of Housing*. Palgrave Macmillan.

Ronald, R. (2018). 'Generation rent' and intergenerational relations in the era of housing financialisation. *Critical Housing Analysis*, 5(2), 14–26.

Ronald, R. & Kadi, J. (2017). The revival of private landlords in Britain's post-homeownership society. *New Political Economy*, 23(6), 786–803.

Rosenthal, T. & Vilchis, L. (2024). *Abolish Rent: How Tenants Can End the Housing Crisis*. Haymarket Books.

Ruggie, J. G. (1982). International regimes, transactions, and change: Embedded

liberalism in the postwar economic order. *International Organization*, 36(2), 379–415.
Ruiz Cayuela, S. & García-Lamarca, M. (2023). From the squat to the neighbourhood: Popular infrastructures as reproductive urban commons. *Geoforum*, 144, 103807.
Ryan-Collins, J., Lloyd, T. & Macfarlane, L. (2017). *Rethinking the Economics of Land and Housing*. Zed Books.
Ryan-Collins, J. & Murray, C. (2023). When homes earn more than jobs: The rentierization of the Australian housing market. *Housing Studies*, 38(10), 1888–917.
Sandel, M., Sheward, R., Ettinger de Cuba, S., Coleman, S. M., Frank, D. A., Chilton, M., Black, M., Heeren, T., Pasquariello, J. & Casey, P. (2018). Unstable housing and caregiver and child health in renter families. *Pediatrics*, 141(2).
Santos, A. C. (2025). Peripheral housing rentierisation in Southern Europe: reflections from the Portuguese case. *Housing Studies*, 40(3), 696–721.
Saunders, E., Samuels, K. & Statham, D. (2018). Rebuilding a shattered housing movement: Living rent and contemporary private tenant struggles in Scotland. In N. Gray (ed.), *Rent and its Discontents: A Century of Housing Struggle* (pp. 85–101). Rowman & Littlefield.
Saunders, P. (1984). Beyond housing classes: The sociological significance of private property rights in means of consumption. *International Journal of Urban and Regional Research*, 8(2), 202–27.
Saunders, P. (2021). *A Nation of Home Owners*. Routledge.
Saunders, P. & and Williams, P. (1988). The constitution of the home: Towards a research agenda. *Housing Studies*, 3(2), 81–93.
Schipper, S. (2015). Towards a 'post-neoliberal' mode of housing regulation. The Israeli social protest of summer 2011. *International Journal of Urban and Regional Research*, 39(6), 1137–54.
Schwartz, A. (2023). Rental housing dynamics and their affordability impact in the United States. In P. Kemp (ed.), *Private Renting in the Advanced Economies* (pp. 42–68). Policy Press.
Schwartz, H. & Seabrooke, L. (2008). Varieties of residential capitalism in the international political economy: Old welfare states and the new politics of housing. *Comparative European Politics*, 6(3), 237–61.
Shelter. (2014). *Safe and Decent Homes*. Shelter.
Shiffer-Sebba, D. (2020). Understanding the divergent logics of landlords: Circumstantial versus deliberate pathways. *City & Community*, 19(4), 1011–37.
Smith, N. (1979). Toward a theory of gentrification: A back to the city movement by capital, not people. *Journal of the American Planning Association*, 45(4), 538–48.
Smith, S. J., Clark, W. A., Ong ViforJ, R., Wood, G. A., Lisowski, W. & Truong, N. K. (2022). Housing and economic inequality in the long run: The retreat of owner occupation. *Economy and Society*, 51(2), 161–86.
Soaita, A. M. (2021). *Renting During the Covid-19 Pandemic in Great Britain: The Experiences of Private Tenants*. Urban Studies.
Soaita, A. M. (2024). The social vibe of the tenant/landlord relationship in a 'tenant market': The case of Romania. *Housing Studies*, 1–22 (early online).

Soaita, A. & McKee, K. (2019). Assembling a 'kind of' home in the UK private renting sector. *Geoforum*, 103, 148–57.
Soaita, A. & McKee, K. (2020). *Private Renters' Housing Experiences in Lightly Regulated Markets*. UK Centre for Colloborative Housing Evidence.
Soederberg, S. (2018). The rental housing question: Exploitation, eviction and erasures. *Geoforum*, 89, 114–23.
Spratt, V. (2022). *Tenants*. Profile Books Ltd.
Standing, G. (2011). *The Precariat: The New Dangerous Class*. Bloomsbury academic.
Stringer, J. (2025). *Renters Unite! How Tenant Unions are Fighting the Housing Crisis*. Pluto Press. https://www.plutobooks.com/9780745350011/renters-unite/
Tester, G. (2008). An intersectional analysis of sexual harassment in housing. *Gender & Society*, 22(3), 349–66.
Van Lanen, S. (2017). Living austerity urbanism: Space–time expansion and deepening socio-spatial inequalities for disadvantaged urban youth in Ireland. *Urban Geography*, 38(10), 1603–13.
Van Lanen, S. (2022). 'My room is the kitchen': Lived experience of home-making, home-unmaking and emerging housing strategies of disadvantaged urban youth in austerity Ireland. *Social & Cultural Geography*, 23(4), 598–619.
Vasudevan, A. (2015). The makeshift city: Towards a global geography of squatting. *Progress in Human Geography*, 39(3), 338–59.
Verstraete, J. & Moris, M. (2019). Action–reaction. Survival strategies of tenants and landlords in the private rental sector in Belgium. *Housing Studies*, 34(4), 588–608.
Vollmer, L. & Gutiérrez, D. (2022). Organizing for expropriation: How a tenants campaign convinced Berliners to vote for expropriating big landlords. *Radical Housing Journal*, 4(2), 47–66.
Waldron, R. (2018). Capitalizing on the state: The political economy of real estate investment trusts and the 'resolution' of the crisis. *Geoforum*, 90, 206–18.
Waldron, R. (2022). Experiencing housing precarity in the private rental sector during the covid-19 pandemic: The case of Ireland. *Housing Studies*, 38(1), 84–106.
Waldron, R. (2023). Generation rent and housing precarity in 'post crisis' Ireland. *Housing Studies*, 38(2), 181–205.
Waldron, R. (2024). Responding to housing precarity: The coping strategies of generation rent. *Housing Studies*, 39(1), 124–45.
Waldron, R. & Redmond, D. (2014) The extent of the mortgage crisis in Ireland and policy responses. *Housing Studies*, 29(1), 149–65.
Watson, D. (2019). 'Theirs Was the Crisis. Ours Was the Remedy': The Squatting Movements of 1946 in Britain, Canada, and Australia, 84(3), 241–65.
Watson, M. (2009). Planning for a future of asset-based welfare? New labour, financialized economic agency and the housing market. *Planning, Practice and Research*, 24(1), 41–56.
Watson, S. (2023). *Accommodating Inequality: Gender and Housing*. Routledge.
Whitehead, C. M. E. & Williams, P. (2011). Causes and consequences? Exploring the shape and direction of the housing system in the UK post the financial crisis. *Housing Studies*, 26 (August), 1157–69.

Wijburg, G. & Aalbers, M. B. (2017). The alternative financialization of the German housing market. *Housing Studies*, 32(7), 968–89.

Wijburg, G., Aalbers, M. B. & Heeg, S. (2018). The financialisation of rental housing 2.0: Releasing housing into the privatised mainstream of capital accumulation. *Antipode*, 50(4), 1098–119.

Wilde, M. (2022). Eviction, gatekeeping and militant care: Moral economies of housing in austerity London. *Ethnos*, 87(1), 22–41.

Yrigoy, I. (2021). The political economy of rental housing in Spain: The dialectics of exploitation (s) and regulations. *New Political Economy*, 26(1), 186–202.

Zavisca, J. R. & Gerber, T. P. (2016). The socioeconomic, demographic, and political effects of housing in comparative perspective. *Annual Review of Sociology*, 42, 347–67.

Zhang, B. (2023). Re-conceptualizing housing tenure beyond the owning-renting dichotomy: Insights from housing and financialization. *Housing Studies*, 38(8), 1512–35.

Ziegelmeyer, M. (2015). *Other Real Estate Property in Selected Euro Area Countries*. Banque centrale du Luxembourg.

Index

Aalbers, M. B.
  GDP and housing 68
  national/local policies 87
Adkins, L.
  concentration of wealth 97
  generational inequalities 107
  residential assets 96
affordability 13
  Berlin 148
  contexts of 165
  demarketizing housing 161
  first-time buyers locked out 54
  gentrification and 66
  inequalities and 91, 92–5, 157–8
  reduction of 88–9
  Spain 51
  tenants' organizations and 130
  tenure and 103
  wage gaps 46–7, 48, 94
age
  abuse of elder tenants 116
  see also generations
agency of tenants 27, 118
  analysis of PRS 13
  relationship with landlord 120–1
Ahmari, S. 33
Airbnb 162
amenities, local
  Berlin's DWE 151
  Built to Rent 83–4
American Homes 4 Rent 72
Amundi 86

Argentina
  Buenos Aires rent strikes 133–4, 135, 136
  tenant unions 135
art, cultural investment 25
Arundel, R.
  inequality and insecurity 61
  promise of homeownership 2, 59
August, M. 81–2
Australia
  affordability gap 49
  decline of homeownership 96
  increase in PRS 43–4
  indigenous people 95
  unimpacted by financial crisis 55–6
Austria
  homeownership 43
  social housing 78, 161
  tenant insecurity 113
  Vienna rent strike 134

Baer, J. A. 134–5
Ball, M. 37
banks *see* financial systems; mortgages
Barbour, Mary 134
Barcelona Tenants' Union (TUB) 131, 139
  media and 141–2
  PAH and evictions 143
  rent reductions 147
Bartels, C. 99

Bate, B. 115
Biden, Joseph 129
Blackstone
  campaigns against 65
  concentration of ownership 86
  Denmark 81
  Invitation Homes 72
  SFRs 73
  Spanish real estate 76
  tenancy management 77
Brazil, Covid upheavals and 117–18
Brickell, K. 126
Buddhism 15
Buela *see* Rouco Buela, Juana
build to rent 69, 70, 71
  financialization 82–4
  hotel-like amenities 83–4
Bulgarian homeownership 43
Bush, George W. 39, 160
Buy to Let mortgages
  Ireland 47
  post financial crisis 54
  products 44
  restricting 162
  UK 48–9, 54
  see also mortgages
Byrne, M. 120

Caixabank, Rosa's case and 144–5
Canada
  institutional landlords and renovation 81–2
  largest landlord 76
capitalism
  accumulation of wealth 7, 9, 96
  worker relations 128
  working class ownership and 59–62
  see also financialization; markets; neoliberalism; property ownership
CAPREIT 76
Card, K. 141
care homes
  Built to Rent 84
  niche market 88
care work
  created by tenant 32

home infrastructure 29–30, 31
Castells, M. 26
Catalunya *see* Barcelona
Catalunya Caixa bank 76
Cerberus: Berlin housing 76, 78
Christophers, B.
  age and resources 106
  financialization of SFRs 73
  large-scale renovation 80
  locked out of ownership 101
Clair, A. 112
class
  hierarchies and eviction 126
  housing and 104
  inequalities and 10, 12, 101
  middle-class identity 58–60, 63
  post-war compromise 113
  post-war Europe 59–60
  PRS and 107
  Spanish Mortgage Bank and 49
  tenure and 104–5
  working class capitalism 59–62
Colburn, G. 72
communities
  Berlin's DWE 151
  neighbour networks 29
  private investment 23
  resident associations 26
  residential stability and 2
  social construction of 18
  urban spaces 18
  see also place
Community Action Tenants' Union (CATU), Ireland 138
Community Land Trusts 160
The Condition of the Working Class in England (Engels) 90
consumption
  collective 26
  home making 127–8
  mass 58
corporate landlords *see* landlords; private rental sector (PRS)
Covid-19 pandemic
  Brazil and housing 117–18
  eviction moratorium 116–17
  living conditions and 91

culture
  investment in 23
  middle-class identity 58–60
  recognizing tenant's home 34
currency values 85
Czech Republic, revival of PRS and 42–3

Dancygier, R.
  DWE campaign 148, 149, 152
  market power of investors 152–3
Davitt, Michael 21
Denmark
  Blackstone Law 81
  institutional landlords 70
  limits financial landlords 65
  renovating social housing 80, 81–2
  revival of PRS 42–3
  social housing 161
  state intervention 87
  student housing 161–2
deregulation
  of PRS 44
  Spain 52
Desmond, M.
  conflict with landlords 120
  eviction and social hierarchy 126
Deutsche (DWE) 163
Deutsche Wohnen und Co Enteignen (DWE) 132
  campaign of 148–52
Dewilde, C. 98
Digs, London 138
distribution of housing 109
  Berlin housing 149
  direct redistribution 159, 163–4
Dublin Tenants' Association
  living in someone's asset 3
  tenants' grievances 1
Du Bois Review 90
Dustmann, C. 99

Easthope, H. 114
economics
  global system 68
  linked to housing 8, 67
  polarization 63
  structural transformation of 41–2
education, housing insecurity and 4
employment and incomes
  declining pay 44
  Engels on 90
  housebuilding workers 165
  inequality 6
  insecurity 44
  low incomes 94
  post-war men 58
  women 58
  workers and rent control 136–7
  see also affordability; poverty; precarity
energy, public investment in 23
Engels, Friedrich
  The Condition of the Working Class in England 90
  on housing conditions 132
Estonian homeownership 43
Europe
  affordability 92–4
  EU monetary union 56
  protections for tenants 112–13
  revival of PRS 42–3
  tenant protections 122
European Central Bank 85
Eurostat 93
eviction and displacement 2
  agreements and conditions 33
  Berlin and DWE 149
  destruction of home 31
  emotional loss 32
  inequalities and PRS 4–5
  Irish ban on 116–17, 129
  landlords' powers 118–19
  large and small landlords 123–4
  legal protections 33, 35
  Living Rent campaign and 140–1
  'no fault' 140
  NYC rent strike 133
  renovation and 82, 138
  return of 40
  revenge 119, 120, 158
  rise with institutional landlords 89

SFR landlords and 74
social hierarchies and 126
Spanish mortgages and 143–4. 145
tenant unions 139
tenure and 102
women protest 134–5
WWI and rent strikes 132

families
  having children late 44
  insecurity of home 115
  lone parents 1, 2, 117
  of soldiers 132
  young adults at home 161–2
Federal Housing Finance Agency 73
Federal Reserve 85
Feldman, M. M. A. 60
feminist studies, domestic violence and 30
Fields, D.
  financialization of SFRs 73
  SFR operators 72
financial institutions
  'bad banks' 75, 87
  economics and mortgage provision 68
  financialization of housing and 85
  international 57
  markets of 40
  post-crisis investments 123
  structural transformation of economies 41–2
  see also mortgages
financialization 11, 12
  apartment buildings 76
  Build to Rent 82–4
  concentration of ownership 86
  definition of 41–2
  former social housing 77–9
  of housing 7, 67–9
  impact on cities 87–8
  internationalization 85–6
  intersection with neoliberalism 54–5
  Ireland case study 44–7
  market bubbles 43–4

PRS and 14, 39–40, 69–75, 84–9
  resistance against 138
  rise of financial landlords 65–7
  sources of 85
  state intervention 86–7
  US institutions of 76
financial systems
  housing markets and 17
  PRS and 12
Finland, homeownership in 43
Florida, R. L. 60
Fordist housing regimes 68–9
Forrest, R.
  'generation own' 58
  late-homeownership society 61–2
France
  low institutional investment 88
  Paris rent strike 134
  rent regulation 130
  revival of PRS 42
Frederiksen, Mette 81
Fuller, G. W. 142

generations 3, 10
  boomers and younger 4, 12, 58, 91–2, 105–7
  decline of homeownership 105
  Dublin and 1–2
  inequalities and 63–4, 101
  millennials 3–4
  Netherlands 56
  tenure and class 92
  transfer of assets 62, 105–6
  UK and 62
gentrification see renovation and gentrification
George, Henry
  critique of land rent 20–2
  The Irish Land Question 21–2
  politics of 25–6
  Progress and Poverty 21
  theories of rent 10
  unearned income from land 20–1
Gerber, T. P.
  inequality and homeownership 97
  on tenure 101

Germany
  Article 15 on public ownership 150–1
  Build to Rent 82
  concentration of ownership 86
  DWE campaign 132, 138, 148–52, 163
  East-West housing differences 78
  expropriation of PRS housing 129
  homeownership 43, 148
  impacts of financial landlords 87–8
  inequalities in 99–100
  landlords and 70, 123
  radical housing redistribution 149
  reduced affordability 88
  rent regulation 130
  sale of social housing 77–9, 164
  state intervention 86
  tenant insecurity 3, 113, 148–52
  tenant unions 35, 137, 158
Gil García, J.
  politicization of PRS 129
  post-crisis tenants 153–4
  state intervention 86–7
  on Stay Put 146–7
Glasgow Women's Housing Association 132
Global Financial Crisis
  acquisition of SFRs and 75–7
  Australia and Netherlands case 55–6
  as catalyst 54
  creates homeowner decline 51–7
  evictions 102
  financial institutions investment 123
  financial landlords 9
  local housing markets and 56–7
  mortgages and 3, 8, 67, 69
  political economy literature and 7
  property bubble 44
  SFT stabilizes housing markets 74
  Spain and mortgage victims 143
  state intervention 87
  tenants' organizations and 137–42

Global North and South perspective 14
Goldman Sachs and San Joan Despí apartments 146
Gomory, Henry 123–4
Gray, N.
  impact of rent strikes 136
  women and rent strikes 134
Greece, tenant activism in 138
Gustafsson, J. 80
Gutiérrez, D. 148–9
Guzmán, J.
  Barcelona rent strike 134, 135
  on Spain's PAH and Rosa 144–5

Hands over the City (film) 26
Harris, J.
  landlord flexibility 124
  unequal landlord relationship 125
Harvey, David 18
  investment in built environment 37
  monopoly rent 22, 25
  uniqueness of place 24–5
  urban development and capitalism 69
healthcare facilities 29
  medical institutions and hospitals 23
hedge funds 85
  institutional landlords 70
  SFRs 73
Hick, R.
  low incomes 94
  revival of PRS 42–3
Hirayama, Y.
  'generation own' 58
  late-homeownership society 61–2
holiday rentals 162
Holm, A.
  European cities and PRS 87–8
  niche markets 88
home and home making
  basic need to 34–5
  as caring workplace 29–30, 31
  creation of home 12, 28–9, 30–1, 114–18
  every practices of 9

INDEX 189

households and economy 42
inequalities of 109
landlord's control of access 125–8, 157
monopolistic nature of 37
not always stable and secure 34
not just consumption 127–8
as place 27–31
precarity 112–13
produced by tenants 10
in the PRS 4
qualities and practices of 115
rights of home 157
social recognition of 34
social relation of 165
social reproduction 111–12
in someone's asset 9, 109, 118–22
stability 2
subjective experience of 5, 10, 12–13, 30–1, 36, 114–15, 121, 127–8, 156
tenant agency and 130–1
uniqueness 32–3
unmaking home 118
value of 31–2
homelessness 93
homemaking urge 34–5
inequalities and 91
Los Angeles 141
homeownership
accumulation of value 106–7
being locked out of 101
as capital asset 96
concentration of 11
decline of 7
expansion of 57–8
financialization of 85
focus of political debate 141
Germany 99–100, 148
global crisis decline 51–7
ideology of 159–60
indebted 102
Ireland 1–2, 44–7
low levels of 96
middle-class identity 58–62
Netherlands 55–6
policies advantaging 162

politics of housing 36
post-war promise of 138
promise of 2, 59
repossessions 75
revival of PRS and 57–64
rights of 126
SFRs dampen access to 75
soaring prices 2
social cement of 61
society and geography 14
stability of 2
state support 2
UK 54
undermined 39–40
unequal distribution of 95–7
wealth distribution 58
see also generations; property ownership
home rent
concept of 10, 11, 16, 18, 18n, 31–4
within economic system 37
form of commodity 35–6
form of conflict 36
political economy 18
as social relationship 33–4
utility of approach 35–8
value for tenants 27
Hoolachan, J. 119–20
Horton, A. 84
housebuilding
Build to rent 69, 70, 71, 82–4
Ireland 52–3
housing
apartments 71, 76
commodification of 66–7, 90, 110, 160–1
common administration 151
commons 160, 164
as consumption 9
cooperatives 160
cost of 109
decommodification of 153
deconcentration 156
demarketizing 156–8, 160–1
detached, semi-detached and terraces 72
distribution of stock 40–1

housing (*cont.*)
    financialization of 66, 67–9
    financial systems and 17
    Fordist/Post-Fordist 60, 68–9
    industrialization and urbanization 132
    land scarcity 7–8
    market bubbles 43
    markets 40
    Marxian influence 8
    monopolistic 7–8
    national systems 17
    political economy of 4, 8–9
    prices 67, 98
    public construction of 45
    redistribution 149
    single family rentals 70
    supply of 165
    tenure and 102
    transitory 161
    wide array of influences 44
    see also home and homemaking; housing policies; living conditions and maintenance
housing inequalities *see* affordability; employment and incomes; inequalities
housing policies
    contexts of 54–5, 155–6
    denaturalizing PRS 155–6
    explanatory power of 56
    neoliberalism and 2, 160
    politics of 166
    post-war pact 137–8
    power of landlords 33
    private to public flow 164
    PRS revival 40
    tenant activism and 135–7
housing studies 4–5, 10
    interview and ethnographic data 111
    making a home 109–10
    political economy 6
    tightness of rental markets 121
Howard, A. 55–6
Hulse, K. 114
Hungary, homeownership in 43

Ill-Raga, M. 134, 135
immigration 44
    housing for 161
    Ireland 47
    locked out housing markets 2
    migrant workers and Dublin Tenants 1
    Spain 50
incomes *see* employment and income
In Defence of Housing (Madden and Marcuse) 8
inequalities
    affordability 92–5
    age and generations 55, 105–7
    agency of tenants 130–1
    analysis of PRS 13
    Anglo-centric view 102
    associated with PRS 5–6
    class 10, 55
    commitment to equality 15
    concentration of ownership 138
    distribution of property 95–7
    Engels on 90
    exclusions and 2
    experience of home and 156
    exploitation ownership 21
    factors of 107–8
    future and 166
    of home 109
    housing insecurity and 4–5
    housing systems 8
    income and 6, 12, 158
    polarizing society 63–4
    PRS as an engine of 97–101
    tenure and 101–5
infrastructure
    access to 29
    private/public investment 23
insecurity *see* security/insecurity
international institution
    financial 57
investment
    buy-to-let 162
    cultural 25
    disincentivizing PRS investment 162–3
    forms of 9

international 53
Ireland 46, 47
market power of 152–3
post-crisis institutions 123
profits from housing 7
PRS as opportunity 3
public 23
restrictions of 159
tenants' rights and 126
Invitation Homes 72, 73
Ireland
  austerity budgets 45
  ban on evictions 129
  Built to Rent 83
  case study 40, 44–7
  CATU 138
  Covid eviction moratorium 116–17
  establishing REITs 87
  EU monetary union 46, 56
  global financial crisis and 56, 75–7
  homeownership decline 1–2, 52, 96–7
  house building 52–3
  house price drop 52–3
  Housing Act 1966 45
  institutional landlords 65, 71, 76–7, 88–9
  Land War 21
  lending collapse 53
  politics and corporate landlords 152
  post-colonialism 59, 160
  property bubble 45–7
  public homeownership support 45
  Real-Estate Investment Trusts 53
  redistribution 163
  rent allowances increase 54
  rent regulation 130
  Residential Tenancies Act 47
  revival of PRS 42–3
  sharing tenants' experiences 141
  source of institutional landlords 70
  stamp duty increase 162
  Strategy for the Rental Sector 83
  tenant activism 138
  'wall of credit' 46
IRES REIT 76
The Irish Land Question (George) 21–2
Italy, Milan rent strike in 134

Joint Centre for Housing Studies 94
Jordà, O. 68

Kadekle, P. 99–100
Kadi, J.
  on global financial crisis 54
  post-ownership societies 61, 62–3
Kemp, P. A.
  expansion of homeownership 57–8
  German landlords 123
  house prices UK 48
Kenya, Nairobi housing and 164
Keynesianism: 'asset price' 58
Kholodilin, K. A. 97–8, 100
King Jr, Martin Luther: on change 139
Kofner, S. 123
Kohl, S. 97–8, 100
Kotti and Co. 148

labour *see* employment
Lancee, B. 98
land
  agricultural 18, 19
  charging for access 20
  as local 19
  not a true commodity 18–19
  politics of 18–22
  post-colonial distribution 59
  resources and amenities 20
  scarcity of 7, 18–19
  unearned income from 19–20
  value of 19–20
landlords
  asset security 121
  Berlin public housing 149
  Biden's promise of rent control 129
  Build to Rent 82–4
  buy to let mortgages 44
  concentration of wealth 86, 97–101

landlords (*cont.*)
  deliberate or circumstantial 124
  discrimination 120–1
  economic structures and 41
  as engine of inequality 100–1
  failure of maintenance and care 79
  financial 9, 69
  focus on long-term capital 123
  German compensation payouts 150–1
  Global Financial Crisis and 54
  'home' value and 31
  international and local policies 87
  investment in Ireland 47
  large companies 72
  management and maintenance 31
  'mom and pop' 72, 123
  niche markets 88
  percentages of in PRS 70–1
  political and economic systems 38
  poor tenant relations 77
  powers of 118–22
  proportion of market 72–3
  reduction of affordable housing 88–9
  relations with tenants 123–4
  renovation and upgrading 79–82
  rent as engine of inequality 97–101
  repossession and 54
  rise in evictions and displacement 89
  rise of 65–7, 90
  small-scale owners 71
  sources of 70
  in Spain 142
  types and terminology of 69–70
  tyranny of 2
  see also landlord–tenant relationship; private rental sector (PRS)
landlord–tenant relations 11, 13
  abusive practices 40, 116
  access to home 157
  Anglophone accounts 122
  challenging 119
  contractual terms 16–17
  depersonalized management 124–5
  dimension of coercion 33–4
  economic relationship 35
  eroding landlord power 157
  financial landlords and 66
  geography of 122–3
  homemaking challenges 118–22
  inequality 98–101
  institutional landlords and 123–4
  lone parents and 117
  market and 16–17, 110, 114, 156
  not like capitalist-owner relations 128
  policy problems 121
  power imbalance 4, 35, 36, 125–8, 130, 155
  property inspections 116
  references 119
  retaliation by landlords 157
  revenge evictions 120
  selectivity 119
  sexual exploitation 116
  social 4
  subjective composition of 125–8
  terrain of homemaking 111
  varying forms of 122–5
  withholding deposits 119
landowners
  as monopolists 19
  rise from exploitation 21
laws and legislation
  California's 'Prop 10' campaign 141
  Denmark
    Blackstone law 81
  establishing REITs 87
  eviction/tenancy termination 33
  Germany
    Article 15 150–1
  Ireland
    Housing Act 1966 45
    Residential Tenancies Act 47
  landlords' power 34–5
  limiting financial landlords 65
  non-compliance 158

protection of tenants 1, 35, 157
Real-Estate Investment Trusts 76
rental rights and obligations 16–17
Spain
   Boyer Decree 49–50
   Urban Letting Act 52, 142
   Urban Rent Law 49
United Kingdom
   Housing Act 48
   Rents and Mortgage Interest Bill 132–3
Lawson, R. 133, 135
Limited Profit Housing Associations, Austria 161
living conditions and maintenance 2, 31
   Berlin and DWE 149
   building inspectors 120
   escaping 117–18
   homemaking 157
   institutional landlords and 79–82, 89
   poor quality of 13
   renovation of social housing 70, 79–82
   tenant constraints and 116
   tenant organization and 130
   tenant repairs 119
   tenants' union and 139
   withholding rent and 120
Living Rent, Scotland 139
   establishment of 3
   media and 141–2
   successful campaign 140–1
Logan, John R. 18
   land not a commodity 19
   land value 20
   place as commodity 22–4
   place as network of people 29
   politics of place 26–7
   value of home 30
London Renters' Union 3
London Tenants' Union 138
Lone Star Capital 76
Lone Star Fund 86
Los Angeles Tenants' Union 141
Loughran, K. 90

McCardle, R. 120
McKee, K.
   landlord flexibility 124
   landlord–tenant relations 119–20, 125, 126
   renting as transitory 115
Madden, D.
   collective power 131
   In Defence of Housing (with Marcuse) 8
   distribution of housing 109
   residential alienation 10, 32
Madrid Tenants' Union 131, 139, 143, 146
maintenance *see* living conditions and maintenance
Malta, PRS and 42–3
Marcuse, P.
   collective power 131
   In Defence of Housing (with Madden) 8
   distribution of housing 109
   residential alienation 10, 32
markets
   'black' 158
   financialized bubbles 43–4
   global housing assets 67
   housing demarketization 156–8
   housing relationships 103–4
   inelasticity of land 19
   landlord–tenant relations 16–17, 110
   private rental sector 5–6
   property bubbles 45–7, 48–9, 50–1
   securities 85
   SFRs stabilize 74
   tyranny of 156
   UK liberalization 48–9
Martin, C. 42
Martínez López, M. A. 86–7
Marx, Karl on distribution 41
Mee, K. 29
mental health, housing insecurity and 4
Michener, J. 138
Milligan, V. 114

Molotch, Harvey 18
  land not a commodity 19
  land value 20
  place as commodity 22–4
  place as network of people 29
  politics of place 26–7
  value of home 30
monetary policy explaining PRS 40, 56
Moreno, M. 9
Moreno Zacarés, J. 37
Morris, A. 115
mortgages
  'deleveraged' 75
  economic development and 67
  expansion of credit 39
  financial institutions and 58
  foreclosures and repossessions 8, 51–7, 53, 75, 143
  global market assets 67
  interest rates 85
  Ireland 45–7
  Irish public provision for 45
  Loan to Value ratio (LTV) 53
  long-term incentives (LTI) 53
  securities market 85
  Spanish securitization 51
  state intervention in crisis 87
  subprime 50–1, 67, 74
  tightening of 74
  see also affordability; Buy to Let mortgages; Global Financial Crisis; homeownership
Murray, C. 96

National Asset Management Agency (NAMA) 75–6
neighbours *see* communities
neoliberalism 11
  explaining PRS revival 39–40
  global process of 57
  housing policy 2, 160
  intersection with financialization 54–5
  Ireland case study 44–7
  private rental sector 5
Nethercote, M. 83

Netherlands
  institutional landlords 70
  low ownership, high rental 55–6
  post-crisis mortgages 53
  rent regulation 130
  restricting Buy-to-Let 162
  tenant insecurity 113
  tenant unions 35
New Zealand
  indigenous people 95
  revival of PRS 42–3
Nigeria: Lagos housing 164
Norris, M. 45

Oaktree 76
ontological security *see under* security
Organization for Economic Cooperation and Development (OECD)
  housing affordability 92–3, 93
Otto-Suhr tenant organization 148–9
overcrowding 2

Palomera, J.
  politicization of PRS 129
  post-crisis tenants 153–4
  on Stay Put 146–7
Patrizia 86
pension and insurance funds 70, 85
place 18
  attachments 24
  as a commodity 22–4
  created by tenants 32
  entrepreneurs 26–7
  home as 27–31
  network of people 29–30
  private investment 23
  social construction of 23
  social produced value 25
  subjective dimension 24
  uniqueness 24–5
  see also communities; home and homemaking
planning system 162
Platforma de Afectados por la Hipoteca (PAH) 143–4, 145

Poland
  revival of PRS 42–3
  tenant organizations 138
Polanyi, K. 18
political economy
  explaining PRS 11, 39–40
  financial landlords 66
  of housing 4, 38
  landlord–tenant relationship 121–2
  literature of 7–10, 10–12
  macro- and micro- 10, 11
  national housing systems and 17
  precarity and 126–7
  private rental sector 17, 152–3
  see also financialization; markets; neoliberalism
politics
  critique of land rent 21–2
  focus on homeowners 141
  housing and 7
  inequality and 91
  institutional landlords and 89
  tenants and 131
Portugal, tenant activism in 138
post-colonial land redistribution 59
poverty
  Dublin Tenants 1
  failing to pay the rent 91
  intensification of 93
  landlord–tenant relationships 124
  rent costs 100–1
Power, E. 29
power relations
  agency of tenants 130–1
  bureaucracy and 161
  collective power 131
  disempowerment of tenants 1, 3
  inequalities 12
  landlord–tenant imbalance 4, 125–8, 155
  tenants and 'home' 33–4
  tenant unions and 140
  see also landlord–tenant relationship
precarity
  concept of 5
  a fact of life for tenants 112–13
  indebted homeowners 102
  locked out housing markets 2
  political economy and 126–7
  see also security/insecurity
privacy 12
  eroding landlord power 157
  experience of home 116
  hidden abuse and 30
  home making 28
private equity funds 164
  housing and 85
  institutional landlords 70
  SFRs 73
  Spain 142
private rental sector (PRS)
  Anglophone countries 113
  apartments 65
  assets of 32
  availability of 165
  beyond homeownership society 57–64
  Build to Rent 70
  commodification of housing 132
  concentration of ownership 63, 86
  decline of 2, 137
  deconcentration 165
  demarketization 165
  denaturalizing 155
  deregulation of 44
  disincentivizing investment in 162–3
  distribution of housing stock 40–1
  Dublin Tenants 1
  explaining revival of 39–40
  financialization of 5, 11, 12, 38, 69–75, 84–9
  five-point analysis of 13–14
  flexibility and informality 125
  forces acting upon 85
  form of conflict 36
  future of 166
  geography of 14, 38
  global financial crisis and 137–42
  growth in Spain 142
  home and homemaking 109–10
  housing demarketization 156–8

private rental sector (PRS) (*cont.*)
　housing insecurity 92
　immigration and 47
　impacts on cities 87–8
　inequality engine 91, 97–101, 107–8
　international investment 53, 85–6
　investment opportunity 3
　Ireland 46–7
　lightly regulated 5
　market forces 5–6
　micro-political economy 17
　neoliberalism 5
　niche markets 88
　only option for some 53
　political economy and 18, 36, 110, 129, 152
　property owners 103
　recognizing tenants' 'home' 34
　renovation and displacement 81–2
　rent control policies and 136–7
　renting as transitory 115
　re-politicization of 3
　residential accumulation 37
　restricting investment 159
　resurgence of 5–6, 11, 42–4, 65–7, 113
　right to obtain rent 33
　short-term renting 88
　small-scale landlords 71
　state intervention 86–7
　subjective experience of 4, 9–10, 125–8
　subletting and informal arrangements 103
　supply and demand 17
　tenant activism and 153–4
　tenants' experience of 12–13, 36
　tenure 103
　transforming 155, 165–6
　unequal distribution 95–7
　see also financialization; landlords; landlord–tenant relations; property ownership
Progress and Poverty (George) 21
property management/estate agents 124–5

property ownership
　collective 161, 162
　concentration of 3, 40–1
　deconcentration of 156, 159, 165
　Germany's Article 15 150–1
　inequality 155
　non-ownership 164
　non-residential 96
　power relations and 33–4
　PRS relationships 103–4
　unequal distribution 17
　unequal relationships 140
　varied forms of 160
　wider distribution 159–62
　see also homeownership; landlords; private rental sector (PRS)

race and ethnicity
　avoiding confrontation 120
　evictions after renovations 82
　housing inequality 90
　rent affordability 95
Raymond, E. L.
　on renovation and eviction 82
　SFR landlords 74
Real-Estate Investment Trusts (REITs) 53
　Built to Rent 83
　expanding financialization 76
　housing and 85
　institutional landlords 70
　legislation for 87
　SFRs 73
　Spain 142
regulations
　private rental sector and 5–6
　see also rent regulation
renovation and gentrification 71
　Berlin and DWE 149
　evictions and 138
　financialization and 66
　rent increase and 80
　social housing 79–82
　see also living conditions and maintenance
rent
　after renovations 80
　Built to 69

critique of 16
defining 18n
deregulation of 48
freezes 129
increases 3, 74, 79, 139
linked to house prices 98
monopoly 18, 18n, 22–7, 25, 32–3
non-payment of 35, 91
politics of land 18–22
profits for investment 7
rack renting 2, 40
redistribution 98
regulation 91, 113
rights and obligations 16–17
Schwaube's Law 98
subsidies 13, 158
theory of 10
tightness of market 121
as 'unearned income' 19–20
see also home rent; rent regulation; rent strikes
rental markets *see* home rent; private rental sector (PRS); tenants
rent regulation 36, 81, 91, 97–8
　aspects of 157–8
　Berlin and DWE 149
　California 141
　contexts of 165
　impact of strikes 136–7
　increase of 129–30
　Living Rent campaign and 140–1
　New York City 136
　Spain 142, 146, 147
　Sweden 129
　tenant demands 155
　United States 129
rent strikes 13
　agency of tenants 131
　historical view of 132–7
　impacts of 136–7
　WWI context 132
residents *see* tenants
Rinaldi, Antonio 133–4
Romanian housing 165
Ronald, R.
　on global financial crisis 54

inequality and insecurity 61
post-ownership societies 61, 62–3
promise of homeownership 2, 59
Rose, D. 104
Rosenthal, T.
　housing inequality 90–1
　rent as engine of inequality 101
　tenant organization 131–2
Rouco Buela, Juana 134
Russia
　Cold War and 60
　sale of social housing 164
　Soviet housing nationalization 163
Ryan-Collins, J.
　landowners not in normal market 19
　residential assets 96

Safe and Decent Homes (Shelter) 119
safety
　home making 28
　investment in 23
SAREB 75
Sassi, Juliana 116–17
Saunders, P. 27, 28, 29, 104
schools
　access to 29
　public investment 23
Schröder, C. 99
Schwartz, A.
　affordability 95
　global crisis and foreclosures 52
Schwaube's Law 98
Scottish Federation of Tenants' Associations 132
security/insecurity 4, 12–13, 112
　analysis of PRS 14
　chronic 122
　created by tenant 32
　demarketization of housing 156–8
　housing market 61
　inequality at home 34
　landlord's controls and 125–8
　ontological 27–9, 30, 34, 111–12
　concept of 114
　practices of homemaking 115
　renting as transitory 115

security/insecurity (*cont.*)
  return of 40
  shorter tenancy periods 48
  tenant organization and 130
  tenure and 102, 103
  see also eviction and displacement
Shelter
  Safe and Decent Homes 119
Shiffer-Sebba, D. 124
Singapore
  decolonization of housing 160
  post-colonialism 59
  redistribution 163
single family rental (SFR)
  dampen access to homeownership 75
  geography of 72–3
  institution investors in 70
  post-crisis acquisition of 75–7
  proportion of market 72
  stabilizes local markets 74
  state role in financialization 73
  in the US 71–5
Slovakia, PRS revival in 42–3
Slovenian homeownership 43
Smith, S. J. 96
Soaita, A.
  landlord's power 126
  renting as transitory 115
social action, construction of place and 23
social housing
  Berlin tenants and 148–9
  distribution of ownership 160–1
  financialization of 77–9
  historical access to 106
  mass acquisition of 70
  mass construction of 57
  Netherlands 56
  privatization of 68
  renovating property 79–82
  residualization of 44
  right to buy 39, 45
  stability of 2
  state intervention 2
  tenants' unions and 36
  UK 47–8

socialism
  Cold War and 60
  German E. and W. housing 78
  Germany's Article 15 150–1
  homeowning antidote 59
  labour organizations and 135
  Marxism and urban development 69
social movements: Spain and housing 143
social reproduction 29–31, 34
society
  exclusion and alienation 63
  homeownership as social cement 61
  late-/post-ownership 61–2
  for ownership and rental 41
SoRelle, M. 138
Spain
  after Global Financial Crisis 143
  Barcelona Tenants 145–8
  Blackstone and 65
  Boyer Decree 49–50
  case study 40, 49–51
  coalition government 147
  decline of rental sector 49–50
  deregulation 52
  establishing REITs 87
  EU monetary union 56
  foreclosures 51–2
  global financial crisis and 56
  homeownership in 50, 52
  housing system 131, 142
  increased rent in Barcelona 142
  PAH and mortgage victims 143–4, 145
  politics and corporate landlords 152
  post-crisis SFR acquit ions 75–7
  property bubble 50–1
  rent regulation 130
  revival of PRS 42–3
  San Joan Despí 146
  source of institutional landlords 70
  state intervention 86–7
  Stay Put/*Ens Quedem* 145–7

INDEX 199

tax incentives 49
tenant activism 138
tenants' unions campaigns 145–8
Urban Letting Act 52, 142
Urban Rent Law, 1946 49
Spanish Mortgage Bank 49
Spratt, Vicki
  Tenants 112
squatting 138
  as redistribution 163–4
states and governments
  economic structural
    transformation 41–2
  expropriation of PRS housing 129
  intervention in financialization
    86–7
  national/local policies 87
  PRS policy changes 129–30
  welfare systems 58
Stay Put/*Ens Quedem,* Spain 145–7
Strategy for the Rental Sector (Irish
  government report) 83
student housing
  Built to Rent 84
  Denmark 161–2
  niche market 88
supply and demand 35
Sweden
  concentration of ownership 86
  fall of homeownership 43
  impacts of financial landlords 87–8
  institutional landlords 70
  large-scale renovation 80–1
  Municipal Housing Companies
    78–9
  reduced affordability 88
  rent regulation 129
  sale of social housing 164
  state intervention 86
  Stockholm rent strike 134
  tenants' unions 35, 137, 158
Swedish Union of Tenants 131, 137

taxes
  relief for homeowners 45
  Spanish incentives 49
  stamp duty 162

telecommunications 23
tenancy
  displacement 79
  period of 48
tenants
  activism 91
  agency of 114, 130
  BTR amenities and 83–4
  constraints on maintenance 116
  creating meaningful homes
    113–14
  displacement 81, 89
  every day experiences 109–10
  informal subletting 35
  institutional landlords and 77
  legal protections 34, 35, 157
  mobility in Brazil 117–18
  politics of 111
  privacy and control 12
  produce 'home' 10
  protections for 112–13
  resurgence of activism 138
  rights of 118–22, 126
  social recognition of home 34
  social relations with landlord 4
  someone else's asset 32
  strategies 120–1
  subjective experience of 13, 104–5,
    110
  subletting arrangements 103
  weakness of rights 74, 123
  see also home and homemaking;
    landlord–tenant relations;
    tenants' organizations
Tenants (Spratt) 112
tenants' organizations
  activism 13, 36
  after the Global Financial Crisis
    137–42
  ambitious forms 153
  campaign for legislative change
    142–8
  countries with strong protections
    and 35
  demands for reform 155
  direct action and civil
    disobedience 139, 144

tenants' organizations (cont.)
  direction action and civil
    disobedience 145
  mass strikes 2
  politics of 10, 141
  rent strikes 13, 132–7
  revival of activism 153–4
  Spain 144–8
  squatters and 138
  tenant identity 154
Tenants Together 141
Tenants' Union of Barcelona (TUB) 145–8
tenants' unions 2, 35, 135, 139–40
  challenging landlords 158
  Germany 35, 137, 158
  Ireland 46, 56
  Spain 131, 145–8
  Sweden 35, 131, 135, 137
  United Kingdom 3, 138–9
tenure
  alternative 159–62
  generation and class 10, 92
  importance of 101–5
  inequalities 12, 101–5
  patterns of 11
  PRS and 107
  security of 113, 155, 157
Thatcher, Margaret 39, 160
transport, public investment in 23

United Kingdom
  Built to Rent 83
  case study of 40, 47–9
  Digs 138
  economic development and housing 67
  establishing REITs 87
  generation rent 62
  Glasgow rent strikes 132, 132–6, 134, 135, 136
  Glasgow trade unions 135
  global financial crisis and 56
  homeownership rises and falls 52, 54, 96
  institutional landlords 88–9
  labour organizations 135
  Leeds rent strike 134
  Living Rent 139, 140–1
  London Tenants' Union 138–39
  post-ownership society 62
  Rents and Mortgage Interest (War Restrictions) Bill 132–3
  revival of PRS 42–4
  sale of social housing 164
  Scottish rent regulation 130
  sharing tenants' experiences 141
  social housing 47–8
  tenant activism 136, 138–9
  Thatcher's right to buy policy 48
  transfer of housing stock 163
United States
  Blackstone and 65
  build to rent 82
  California's 'Prop 10' campaign 141
  decline of homeownership 96
  economic development and housing 67
  economic inequality 98
  financialization institutions 76
  global crisis foreclosures 52
  institutional landlords 71
  migrants and ethnic minorities 95
  New York
    labour unions 135
    rent controls 136
    rent strikes 133, 136
  post-crisis mortgages 53
  revival of PRS 42–3
  single family rentals 70, 71–5
  state intervention 87
  Sunbelt properties 73
  tenant activism 138
urban development 132
  gentrification 71
  Harvey's view 69
  surplus value 69
  see also communities
Urban Letting Act, Spain 142

value, subjective 30–1
Vergerio, M. 72

Vilchis, L.
　housing inequality 90–1
　rent as engine of inequality 101
　tenant organization 131–2
violence and abuse
　at home 34
　privacy of home and 30
Vollmer, L. 148–9
Vonovia 80

Waldron, R. 112
Walks, A. 81–2
Walshaw, A. 28, 114
water, public investment and 23
welfare inequalities 6

wellbeing, impacts of tenure and 102
Wiedemann, A.
　DWE campaign 148, 149, 152
　market power of investors 152–3
Williams, P. 27, 28, 29
women
　'broom parades' 134
　rent strikes 132, 134–5
work *see* employment and incomes

Zavisca, J. R.
　inequality and homeownership 97
　on tenure 101
Zhang, B. 102